INFRARED AND MILLIMETER WAVES

VOLUME 4 MILLIMETER SYSTEMS

CONTRIBUTORS

George W. Ewell
Gary A. Gordon
Richard L. Hartman
Tatsuo Itoh
N. Bruce Kramer
Paul W. Kruse
Edward K. Reedy
Charles R. Seashore
H. Shigesawa
K. Takiyama
M. Tsuji
James C. Wiltse

INFRARED AND MILLIMETER WAVES

VOLUME 4 MILLIMETER SYSTEMS

Edited by *KENNETH J. BUTTON*

NATIONAL MAGNET LABORATORY
MASSACHUSETTS INSTITUTE OF TECHNOLOGY
CAMBRIDGE, MASSACHUSETTS

JAMES C. WILTSE

ENGINEERING EXPERIMENT STATION
GEORGIA INSTITUTE OF TECHNOLOGY
ATLANTA, GEORGIA

1981

ACADEMIC PRESS

A Subsidiary of Harcourt Brace Jovanovich, Publishers

New York London Toronto Sydney San Francisco

ACADEMIC PRESS, INC.
111 Fifth Avenue, New York, New York 10003

United Kingdom Edition published by
ACADEMIC PRESS, INC. (LONDON) LTD.
24/28 Oval Road, London NW1 7DX

Library of Congress Cataloging in Publication Data
Main entry under title:

Infrared and millimeter waves.

Vol. 4- edited by K. J. Button, J. C. Wiltse.
Includes bibliographies and indexes.
CONTENTS: v. 1. Sources of radiation.--v. 2. Instrumentation.-- --v. 4. Millimeter systems.
1. Infra-red apparatus and appliances.
2. Millimeter wave devices. I. Button, Kenneth J.
II. Wiltse, James C.
TA1570.I52 621.36'2 79-6949
ISBN 0-12-147704-5 AACR1

PRINTED IN THE UNITED STATES OF AMERICA

81 82 83 84 9 8 7 6 5 4 3 2 1

CONTENTS

List of Contributors	vii
Preface	ix
Contents of Other Volumes	xi

Chapter 1 **Introduction and Overview of Millimeter Waves**
James C. Wiltse

I.	Background	1
II.	New Technology	5
III.	System Applications	11
	References	17

Chapter 2 **Millimeter Radar**
Edward K. Reedy and George W. Ewell

I.	Fundamentals of Millimeter-Wave Radar	23
II.	Propagation Effects	36
III.	Clutter Characteristics	44
IV.	Technology Base	56
V.	Millimeter-Wave Radar Example	76
VI.	Millimeter-Wave Radar Applications	80
	References	91

Chapter 3 **Missile Guidance**
Charles R. Seashore

I.	Introduction	95
II.	Propagation and Targets	98
III.	Range Equations	112
IV.	Millimeter-Wave Seeker Design	122
V.	Countermeasures	148
	References	150

Chapter 4 **Sources of Millimeter-Wave Radiation: Traveling-Wave Tube and Solid-State Sources**
N. Bruce Kramer

I.	Introduction	151
II.	Traveling-Wave Tubes	152

III. Solid-State Sources 168
IV. Summary and Conclusions 196
References 197

Chapter 5 **Dielectric Waveguide-Type Millimeter-Wave Integrated Circuits**
Tatsuo Itoh

I. Introduction 199
II. Dielectric Waveguides for Integrated Circuits 201
III. Propagation Characteristics of Dielectric Waveguides 207
IV. Passive Components Made of Dielectric Waveguides 243
V. Active Components 253
VI. Antennas for Dielectric Millimeter-Wave Integrated Circuits 262
VII. Subsystems 269
VIII. Conclusions 271
References 271

Chapter 6 **Submillimeter Guided Wave Experiments with Dielectric Waveguides**
M. Tsuji, H. Shigesawa, and K. Takiyama

I. Introduction 275
II. Dielectric Slab Waveguides 277
III. Dielectric Rib Waveguides 297
IV. Dielectric Cylindrical Waveguides 310
V. Conclusion 319
Appendix I 322
Appendix II 324
Appendix III 325
References 325

Chapter 7 **Imaging-Mode Operation of Active NMMW Systems**
Gary A. Gordon, Richard L. Hartman, and Paul W. Kruse

I. Introduction 327
II. Propagation of NMMW Radiation 330
III. System Design Considerations 333
IV. System Limits and Wavelength Tradeoffs 340
V. Image Quality Considerations 344
References 351

INDEX 353

LIST OF CONTRIBUTORS

Numbers in parentheses indicate the pages on which the authors' contributions begin.

GEORGE W. EWELL (23), *Radar and Instrumentation Laboratory, Engineering Experiment Station, Georgia Institute of Technology, Atlanta, Georgia 30332*

GARY A. GORDON (327), *R & D Associates, Marina del Rey, California 90291*

RICHARD L. HARTMAN (327), *DRSMI-RR, U.S. Army Missile Command, Redstone Arsenal, Alabama 35898*

TATSUO ITOH (199), *Department of Electrical Engineering and Electronics Research Center, The University of Texas at Austin, Austin, Texas 78712*

N. BRUCE KRAMER (151), *Electron Dynamics Division, Hughes Aircraft Company, Torrance, California 90509*

PAUL W. KRUSE (327), *Honeywell Corporate Technology Center, Bloomington, Minnesota 55420*

EDWARD K. REEDY (23), *Radar and Instrumentation Laboratory, Engineering Experiment Station, Georgia Institute of Technology, Atlanta, Georgia 30332*

CHARLES R. SEASHORE (95), *Honeywell, Inc., Millimeter-Wave Technology Center, Bloomington, Minnesota 55430*

H. SHIGESAWA (275), *Department of Electronics, Doshisha University, Kamikyo-ku, Kyoto, 602 Japan*

K. TAKIYAMA (275), *Department of Electronics, Doshisha University, Kamikyo-ku, Kyoto, 602 Japan*

M. TSUJI (275), *Department of Electronics, Doshisha University, Kamikyo-ku, Kyoto, 602 Japan*

JAMES C. WILTSE (1), *Engineering Experiment Station, Georgia Institute of Technology, Atlanta, Georgia 30332*

PREFACE

For the first time in this treatise, we deal with millimeter wave systems, but there will be further volumes on the subject. Recent progress toward commercialization in this range of the spectrum assures that books on millimeter waves will be received with some interest. Larger numbers of sessions on this subject appear in the programs of the Annual International Conference on Infrared and Millimeter Waves.

The chapters on radar, on missile guidance, and on imaging immediately define the nature of the subject matter in this volume. The thrust of the subject matter, however, is the extension of systems from the microwave toward the shorter millimeter wavelengths. This thrust is indicated by the chapters on sources of millimeter wave radiation and by the two chapters devoted to the fundamentals of propagation in dielectric waveguides.

Chapter 1 is the introduction and overview of millimeter waves and is comprehensive. It was written by one of us who served the IEEE Microwave Theory and Techniques Society as the Annual Lecturer on the subject of millimeter waves during 1979–1980. Subsequent chapters in this volume cover only a few of the topics mentioned in Chapter 1. Radar, missile guidance, imaging, and propagation in dielectric waveguides have been treated in detail. However, chapters on much of the new technology on components and the chapter on communications have been omitted: an additional volume on millimeter waves and their applications will be planned, eventually.

We have expressed our gratitude in previous volumes to the authors and to the institutions that support their research. We have also expressed our appreciation to the IEEE Microwave Theory and Techniques Society for supporting the Annual International Conference on Infrared and Millimeter Waves from which some of this material is derived. The time has come now to thank the staff of Academic Press who have solved some most difficult production problems in order to bring these books to fruition.

CONTENTS OF OTHER VOLUMES

Volume 1: Sources of Radiation

J. L. Hirshfield, Gyrotrons
H. J. Kuno, IMPATT Devices for Generation of Millimeter Waves
Thomas A. DeTemple, Pulsed Optically Pumped Far Infrared Lasers
G. Kantorowicz and P. Palluel, Backward Wave Oscillators
K. Mizuno and S. Ono, The Ledatron
F. K. Kneubühl and E. Affolter, Infrared and Submillimeter-Wave Waveguides
P. Sprangle, Robert A. Smith, and V. L. Granatstein, Free Electron Lasers and Stimulated Scattering from Relativistic Electron Beams

Volume 2: Instrumentation

N. C. Luhmann, Jr., Instrumentation and Techniques for Plasma Diagnostics: An Overview
D. Véron, Submillimeter Interferometry of High-Density Plasmas
J. R. Birch and T. J. Parker, Dispersive Fourier Transform Spectroscopy
B. L. Bean and S. Perkowitz, Far Infrared Submillimeter Spectroscopy with an Optically Pumped Laser
Wallace M. Manheimer, Electron Cyclotron Heating of Tokamaks

Volume 3: Submillimeter Techniques

T. G. Blaney, Detection Techniques at Short Millimeter and Submillimeter Wavelengths: An Overview
W. M. Kelley and G. T. Wrixon, Optimization of Schottky-Barrier Diodes for Low-Noise, Low-Conversion Loss Operation at Near-Millimeter Wavelengths
A. Hadni, Pyroelectricity and Pyroelectric Detectors
A. F. Gibson and M. F. Kimmitt, Photon Drag Detection
F. W. Kneubühl and Ch. Sturzenegger, Electrically Excited Submillimeter-Wave Lasers
Michael von Ortenberg, Submillimeter Magnetospectroscopy of Charge Carriers in Semiconductors by Use of the Strip-Line Technique

Eizo Otsuka, Cyclotron Resonance and Related Studies of Semiconductors in Off-Thermal Equilibrium

Volume 5: Coherent Sources and Applications

Benjamin Lax, Coherent Sources and Scientific Applications

J. O. Henningsen, Spectroscopy of Molecules by Far Infrared Laser Emission

F. Strumia and M. Inguscio, Stark Spectroscopy and Frequency Tuning in Optically Pumped Far Infrared Molecular Lasers

Jun-ichi Nishizawa, The GaAs TUNNETT Diodes

V. L. Granatstein, M. E. Read, and L. R. Barnett, Measured Performance of Gyrotron Oscillators and Amplifiers

F. K. Kneubühl and E. Affolter, Distributed Feedback Gas Lasers

INFRARED AND MILLIMETER WAVES

VOLUME 4 MILLIMETER SYSTEMS

CHAPTER 1

Introduction and Overview of Millimeter Waves

James C. Wiltse

Engineering Experiment Station
Georgia Institute of Technology
Atlanta, Georgia

I. BACKGROUND 1
II. NEW TECHNOLOGY 5
A. *Sources* 6
B. *Mixers* 8
C. *Components and Devices* 10
D. *Antennas* 10
III. SYSTEM APPLICATIONS 11
A. *Missile Guidance* 11
B. *Radar* 12
C. *Radiometry, Remote Sensing, and Radio Astronomy* 12
D. *Communications* 14
E. *Electronic Warfare* 16
F. *Biological Applications* 17
REFERENCES 17

I. Background

The millimeter-wave region has seen significant advances in recent years in the development of transmitters, receivers, devices, and components. A considerable amount of progress is now occurring in the evolution of systems applications in such fields as communications, radar, radiometry, remote sensing, missile guidance, radio astronomy, and spectroscopy. This chapter provides a summary of activities in the field and includes an extensive list of relevant references. The succeeding chapters provide further details on new technologies and systems.

The millimeter-wave range is frequently defined as the frequency region from 30 to 300 GHz or 1-cm to 1-mm wavelength, although a recent standard promulgated by the Institute of Electrical and Electronics Engineers (IEEE, 1976) gives 40–300 GHz as the nominal frequency range, identifying 27–40 GHz as the K_a-band range. Other current terminology includes *near-millimeter waves* for frequencies from approximately 100 GHz (sometimes from 90 GHz) to 1000 GHz, and *submillimeter waves*

ISBN 0-12-147704-5

from about 150 to 3000 GHz (3 THz). Figure 1 illustrates these and other nomenclature, while Table I shows various letter-band designations. From the optical point of view, "extreme far infrared" extends from the far infrared down in frequency to 150 GHz.

A characteristic of the millimeter-wave range is that for a given available antenna size (aperture) the beamwidth is smaller and the gain is higher than at microwave frequencies, and, conversely, to obtain a specified gain or narrow beamwidth, a smaller antenna may be used. This is important in several applications, such as missile terminal guidance seekers and airborne sensors. For a missile seeker, a narrow beamwidth reduces the unwanted radar or radiometric return from the terrain "clutter patch" around a target.

In general, atmospheric propagation effects dominate considerations relating to many applications. This is true even for satellite communications outside the atmosphere, since frequencies may be chosen for which the atmosphere is opaque, thus preventing detection of satellite-to-satellite communications by ground-based receivers. Terrestrial systems desiring to prevent signal "overshoot" in range may similarly operate at a frequency of high atmospheric absorption to gain a degree of covertness. Typical values of atmospheric attenuation are shown in Fig. 2 for sea level and for 4-km altitude. Noteworthy are the absorption peaks due to such atmospheric constituents as oxygen and water vapor, and the minima regions called *windows*. Also characteristic is the fact that the window minima increase monotonically with frequency. Additional attenuation (and backscatter) is produced by rainfall, clouds, and fog (see Fig. 3), although fog losses are small unless visibility is very limited (e.g., 100 m or less). Depending on liquid water content, cloud attenuation rates may not be insignificant, but total attenuation is usually low because of limited range extent. The attenuation of cirrus ice clouds is negligibly small, since the dielectric constant of ice is much lower than that of liquid water. Values of attenuation and backscatter due to atmospheric effects

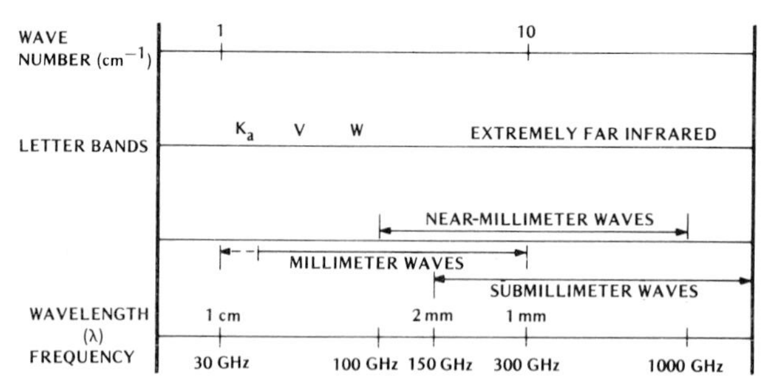

FIG. 1 Nomenclature for millimeter waves and other frequency ranges.

TABLE I

Frequency-Band Designations

Microwave and radar usage: United Kingdom usage	Microwave and radar usage: United States usage	Official joint chiefs of staff band designation	Official international telecommunications union, Geneva: Band designation	Official international telecommunications union, Geneva: Metric designation	EIA JAN standard waveguide: Staggered series of rectangular guide sizes (no ridges)	
V–BAND 50–75 GHz	W–BAND 56–100 GHz	M 60–100 GHz	BAND NO. 11 EHF 30–300 GHz	MILLIMETRIC	90–140 GHz	WR 10 75–110 GHz
O–BAND 40–70	V–BAND 46–56	L 40–60	BAND NO. 10 SHF 3–30	CENTIMETRIC	WR 12 RG 99 60–90	WR 15 RG 98 50–75
Q–BAND 27–40	Q–BAND 36–46	K 20–40			WR 19 40–60	WR 22 RG 97 33–50
K–BAND 18–27	K_a 33–36				WR 28 RG 96 26.5–40	WR 34 22–33
	K–BAND				WR 42 RG 53 18–26.5	WR 51 15–22

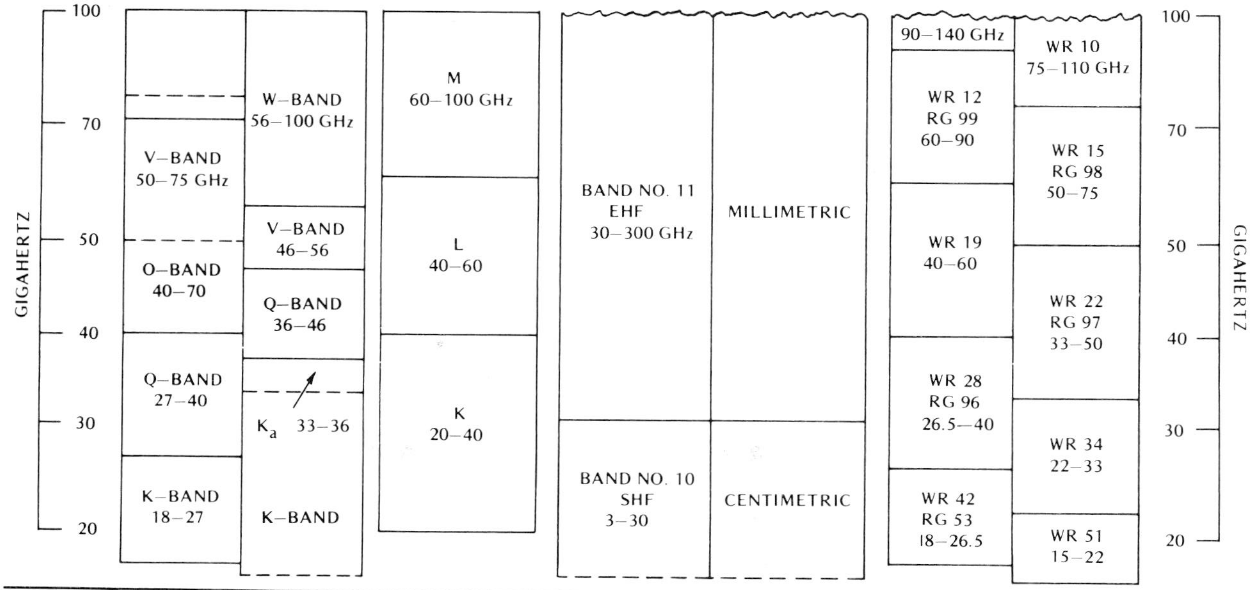

GIGAHERTZ: 100, 70, 50, 40, 30, 20

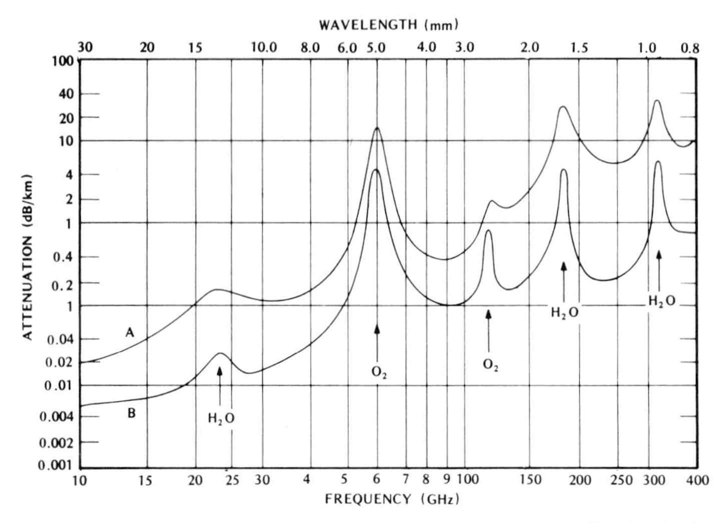

FIG. 2 Average atmospheric absorption of millimeter waves. *A:* Sea level; $T = 20°C$; $P = 760$ mm; $P_{H_2O} = 7.5$ g/m³. *B:* 4 km; $T = 0°C$; $P_{H_2O} = 1$ g/m³.

have been well established for frequencies up to 100 GHz (Chen, 1975; Sander, 1975; Dyer and Currie, 1978; Richard and Kammerer, 1975; Persinger *et al.*, 1980), and programs are now under way to obtain better information about atmospheric effects above 100 GHz (Kulpa and Brown,

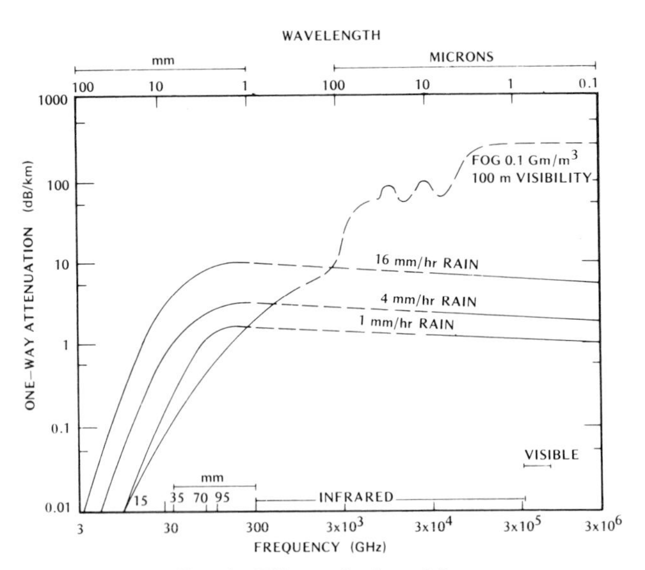

FIG. 3 Effects of rain and fog.

1979; Kobayashi, 1980). Of particular interest are effects near the attenuation minima at 35, 94, 140, and 220 GHz.

Atmospheric turbulence effects due to time-varying localized temperature and humidity variations in clear air produce amplitude scintillations and angle-of-arrival changes, and these are now being investigated in more detail (Andreyev *et al.*, 1977; McMillan *et al.*, 1979a and b). Amplitude fluctuations of several decibels can occur on near-earth paths of moderate length (say, one or a few kilometers) in clear air. The fluctuation rate is slow (spectral width of perhaps a few cycles or tens of cycles per second), so automatic gain control can compensate for this in some applications. The angle of arrival effects may not be so easy to compensate for, however. Since millimeter-wave systems may sometimes be designed to provide very good angular resolution, atmospheric angle-of-arrival fluctuations may contribute unwanted errors for some cases. Peak-to-peak fluctuations up to $\frac{1}{2}$ mrad may occur for clear-air path lengths of one to a few kilometers.

The effects of smoke and/or dust on propagation are also being measured. An example is a test (DIRT-I) conducted in September 1978 at White Sands Proving Ground by the Army Atmospheric Sciences Laboratory. Propagation through a dust and oil smoke cloud was measured at 94 and 140 GHz and at infrared wavelengths (Lindberg, 1979).

In another test near Lake Havasu City, Arizona, the transmission through, and reflection from, a large dust cloud produced by 720 tons of high explosive were measured at several millimeter wavelengths (Martin, 1979). The results of such tests now provide systems engineers with specific information for design purposes. In other programs the reflectivities and radiometric signatures of land and sea (i.e., clutter), as well as vehicles and other targets, are also being obtained at window frequencies (Trebits *et al.*, 1978; Kulpa and Brown, 1979).

II. New Technology

The strong revival during the past several years of research and applications in the millimeter-wave region is attributed to the advent of new technology, the evolution of new requirements for sensors and communication links, and the superiority of millimeter-wave systems over optical and infrared systems for penetration of smoke, fog, haze, dust, and other adverse environments. The improved technology includes better sources (such as IMPATTs, Gunn oscillators, gyrotrons, extended interaction oscillators, and magnetrons), which have higher-power outputs and/or operate at higher frequencies, and in some cases have longer lifetimes than earlier designs. Lower-noise mixers have also been developed, providing noise temperatures below 500 K (uncooled) or 100 K (cooled) near 100 GHz. Component development has progressed, too, particularly in the areas of integrated circuits, image lines, fin line waveguides, and quasi-optical techniques. Nonetheless, there is room for improvement in components and devices, particularly in extending ferrite or other nonlinear devices to frequencies above 100 GHz and in reducing losses in most components. For a variety of reasons the frequency range from 35 to 100 GHz (with some special examples at 140 GHz) has seen the heaviest system development, while the range above 100 GHz (and on into the submillimeter range) is seeing a concentration of research effort on components and techniques.

A. Sources

In the area of sources, both solid-state and vacuum-tube types are available and in use in various ways (details are given in Chapter 4 of this volume). For low-power applications (watts of peak power, tens of milliwatts average), solid-state IMPATTs have seen extensive development and now operate to nearly 300 GHz. Gunn oscillators are available to 100 GHz, klystrons to 200 GHz, and carcinotrons to over 1000 GHz. In medium-power tubes, the magnetron has been improved in lifetime and is available to 95 GHz, and the extended interaction oscillator (EIO) has appeared and is in use up to 280 GHz. A breakthrough in extremely high power has been obtained with the gyrotron tube and related devices, which have already achieved megawatts peak at efficiencies greater than 30%, and hundreds of kilowatts cw at efficiencies near 50%. The gyrotron has seen application in fusion research and undoubtedly will soon see use in communications and radar. Optically pumped lasers can also be used to produce significant power at specific millimeter and submillimeter frequencies, but efficiencies and pulse repetition rates are low (Gallagher *et al.*, 1977; DeTemple, 1979).

The growth in millimeter-wave applications is certainly attributable in part to the availability of improved solid-state sources. While such sources provide relatively low power, they are smaller in overall size and require much lower voltages and prime powers than vacuum-tube types, such as magnetrons, EIOs, and klystrons. The most obvious improvements in solid-state sources during the past several years have been the availability of higher outputs and operation at much higher frequencies. Some of the other improvements are more subtle, but very important for systems; these include development of frequency stabilized or phase-locked sources, injection-locked IMPATT amplifiers, and frequency-doubled Gunn oscillators. (As an aside, frequency stabilization and phase-locking are also being developed for pulsed EIOs.) These types of improvements are permitting the extension of all-solid-state pulse-compression and coherent MTI (moving target indication) radars to as high as 94 GHz (Bernues *et al.*, 1979; Simonutti *et al.*, 1980). Similarly, all-solid-state receivers are being extended to 200 GHz.

IMPATT oscillators are available at frequencies from 3 to above 230 GHz. Typical efficiencies are 5% near 40 GHz and less than 1% above 100 GHz. Continuous-wave power output from single oscillators ranges from about 500 mW at 40 GHz to 10 mW at 230 GHz, and powers greater than a watt at 60 GHz have been obtained by combining the outputs of several diodes. At 94 GHz peak-pulsed power output of 13 W has been obtained from single diodes and 40 W from a four-diode combiner with 6% efficiency (Chang and Ebert, 1980). IMPATT devices have also been used as amplifiers (Ma *et al.*, 1980). Although inherently noisier than Gunn diodes because of their avalanche carrier generation mechanism, IMPATTs may be improved by phase or injection-locking (Kuno and Fong, 1979; Midford and Bernick, 1979; Fong and Kuno, 1979; Fong *et al.*, 1980). A very thorough review of IMPATT technology has been given by Kuno (1979) in Volume 1 of this treatise.

Gunn oscillators are available at frequencies from the microwave region to as high as 100 GHz. Although lower in power than IMPATTs, Gunn devices are very useful because of their clean spectral characteristics. As a result, they find greatest use as receiver local oscillators and laboratory signal sources. Efficiencies for gallium-arsenide (GaAs) diodes range from 3% at 40 GHz to less than 1% at 100 GHz. Continuous-wave power outputs vary from 200 mW at 40 GHz to 20 mW at 100 GHz. Above 100 GHz, because of material properties ("mobility cutoff" for GaAs), the power output dramatically drops. To produce higher powers and efficiencies and to provide useful power at frequencies above 100 GHz, research has been conducted on indium phosphide Gunn diodes (Fank *et al.*, 1979). Alternatively, varactor multipliers can be used to obtain signals at these higher frequencies. Typical commercially available doublers give 35% efficiency with a 45-mW input at 70 GHz.

Over the years more conventional vacuum tube sources such as klystrons and magnetrons have seen gradual improvement. The magnetron in particular has at least two limitations: the maximum frequency is around 100 GHz (where peak power output is about 1 kW), and the lifetime is often less than 100 to 200 hr. These limitations have led to the use of the extended interaction oscillator, with a frequency coverage from 30 to 280 GHz (EIO, 1975). Because the electron beam is not collected by the frequency-determining structure, the lifetime is predicted to be an order-of-magnitude longer than for the magnetron. Pulsed-power output is approximately 5 kW at 35 GHz, 1 kW at 95 GHz, and 40 W at 280 GHz. In addition, cw tubes are available that give hundreds of watts at K_a-band, tens of watts near 100 GHz, and more than 10 W at 140 GHz. Extended interaction amplifiers (EIAs) are also under development.

Backward-wave oscillators (BWOs) or Carcinotrons† are also in current use, and these tubes extend in frequency coverage from the middle-microwave- through the millimeter-wave- and well into the submillimeter-wave range (Kantorowicz *et al.*, 1979; Kantorowicz and Palluel, 1979). In the Soviet Union frequency coverage has been extended to beyond 1100 GHz. These tubes, normally in the tens or hundreds of milliwatts power regime, have the particular advantage of wide tuning range (sometimes a full waveguide band) and are useful as general signal sources.

The gyrotron has received extensive publicity recently because of its capability for producing extremely high pulse or cw powers. Research on this tube has been under way for many years, with particular emphasis in the Soviet Union for the past two decades and more recently in the United States (Jory *et al.*, 1978; Chu *et al.*, 1979; Zaytsev *et al.*, 1974). Detailed information on this and related relativistic electron beam tubes is given in Volume 1 of this treatise (Hirshfield, 1979). Briefly, gyrotrons have been operated throughout the entire millimeter-wave range, producing megawatts of peak or tens to hundreds of kilowatts cw power, usually with extremely good efficiencies (>30%) (Chu *et al.*, 1980). Specified tubes are now available commercially. In general, gyrotrons require very high voltages (60–100 kV) and large magnetic fields, which often means that superconducting magnets are needed. Thus the packaging and power supply problems are severe. Another problem occurs because the output is usually produced in higher-order waveguide modes.

B. Mixers

In the mixer area extensive investigations have been conducted in both the microwave structures (Kerr, 1979a,b; Meier, 1979; Swanberg and Paul, 1979; Cardiasmenos, 1979, 1980; Kawasaki and Yamamoto, 1979; Keen *et al.*, 1979) and the diode materials (Schneider, 1978; Keen, 1978; Smirnova and Cherpak, 1975; Mirovskiy *et al.*, 1975; Strukov *et al.*, 1975). In a decade of effort, mixer noise temperatures have been lowered by an order of magnitude. Figure 4 shows typical measured mixer noise figure performance for frequencies from 30 to 140 GHz (Cardiasmenos, 1979). Another innovation has been the use of subharmonic mixing,

† Carcinotron is the trade name for BWOs manufactured by the Electron Tube Division of Thomson-CSF.

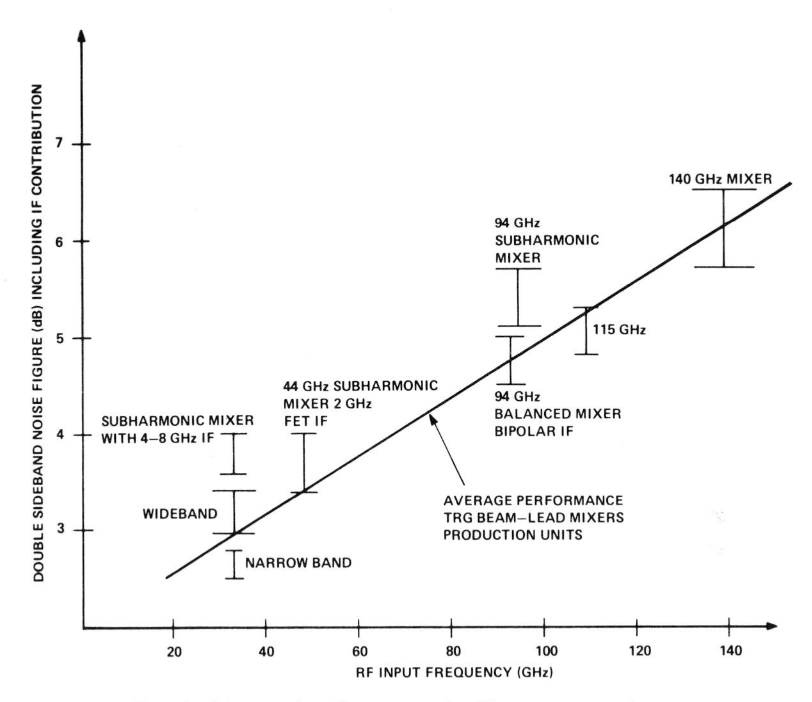

FIG. 4 Measured performance of millimeter-wave mixers.

wherein employing two back-to-back diodes enables one to use a local oscillator frequency of $\frac{1}{2}$ or $\frac{1}{4}$ the normal fundamental local oscillator with excellent performance. This permits building all-solid-state receivers and radiometers at frequencies much higher (2–4 times) than those available from solid-state sources. Of course, another approach is to use harmonic generators as sources for fundamental mixing (Calviello, 1979; Takada and Ohmori, 1979). The various types of diodes investigated in recent years have included superconducting tunnel junctions, point-contact Josephson junctions, Schottky devices, Mott diodes (a low-doped Schottky with an epitaxial layer that is completely depleted at zero bias), and Mottky diodes (Schottky barrier diode with a thin epitaxial layer, approximating a Mott). An additional advantage for some of these mixers is a lower requirement for local oscillator power (as low as the order of 0.1 mW). Some of the best research has been done for radio astronomy applications, but the results carry over to other fields. Important considerations for some of these mixers are the needs for being ruggedized and being made more producible. Nonetheless, the improvements have led to better receivers for radar, communications, and radiometry (passive sensing and radio astronomy).

C. Components and Devices

In addition to the work on sources and mixers mentioned above, the literature on component and device development is extensive (Kramer, 1978; Fank *et al.*, 1979; Calviello, 1979; Valkenburg and Khandelwal, 1979). In general, standard rectangular waveguide components such as couplers, tees, tuners, and hybrid rings are available to 220 GHz. Calibrated attenuators, phase shifters, and wavemeters are available to about 110 GHz (direct reading) and to 220 GHz with calibration curves. Insertion loss for all these components is undesirably high at the upper frequencies. Ferrite devices such as isolators, switches, circulators, and phase shifters are available to 140 GHz with isolations of 15–20 dB; extension to higher frequencies is very severely limited by small sizes and losses in ferrite materials.

Limitations in performance and difficulties in manufacture of rectangular waveguide components has spurred the development of alternate waveguide structures (such as dielectric image line, insular guide, H-guide, and fin line) (Deo and Mittra, 1979; Cohen and Meier, 1978) and the use of quasi-optical components such as attenuators, diplexers, beam waveguides, duplexers, interferometers, and polarization rotators (Rainwater and McMillan, 1980; Sobel *et al.*, 1961). In addition, microstrip techniques have been extended to as high as 140 GHz (Oxley, 1979). Extension of integrated circuit techniques to millimeter wavelengths was the subject of a special issue of the IEEE Transactions (Special Issue, 1978; Mittra and Bhooshan, 1979).

D. Antennas

Antenna technology has been extended upward from the microwave region and downward from the optical region; thus many millimeter-wave designs have familiar counterparts (Chu *et al.*, 1978; Kay, 1966; Kott, 1964; Tolbert *et al.*, 1965). Some do not, however, being based on techniques such as radiation from surface waveguides (Klohn *et al.*, 1978; Shiau, 1976) or the use of phase-correcting Fresnel zone plates instead of lenses (Sobel *et al.*, 1961). Monopulse designs in either rectangular waveguide or dielectric image line have been used up to 95 GHz (Waineo and Koniexzny, 1979; Cahill *et al.*, 1973). A folded geodesic Luneberg lens antenna has been constructed for operation at 70 GHz (Goodman *et al.*, 1967). Other antennas have been developed to provide scan capability (Kesler *et al.*, 1979; Kay, 1966) and/or special pattern coverage (Ore, 1972).

Two areas that have seen significant advancement are the development of large, precision surface antennas (>15-ft diameter) for radio astronomy applications (Chu *et al.*, 1978; Tolbert *et al.*, 1965) and of small antennas for tactical missile-seeker terminal guidance (Britt, 1977; Houseman, 1979; Waineo and Koniexzny, 1979). In the latter case, the electrical design often requires wide bandwidth, monopulse or conical scan plus mechanical scan for wide field of view coverage, and low sidelobes; and all of this must be built to stand an adverse environment with severe vibration, acceleration, and temperature conditions.

III. System Applications

A. Missile Guidance

Guidance for missiles and projectiles is one of the areas of greatest activity in millimeter waves. Present-day precision-guided weapons are often based on the use of laser or TV seekers, which are not satisfactory in adverse weather or nighttime use. In general, a high-accuracy tracking or weapon-guidance capability is needed that will give good performance in smoke, fog, dust, or rain, either day or night.

Particular emphasis is being placed on terminal guidance of tactical air-to-surface missiles (Seashore *et al.*, 1979; Sundaram, 1979; Williams and Boykin, 1979; Wiltse, 1978, 1979a). The apertures available for antennas in such missiles are in the 4–8-in. range. In order to obtain the desired narrow beamwidths from these small antennas, frequencies between 30 and 140 GHz are being used, with the preference usually for 35 or 94 GHz. The 45–65-GHz regime has such high atmospheric attenuation, even in clear weather (oxygen absorption), that it is normally not chosen for seekers.

Extensive design and measurement programs have been carried out to develop "terminal seekers" optimized for various military scenarios. Consistent with the need for small size, low voltage, and eventual low cost, all solid-state design has been emphasized. Since the range needed is small, typically one to several kilometers, low transmitter power is satisfactory. Both active radar and passive radiometer sensors have been investigated (Seashore *et al.*, 1979). Combined active–passive systems have also been studied. The active portion gives greater initial acquisition range but suffers from aimpoint wander due to target glint in the terminal phase, whereas the passive sensor does not have this problem.

Other semiactive systems have also been investigated (with obvious analogy to laser seekers and designators) and, in addition, beam rider guidance is under development for short-range point-to-point missiles, such as those used for antitank weapons. For these cases higher transmitter powers are required, which generally rules out solid-state sources.

B. Radar

Considering radars other than those used in seekers, several types have been worked on. These include radar systems for short-range high-resolution air defense, helicopter-borne target acquisition, airborne imaging, tank fire control, or instrumentation purposes (Currie *et al.*, 1978). The latter have seen considerable use for measuring target and terrain (or sea) clutter signatures, often with simultaneous dual polarization and sometimes dual frequencies (such as 35 and 94 GHz).

An example of a recently developed radar is the U.S. Army tank fire control system called STARTLE (an acronym for Surveillance and Target Acquisition Radar for Tank Location and Engagement). Mounted on tanks, it provides the capability to detect and track targets such as other tanks or armored vehicles in both clear and adverse environmental conditions such as fog, smoke, dust, or rain (Backus, 1979; Balcerak *et al.*, 1977; Fawcette, 1978; Sundaram, 1979; Wiltse, 1978, 1979a,b). Two competing contractors have built all-solid-state models that meet the system parameters given in Table II (Balcerak *et al.*, 1977). One model employs a dual-mode radar that uses coherent MTI (moving target indication) to detect radially moving targets and an area MTI to acquire stationary, creeping, and tangentially moving targets. For the area MTI a wideband, spread-spectrum transmitted waveform provides smoothing of glint, clutter fluctuation, and multipath effects (Sundaram, 1979).

Other radar applications include scale modeling (e.g., measuring radar returns from scale models of aircraft, missiles, or tanks with millimeter or submillimeter waves to simulate returns at microwave frequencies) (Cram and Woolcock, 1978; Gabsdil and Jacobi, 1978), Coast Guard obstacle avoidance (Bearse, 1975), auto collision avoidance (Moncrief, 1978; Wollins, 1978; Zur Heiden and Oehlen, 1977), and missile fuzing.

C. Radiometry, Remote Sensing, and Radio Astronomy

As indicated earlier, receiver technology has improved greatly in recent years. Noise figures are lower, in some cases because of the use of cooled mixers, and operating frequencies are being pushed higher. Very wide bandwidth receiver front ends have been achieved, and of course the combination of wider bandwidths and lower noise figures results in significantly improved sensitivity for detection of thermal-type or continuum radiation.

As mentioned before, radiometric receivers have been used in missile terminal guidance. Instrumentation radiometers have been used for measuring target and terrain signatures, often at simultaneous dual polarizations and on occasion at dual frequencies (e.g., 35 and 94 GHz). Other ra-

TABLE II

PARAMETERS FOR STARTLE RADAR SYSTEMS[a]

Frequency: 94 GHz (3.2 mm)
Beamwidth: 11 mrad
Average power: 0.1–0.5 W
Antenna aperture: 14 in.
Field of view: 15° × 7.5° wide; 5° × 2.5° narrow
Frame time: <2 sec
Target detection: 3000 m (100-m visibility)
Target tracking: 0.5 mrad at 2 km (100-m visibility)

[a] Balcerak *et al.* (1977).

diometers are in use by NASA for remote sensing (Rainwater, 1978; Schuchardt and Stratigos, 1978; Schuchardt *et al.*, 1979; Blue, 1979) and by the U.S. Navy for navigation and imaging (Hollinger *et al.*, 1976). The NASA Nimbus 6 satellite carried five superheterodyne radiometers with center frequencies between 22 and 60 GHz. Temperature profiles were successfully measured for a variety of geographical locations and atmospheric and climatic conditions. Recently, radiometer receivers centered at frequencies between 90 and 183 GHz were flown operationally in a Convair 990 aircraft. At 183 GHz, which is the frequency of an atmospheric water vapor transition, the radiometer provides atmospheric temperature data related to water content, while a 93-GHz channel obtains surface or lower atmosphere temperature information. A dual-channel (93/183 GHz) system has been operated at high altitudes over storms in an RB-57 aircraft. Typical state-of-the-art minimum detectable temperature sensitivities (rms noise temperature fluctuations) are in the 0.1°–0.4°-K range (1 sec integration time) between 35 and 100 GHz for ambient temperature systems. With the use of cryogenic cooling, this has been improved to 0.01° K in a 90-GHz receiver for airborne radiometry (Vowinkel *et al.*, 1980).

Other radiometer applications include thermography or biological imaging (Edrich, 1979; Gautherie *et al.*, 1979; Robert *et al.*, 1979) and a heavy activity in the field of radio astronomy (Cong *et al.*, 1979; Kislyakov, 1971; Penzias and Burrus, 1973; Ulich, 1977; Ulich *et al.*, 1980). Research in radio astronomy has been helped by the availability in recent years of large, accurate antennas, good to frequencies above 140 GHz. Examples include the 45-ft diameter antenna at the University of Massachusetts (near Amherst) and the 36-ft diameter antenna of the National Radio Astronomy Observatory at Kitt Peak, Arizona. Another is the new facility at Pico Veleta, near Granada, Spain, established in a joint effort by German and French organizations.

Radio astronomy efforts have included measurements of the brightness temperatures of the sun, moon, and planets, and the opening up of a completely new field of molecular spectroscopy of interstellar clouds (several dozen different molecules have been detected). For their research in the measurement of cosmic background radiation at microwave and millimeter-wave frequencies and their interpretation of the results in terms of cosmology and the "big bang" theory of the creation of the universe, Dr. A. A. Penzias and Dr. R. W. Wilson of Bell Laboratories received the 1978 Nobel Prize in physics.

D. Communications

1. *Terrestrial*

Terrestrial systems generally fit into one of two categories: short-range point-to-point links where small antenna size and/or narrow beamwidth are desired, or enclosed, low-loss, circular waveguide for use in wide-band, long-haul, heavy-route communication systems. The point-to-point (or ship-to-ship) systems further divide into cases where the carrier frequency is chosen to be in a range of low atmospheric attenuation (e.g., the 30–40 GHz window) to minimize losses or in a high-attenuation region (e.g., near 60 GHz) for privacy or security reasons (Chang *et al.*, 1978; Chang and Yuan, 1980; Davis, 1974; Dudzinsky, 1975; Keitzer *et al.*, 1976; Matsuo *et al.*, 1976).

A straightforward system that has been developed for point-to-point communications is a short-range (5–20 miles) duplex unit operating at 38 GHz (Hughes, 1976). A 35-GHz railway communication link has also been developed (Meinel *et al.*, 1979). Advantages inherently available from millimeter-wave systems include wide bandwidth (possibility of very high data rates or frequency hopping) as well as link privacy and reduced interference by virtue of the narrow beamwidths and low density of other radios and radars.

When a frequency is chosen at which atmospheric attenuation is high, signal overshoot is low and covert operation is possible. Of course, for a given range more system margin is needed, which implies higher power, more gain, and/or better receiver sensitivity to compensate for the attenuation. The combination of narrow beam radiation and high attenuation can provide reliable communications that are significantly free from both signal interception and jamming. Several types of portable, solid-state, battery-powered units have been built. Chang and Yuan (1980) have described a 70-GHz binocular radio in which one half of the binocular remains as a monocular, while the other half is taken up by a radio transceiver suitable for voice communication. Range is said to be 7 km in clear weather and 3 km in 4 mm/hr rain, and the battery will provide continuous operation for 9 hr. Dielectric waveguide millimeter-wave integrated circuits are utilized in this and other designs (Chang *et al.*, 1978; Keitzer *et al.*, 1976). Several systems have been devised that emphasize low cost or civilian applications (Matsuo *et al.*, 1976; Becker, 1979).

Research and development on low-loss hollow metal waveguide as a wide-band communications medium have been conducted for more than two decades. The so-called circular-electric mode (TE_{01}) has been shown to provide extremely low loss (about 2 dB/km) over the frequency range from 33 to 125 GHz (Steier, 1965). Since an inert gas like nitrogen can be used in the tube, oxygen absorption is not a problem. Complete systems have been developed and installed in the United States, Japan, and England. A 14-km long test route has been operated by the Bell System in New Jersey, and a 22.7-km length of experimental guide has been operated in Japan (Abele *et al.*, 1975; Bernardi *et al.*, 1974; Mahieu *et al.*, 1974; Millimeter, 1977; Miyauchi *et al.*, 1975; Miyauchi, 1976; Warters, 1975; White *et al.*, 1974). A special issue of the Bell System Technical Journal was used to describe the 14-km test system. Capable of handling 475,000 two-way voice circuits, high-speed (274 Mbits/sec) digital bit streams can carry voice, data, or video communications at a bit error rate designed to be 10^{-7} or better on a coast-to-coast circuit. For a fully loaded system the cost per circuit mile would be significantly below that of any long haul system existing in 1977 [see Warters (1977), introduction to special issue on the WT4 millimeter waveguide system].

2. *Satellite*

An area with a surprising amount of technical activity in recent years has been the consideration of millimeter-wave satellite links. NASA, the Department of Defense, and the Armed Services have supported numerous analyses and hardware investigations.

Early satellite communications experiments between 30 and 33 GHz were carried out via the NASA Applications Technology Satellites (ATS-5 and ATS-6) and ground stations (Ippolito, 1971; King *et al.*, 1968). Satellite-to-satellite relay links were also designed (Dees *et al.*, 1970), and most of these space links would have taken advantage of the very large atmospheric attenuation near 60 GHz to provide intercept isolation from ground receivers. Somewhat later a satellite-to-satellite link operating at 37 GHz was built for Lincoln Laboratories and successfully tested in orbit (Snider and Coomber, 1978). Identified as Lincoln Experimental Satellites (LES-8 and LES-9), they were launched into geosynchronous orbit in

March 1976 and have been operating for several years. Each satellite has a steerable antenna beam of about 1.2° beamwidth and a solid-state IMPATT amplifier that provides $\frac{1}{2}$ W output. This permits operation of a 100-kbs (kilobits per second) link at a satellite separation of 40,000 km.

More recently a variety of satellite links has been considered, partly for reasons of spectrum congestion, cost benefits, and/or defense needs (Castro and Healy, 1975; Eaves, 1979; Feldman and Dudzinsky, 1977; Holland *et al.*, 1980; Kaul *et al.*, 1980; Mundie and Feldman, 1978; Ricardi, 1979; Tsao *et al.*, 1975; Waylan and Yowell, 1979). It seems highly probable that further satellite communications activity will evolve in the near future.

E. Electronic Warfare

Currently in the United States and other countries major radar, seeker, and communications millimeter technology programs are under way that are designed to provide unique solutions to critical operational problems. Important improvements in sources and components have occurred in the past few years. For example, both extremely compact and light-weight solid-state sources and extremely high-power sources have been developed, leading to the possibility of novel operational concepts that may be uniquely difficult to counter with conventional ECM approaches. On the other hand, advances in components and other technology such as low-noise receivers, together with new signal processing concepts, can provide significant improvements in wide-beam reconnaissance–surveillance–warning receivers.

Several wideband millimeter-wave receivers have been described at the IEEE Microwave Theory and Techniques Symposia (e.g., Hislop, 1979) and in the literature (Hartman, 1979; Rubin and Saul, 1980). Hislop's design covers 88–100 GHz, while others cover variously 26–42 GHz or 18–40 GHz. A unit has been developed by the U.S. Electronics Research and Development Command that covers 18–66 GHz in 12 separate 4-GHz bands.

The narrow mainbeam and low sidelobes of a good millimeter antenna make intercept and jamming difficult. Also the relatively high atmospheric losses at certain millimeter frequencies make some systems intercept-resistant and limit the use of standoff surveillance and/or jamming. Currently proven ECCM techniques such as pulse compression, spread spectrum, and frequency agility can make millimeter-wave systems even more resistant to countermeasures.

Nonetheless both passive and active countermeasures can be effective. Radar camouflage (mesh, pads, blankets) can be used to cover targets such as tanks, making them less vulnerable to active or passive seekers.

Passive decoys, such as corner reflectors, can be dispensed or distributed in the vicinity of a real target, with the intention of misguiding a precision-guided missile or projectile. A target that looks "cold" (relative to the background) when viewed by a radiometer system can be made to look warm by providing a low-power emitter to radiate a cw signal in the radiometer band, which probably is very wide.

The antenna and atmosphere characteristics mentioned above make active jamming difficult, especially since the jammer probably will need an antenna with moderate or small beamwidth, which brings forth requirements for pointing and tracking (and possibly stabilizing) the jammer antenna. It is interesting to speculate, however, what might be done with the enormous powers available from gyrotrons, if the power could be concentrated by a high-gain antenna. A 1-m antenna gives about 57 dB of gain near 100 GHz. Ignoring unsolved engineering problems and possible atmospheric effects, it is nonetheless simple to *calculate* that destructive jamming may be possible. Receiver front-end mixers might be damaged at ranges of kilometers.

F. Biological Applications

The technique of biological imaging or thermography has already been mentioned. A considerable amount of research has been done for several years, with some possibilities for successful use of millimeter imaging for detection of cancerous tumors (notably breast tumors) in humans (Devyatkov, 1974; Edrich, 1979; Fröhlich, 1978; Gautherie *et al.*, 1979; Robert *et al.*, 1979).

Various investigations have also been conducted to determine the absorptive or emissive characteristics of biological material. These include the permittivity and penetration depth of muscle and fat tissues (Edrich and Hardee, 1976); millimeter waves, Raman spectra, and bioeffects (Jaggard and Lords, 1980); the optical constants of liquid H_2O and D_2O; millimeter-wave absorption in biological systems (Hershberger, 1978), and other influences of the radiation on biological objects.

References

Abele, T. E., Alsberg, D. A., and Huchison, P. T. (1975). *IEEE Trans. Microwave Theory Tech.* **23,** 326–333.

Andreyev, G. A., Gulonov, V. A., Ismailov, A. T., Parshikov, A. A., Rozanov, B. A., and Tanyigin, A. A. (1977). *Jt. Anglo-Sov. Semin. Atmos. Propag. Millimetre Submillimetre Wavelengths, Inst. Radioeng. Electron., Moscow* pp. R-1–R-8.

Backus, P. H. (1979). *J. Electron. Def.* **2**(Mar./Apr.), 24–35.

Balcerak, R., Ealy, W., Martino, J., and Hall, J. (1977). *Soc. Photo-Opt. Instrum. Eng. (SPIE) Symp. Eff. Util. of Opt. Radar Syst., Huntsville, Ala.* pp. 172–184.

Bearse, S. V. (1975). *Microwaves* **14**(Sept.), 10.
Becker, S. (1979). *DARPA/Tri-Serv. Millimeter Wave Conf., 8th, Eglin AFB, Fl.* pp. 267–276.
Bernardi, P., Corazza, G. C., Faliciasecca, G., Koch, R., Magni, G., Mastellari, G. A., Stracca, G. B., and Vardoni, F. (1974). *Proc. Eur. Microwave Conf.* pp. 619–623.
Bernues, F. J., Ying, R. S., and Kaswen, M. (1979). *Microwave Syst. News* **9**(May), 79–86.
Blue, M. D. (1979). *IEEE Int. Microwave Symp. Dig.* pp. 545–546.
Britt, P. (1977). *Abstr. USAF Antenna Symp., 23rd, Robert Allerton Park, Ill.* pp. 27–29.
Cahill, T. C., Gill, G., and Syrogos, H. (1973). *Microwave J.* **16**(Nov.), 53–58.
Calviello, J. A. (1979). *Microwave J.* **22**(Aug.), 53–56, 85.
Cardiasmenos, A. G. (1979). *Microwave Syst. News* **9**(May), 46–56.
Cardiasmenos, A. G. (1980). *Microwave Syst. News* **10**(Aug.), 37–51.
Castro, A., and Healy, J. (1975). *Int. Conf. Commun., 11th* pp. 36-9–36-13.
Chang, K., and Ebert, R. L. (1980). *IEEE Int. Solid-State Circuits Conf. Dig.* pp. 126–127.
Chang, Y., and Yuan, L. T. (1980). *Microwave J.* **23**, 31–36.
Chang, Y., Paul, J. A., and Ngan, Y. C. (1978). Final Rep. No. DELET-TR-76-1353-F, Contract DAAB07-76-C-1353 with ERADCOM. Hughes Aircraft, Torrance, California.
Chen, C. C. (1975). Rand Corp. Rep. R-1694-PR (AD A011-642.), Santa Monica, California.
Chu, K. R., Drobot, A. T., Granatstein, V. L., and Seftor, J. L. (1979). *IEEE Trans. Microwave Theory Tech.* **27**, 178–187.
Chu, K. R., Read, M. E., and Ganguly, A. K. (1980). *IEEE Trans. Microwave Theory Tech.* **27**, 318-325.
Chu, T. S., Wilson, R. W., England, R. W., Gray, D. A., and Legg, W. E. (1978). *Bell Syst. Tech. J.* **57**, 1257–1288.
Cohen, L. D., and Meier, P. J. (1978). *Microwave J.* **21**(Aug.), 63–65.
Cong, H., Kerr, A. R., and Mattauch, R. J. (1979). *IEEE Trans. Microwave Theory Tech.* **27**, 245–248.
Cram, L. A., and Woolcock, S. C. (1978). *Symp. Millimeter Submillimeter Wave Propag. Circuits; AGARD Conf. Proc.* No. 245, pp. 6/1–6/5.
Currie, N. C., Scheer, J. A., and Holm, W. A. (1978). *Microwave J.* **21**(Aug.), 35–42.
Davis, R. T. (1974). *Microwaves* **13**(Oct.), 9, 14.
Dees, J. W., Kefalas, G. P., and Wiltse, J. C. (1970). *Proc. IEEE Int. Conf. Commun. San Francisco, Calif.* pp. 22-20–22-26.
Deo, N. C., and Mittra, R. (1979). *Microwaves* **18** (Oct.), 38–42.
DeTemple, T. A. (1979). *In* "Infrared and Millimeter Waves" (K. J. Button, ed.), Vol. 1, Chap. 3. Academic Press, New York.
Devyatkov, N. D. (1974). *Sov. Phys.*—Uspekhi **16, 568**–579.
Dudzinsky, S. J., Jr. (1975). *Microwave J.* **18**(Dec.), 39.
Dyer, F. B. (1980). *Microwave J.* **23**(September), 16–17.
Dyer, F. B., and Currie, N. C. (1978). *Symp. Millimeter and Submillimeter Wave Propag. Circuits AGARD Conf. Proc.* No. 245, pp. 2/1–2/9.
Eaves, R. E. (1979). *IEEE EASCON Proc.* pp. 630–636.
Edrich, J. (1979). *J. Microwave Power* **14**(June), 95–103.
Edrich, J., and Hardee, P. C. (1976). *IEEE Trans. Microwave Theory Tech.* **24**, 273–275.
EIO (1975). Data Sheet No. 3445 5M. Varian Assoc. Can., Georgetown, Ontario, Canada.
Fank, F. B., Crawley, J. D., and Bernz, J. J. (1979). *Microwave J.* **22**(June), 86–91.
Fawcette, J. (1978). *Microwave Syst. News* **8**(July), 23–24.
Feldman, N. E., and Dudzinsky, S. J., Jr. (1977). Rep. R-1936-RC. Rand Corp., Santa Monica, California.
Fong, T. T., and Kuno, H. J. (1979). *IEEE Trans. Microwave Theory Tech.* **27**, 492–499.

Fong, T. T., Midford, T. A., Nakaji, E. M., and Ngan, Y. C. (1980). Contract Rep. No. ARBRL-CR-00414. Hughes Aircraft, Torrance, California.
Fröhlich, H. (1978). *IEEE Trans. Microwave Theory Tech.* **26,** 613–617.
Gabsdil, W., and Jacobi, W. (1978). *Symp. Millimeter Submillimeter Wave Propag. Circuits; AGARD Conf. Proc.* No. 245, pp. 7/1–7/8, September.
Gallagher, J. J., Blue, M. D., Bean, B., and Perkowitz, S. (1977). *Infrared Phys.* **17,** 43–55.
Gautherie, M., Edrich, J., Zimmer, R., Guerguin-Kern, J. L., and Robert, J. (1979). *J. Microwave Power* **14**(June), 123–129.
Goodman, R. M., Jr., Johnson, R. C., Ecker, H. A., and Rivers, W. K., Jr. (1967). *Abstr. Annu. Symp. USAF Antenna Res. Dev., 17th, Robert Allerton Park, Ill.*
Hartman, R. (1979). *J. Def. Electron.* **2**(May), 79–86; see also *Microwave Syst. News* **9**(May), 32 (1979).
Hershberger, W. D. (1978). *IEEE Trans. Microwave Theory Tech.* **26**(Aug.), 618–619.
Hirshfield, J. L. (1979). *In* "Infrared and Millimeter Waves" (K. J. Button, ed.), Vol. 1, Chap. 1. Academic Press, New York.
Hislop, A. (1979). *IEEE MTT-S Int. Microwave Symp. Dig., Orlando, Fl.* pp. 222–223.
Holland, L. D., Hilsen, N. B., Gallagher, J. J., and Stevens, G. (1980). *Microwave J.* **23**(June), 35–43.
Hollinger, J. P., Kinney, J. E., and Troy, B. E., Jr. (1976). *IEEE Trans. Microwave Theory Tech.* **24,** 786–793.
Houseman, E. O., Jr. (1979). *1979 Int. IEEE/AP-S Symp. Dig., Univ. Maryland, College Park* pp. 51–54.
Hughes Unveils New 38 GHz Radio (1976). *Microwave Syst. News* **6**(Feb./Mar.), 13.
IEEE (1976). Standard STD-521-1976, Nov. 30.
Ippolito, L. J. (1971). *Proc. IEEE* **59,** 189–205.
Jaggard, D. L., and Lords, J. L. (1980). *Proc. IEEE* **68,** 114–119.
Jory, H., Heggi, S., Shively, J., and Symons, R. (1978). *Microwave J.* **21**(Aug.), 30–32.
Kantorowicz, G., and Palluel, P. (1979). *In* "Infrared and Millimeter Waves" (K. J. Button, ed.), Vol. 1, Chap. 4. Academic Press, New York.
Kantorowicz, G., Palluel, P., and Pontivianne, J. (1979). *Microwave J.* **22**(Feb.), 57–59.
Kaul, R., Wallace, R., and Kinal, G. (1980). Rep. ORI TR-1679 for NASA. ORI, Silver Spring, Maryland.
Kawasaki, R., and Yamamoto, K. (1979). *IEEE Trans. Microwave Theory Tech.* **27,** 530–533.
Kay, A. F. (1966). *Proc. IEEE* **54,** 641–647.
Keen, N. J. (1978). *Symp. Millimeter Submillimeter Wave Propag. Circuits; AGARD Conf. Proc.* No. 245, pp. 16/1–16/9.
Keen, N. J., Kelly, W. M., and Wrixon, G. T. (1979). *Electron. Lett.* **15,** 689.
Keitzer, J. E., Kaurs, A. R., and Levin, B. J. (1976). *IEEE Trans. Microwave Theory Tech.* **24,** 797–803.
Kerr, A. R. (1979a). *IEEE Trans. Microwave Theory Tech.* **27,** 938–943.
Kerr, A. R. (1979b). *IEEE Trans. Microwave Theory Tech.* **27,** 944–950.
Kesler, O. B., Montgomery, W. F., and Liu, C. C. (1979). *Proc. 1979 Antenna Appl. Symp., Univ. Illinois, Urbana, Ill.*
King, J. L., Dees, J. W., and Wiltse, J. C. (1968). *IEEE Int. Conv. Rec.* p. 248.
Kislyakov, A. G. (1971). *Sov. Phys.—Uspekhi* **13,** 495–521.
Klohn, K. L., Horn, R. E., Jacobs, H., and Freibergs, E. (1978). *IEEE Trans. Microwave Theory Tech.* **26,** 764–773.
Kobayashi, H. K. (1980). Rep. ASL-TR-0049. U.S. Army Atmos. Sci. Lab., White Sands Missile Range, New Mexico.

Kott, M. A. (1964). *IEEE Trans. Antennas Propag.* **12,** 662–667.
Kramer, N. B. (1978). *Microwave J.* **21**(Aug.), 57–61.
Kulpa, K., and Brown, E. (1979). Rep. HDL-SR-79-8. Harry Diamond Lab., Adelphi, Maryland.
Kuno, H. J. (1979). *In* "Infrared and Millimeter Waves" (K. J. Button, ed.), Vol. 1, Chap. 2. Academic Press, New York.
Kuno, H. J., and Fong, T. T. (1979). *Microwave J.* **22**(June), 47–48, 73–75, 85.
Lindberg, J. D. (1979). Rep. ASL-TR-0021. U.S. Army Atmos. Sci. Lab., White Sands Missile Range, New Mexico.
Ma, Y., Sun, C., and Nakaji, E. M. (1980). *IEEE MTT-S Int. Microwave Symp. Dig.* p. 75.
McMillan, R. W., Rogers, R., Platt, R., Guillory, D., Gallagher, J. J., and Snider, D. E. (1979a). Rep. ASL-CR-79-0026-1. U.S. Army Atmos. Sci. Lab., White Sands Missile Range, New Mexico.
McMillan, R. W., Wiltse, J. C., and Snider, D. E. (1979b). *Proc. IEEE Symp. Aerosp. Electron. Syst., Washington, D.C.* pp. 42–47.
Mahieu, J. R., Marchalot, J. N., Boujet, J. P., Coz, G. I., Gibeau, P., and Bouvet, J. V. (1974). *Ann. Telecommun.* **29,** 443–464.
Martin, E. E. (1979). *Proc. U.S. Army MIRADCOM Workshop Millimeter Submillimeter Atmos. Propag. Appl. Radar Missile Syst., Huntsville, Ala.* pp. 109–113.
Matsuo, Y., Adaiwa, Y., and Takase, I. (1976). *IEEE Trans. Microwave Theory Tech.* **24,** 794–797.
Meier, P. J. (1979). *Microwave J.* **22**(Aug.), pp. 66–68.
Meinel, H., Plattner, A., and Breitschadel, R. (1979). 9th *Eur. Microwave Conf., Brighton, Eng.* pp. 259–262.
Midford, T. A., and Bernick, R. L. (1979). *IEEE Trans. Microwave Theory Tech.* **27,** 483–492.
Millimeter Waveguide Systems (1977). *Microwave J.* **20**(Mar.), 24–26.
Mirovskiy, V. G., Strukov, I. A., and Etkin, V. S. (1975). *Radiotekh. Elektron.* **20,** 796.
Mittra, R., and Bhooshan, S. (1979). *IEEE Int. Microwave Symp. Dig., Orlando, Fl.* pp. 211–213.
Miyauchi, K. (1976). *Microwave Syst. News* **6,** 91–97.
Miyauchi, K., Seki, S., Ishida, N., and Izumi, K. (1975). *Rev. Electr. Commun. Lab.* **23,** 707–741.
Moncrief, F. (1978). *Microwave Syst. News* **8**(Apr.), 23–26.
Mundie, L. G., and Feldman, N. E. (1978). Rep. R-2275-DCA. Rand Corp., Santa Monica, California.
Ore, F. R. (1972). *IEEE Trans. Antennas Propag.* **20,** 481–482.
Oxley, T. H. (1979). *Microwave Syst. News* **9**(Sept.), 75–83, 155.
Penzias, A. A., and Burrus, C. A. (1973). *Annu. Rev. Astron. Astrophys.* **11,** 51–72.
Persinger, R. R., Stutzman, W. L., Castle, R. E., and Bostian, C. W. (1980). *IEEE Trans. Antennas Propag.* **28,** 149.
Rainwater, J. H. (1978). *Microwaves* **17**(Sept.), 59–62.
Rainwater, J. H., and McMillan, R. W. (1980). *Microwaves* **19**(June), 76–82.
Ricardi, L. F. (1979). *IEEE EASCON Proc.* pp. 617–722.
Richard, V. W., and Kammerer, J. E. (1975). Rep. BRL-1838. U.S. Army Ballist. Res. Lab., Aberdeen, Maryland. (AD B008173L.)
Robert, J., Edrich, J., Thouvenot, P., Gautherie, M., and Escanyé, J. M. (1979). *J. Microwave Power* **14**(June), 131–134.
Rubin, D., and Saul, D. L. (1980). *Microwave J.* **23**(June), 55–57.
Sander, J. (1975). *IEEE Trans. Antennas Propag.* **23,** 213–220.

Schneider, M. V. (1978). *Microwave J.* **21**(Aug.), 78–83.
Schuchardt, J. M., and Stratigos, J. A. (1978). *Microwaves* **17**(Sept.), 64–74.
Schuchardt, J., Stratigos, J., Galiano, J., and Gallentine, D. (1979). *IEEE Int. Microwave Symp. Dig.* pp. 540–542.
Seashore, C. R., Miley, J. E., and Kearns, B. A. (1979). *Microwave J.* **22**(Aug.), pp. 47–51, 58.
Shiau, Y. (1976). *IEEE Trans. Microwave Theory Tech.* **24,** 869–872.
Simonutti, M. D., English, D. C., and Bernues, F. J. (1980). *IEEE MTT-S Int. Microwave Symp. Dig.* p. 73.
Smirnova, T. A., and Cherpak, N. T. (1975). *Radiotekh. Elektron.* **20,** 348.
Snider, D. M., and Coomber, D. B. (1978). *Proc. AIAA Commun. Conf., 7th, San Diego, Calif.* pp. 457–470.
Sobel, F., Wentworth, F. L., and Wiltse, J. C. (1961). *IEEE Trans. Microwave Theory Tech.* **9,** 512–518.
Special Issue on Microwave and Millimeter-Wave Integrated Circuits (1978). *IEEE Trans. Microwave Theory Tech.* **26,** No. 10.
Steier, W. H. (1965). *Bell Syst. Tech. J.* **44,** 899–906.
Strukov, I. A., Khapin, Y. U. B., and Etkin, V. S. (1975). *Radiotekh. Elektron.* **20,** 1058.
Sundaram, G. S. (1979). *Int. Def. Rev.* **2,** 271–277.
Swanberg, N. E., and Paul, J. A. (1979). *Microwave Syst. News* **9**(May), 58–60.
Takada, T., and Ohmori, M. (1979). *IEEE Trans. Microwave Theory Tech.* **27,** 519–523.
Tolbert, C. W., Straiton, A. W., and Krause, L. C. (1965). *IEEE Trans. Antennas Propag.* **13,** 225–229.
Trebits, R. N., Hayes, R. D., and Bomar, L. C. (1978). *Microwave J.* **21**(Aug.), 49–53, 83.
Tsao, C. K. H., Connor, W. J., and Joyner, T. E. (1975). *Int. Conf. Commun., 11th* pp. 36/4–36/8.
Ulich, B. L. (1977). *IEEE Trans. Antennas Propag.* **25,** 218–223.
Ulich, B. L., Davis, J. H., Rhodes, P. J., and Hollis, J. M. (1980). *IEEE Trans. Antennas Propag.* **28,** 367–376.
Valkenburg, E. P., and Khandelwal, D. D. (1979). *Microwave J.* **22**(Aug.), 59, 60, 63, 64.
Vowinkel, B., Petonen, J. K., and Reinert, W. (1980). *IEEE MTT-S Int. Microwave Symp. Dig., Washington, D.C.* pp. 21–23.
Waineo, D. K., and Koniexzny, J. F. (1979). *1979 Int. IEEE/AP-S Symp. Dig., Univ. Washington, Seattle* pp. 477–480.
Warters, W. D. (1975). *Bell Lab. Rec.* **53,** 401–408.
Warters, W. D. (1977). *Bell Syst. Tech. J.* **56,** 1825–1829.
Waylan, C. J., and Yowell, G. M. (1979). *IEEE EASCON Proc.* pp. 623–629.
White, R. W., Read, M. B., and Moore, A. J. (1974). *IEEE Trans. Commun. Soc.* **22,** 1378–1390.
Williams, D. T., and Boykin, W. H., Jr. (1979). *AIAA J. Guidance Control* **2**(May/June), 196–203.
Wiltse, J. C. (1978). *Microwave J.* **21**(Aug.), 16–18.
Wiltse, J. C. (1979a). *Microwave J.* **22**(Aug.), 39–42.
Wiltse, J. C. (1979b). *Proc. Int. Telemeter. Conf., 15th, San Diego, Calif.* pp. 495–501.
Wollins, B. (1978). *Microwaves* **17**(Jan.), 9–12.
Zaytsev, N. I., Pankrotova, T. B., Petelin, M. I., and Flyagin, V. A. (1974). *Radiotekh. Elektron.* **19,** 1055–1061; Engl. transl., *Radio Eng. Electron. Phys.* **19,** 103–107.
Zur Heiden, D., and Oehlen, H. (1977). *Electr. Commun.* **52**(Feb.), 141–145.

CHAPTER 2

Millimeter Radar

Edward K. Reedy and George W. Ewell

Radar and Instrumentation Laboratory
Engineering Experiment Station
Georgia Institute of Technology
Atlanta, Georgia

I.	FUNDAMENTALS OF MILLIMETER-WAVE RADAR	23
	A. *Millimeter Radar Characteristics*	25
	B. *Radar Range Equation*	31
	C. *Components and Technology*	33
	D. *Applications for Millimeter Radar*	35
II.	PROPAGATION EFFECTS	36
	A. *Attenuation and Reflectivity*	36
	B. *Multipath Effects*	42
III.	CLUTTER CHARACTERISTICS	44
	A. *Land Clutter*	48
	B. *Sea Clutter*	53
IV.	TECHNOLOGY BASE	56
	A. *Receivers*	56
	B. *Transmitters*	58
	C. *Antennas*	68
	D. *Passive Components*	72
V.	MILLIMETER-WAVE RADAR EXAMPLE	76
VI.	MILLIMETER-WAVE RADAR APPLICATIONS	80
	A. *Surveillance and Target Acquisition*	80
	B. *Instrumentation and Measurement*	82
	C. *Guidance and Seekers*	85
	D. *Fire Control and Tracking*	88
	REFERENCES	91

I. Fundamentals of Millimeter-Wave Radar

Historically, the most popular radar operating frequencies have been located in the region of the electromagnetic spectrum between approximately 300 MHz and 35 GHz, which includes, basically, the ultrahigh-frequency (UHF) and superhigh-frequency (SHF) bands. However, with the exception of long-range "over-the-horizon" radars, most modern radars operate in the high end of the UHF band or, more probably, somewhere in the SHF frequency band—typically between 1 and 35 GHz. The

ISBN 0-12-147704-5

electromagnetic frequency spectrum with band designations is shown in Fig. 1. In this figure, the region of the spectrum between 300 MHz and 300 GHz has been expanded to illustrate the region of standard radar operation and assigned radar band nomenclature (Skolnik, 1978; Dyer *et al.*, 1977). Radars operating below 300 MHz normally carry the standard band designations shown at the top of Fig. 1, either HF or VHF.

In this chapter, attention will be focused on the physics and technology associated with radars operating in the upper end of the radar frequency band—the so-called millimeter and, more recently, the near-millimeter region. Strictly, and somewhat arbitrarily defined, millimeter waves range from 1- to 10-mm wavelengths, or from 300 to 30 GHz, respectively. The term ''near-millimeter'' has been recently applied to the frequency region ''bounded by the atmospheric windows at 100 and 1000 GHz (3–0.3 mm)'' (Kulpa and Brown, 1979). Thus significant overlap exists between the designations ''millimeter'' and ''near-millimeter''.

Until recently support for radar research in the millimeter-wave region, especially at frequencies above 35 GHz, tended to be somewhat cyclic in nature with little or no concerted effort or funding applied to the problem by Federal agencies or the civilian radar community. Therefore, the basic continuity required to develop millimeter-radar technology, both components and systems, and to define more completely the basic phenomenological questions associated with propagation and electromagnetic scattering problems in this frequency region was maintained only at a few research organizations. Under these circumstances, it is, therefore, somewhat surprising that millimeter-radar technology is sufficiently viable and advanced to support radar hardware programs.

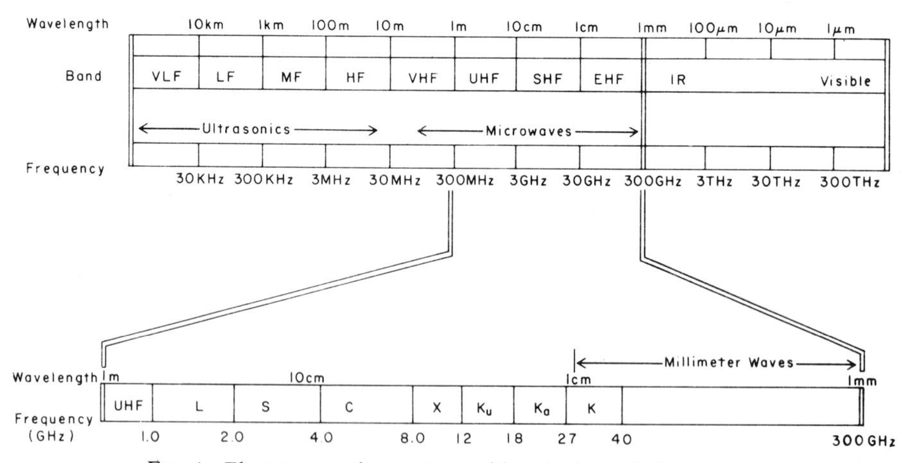

FIG. 1 Electromagnetic spectrum with radar band designations.

A. MILLIMETER RADAR CHARACTERISTICS

At the low end of the millimeter spectrum, millimeter wave radar has many of the same operational characteristics as its microwave counterparts, such as X- or K_u-band, except that the extremely short wavelengths of millimeter radar cause certain fundamental, operation characteristics to be more extreme. For example, practical millimeter-wave radars have narrower antenna beamwidths than X- or K_u-band radars by a factor of 2–20 for a fixed antenna aperture. Certain significant operational advantages result from the narrow beamwidth characteristics of millimeter radar (Richard, 1976): (1) relatively high antenna gains with small antenna aperture, (2) high track and/or guidance accuracy, (3) reduced Electronic Counter Measures (ECM) vulnerability, (4) operation at low elevation angles without significant multipath and ground clutter interference, (5) multiple target discrimination and, possibly, identification, and (6) mapping quality resolution.

Certain other unique operational characteristics make millimeter radar attractive for potential applications in both the defense and civilian arenas:

(1) narrow beamwidth can be achieved with a physically small, lightweight antenna structure;

(2) large radar bandwidths and Doppler frequency shifts are available for spread spectrum Electronic Counter Counter Measures (ECCM) operation, radar target signature processing, high range resolution, increased immunity to electromagnetic interference, and moving target detection;

(3) atmospheric attenuation can be "tuned" to provide transmission security for covert operation;

(4) lower sea return at millimeter wavelengths may provide target-to-clutter advantages for detecting targets in a sea clutter environment (Rivers, 1970).

When compared to higher frequencies in the infrared and visible regions, millimeter wavelength transmission is relatively unaffected by dry contaminants (dust, smoke, etc.), and certain other atmospheric effects such as fog and dry snow.

As with any engineering design situation, certain tradeoffs or disadvantages are associated with operation of radars at millimeter wavelengths:

(1) component costs are currently high, and reliability and availability relatively low;

(2) foliage penetration is very limited but, conversely, foliage reflectivity is high, which can be an advantage in some applications;

(3) operational range may be relatively short, even under "clear" atmospheric conditions (due to both equipment limitations and absorption by water vapor and oxygen molecules); and also

(4) range is reduced in inclement weather due to atmospheric scattering by heavy fog and rain.

The preceding millimeter-wave system tradeoff considerations are summarized in Table I.

As a result of these considerations, the primary potential applications of millimeter wave radar exist in satisfying sensor requirements that dictate physically small, lightweight, high resolution, relatively short-range radar possessing good Doppler characteristics. Millimeter radar is limited under adverse weather conditions when compared with microwave systems, but does exhibit higher resolution characteristics. Again, when compared to microwave radar, millimeter systems are typically much smaller. Although not having the extreme resolution of higher frequency sensors operating in the IR and visible regions, millimeter radar has superior propagation characteristics through smoke, haze, dust, and fog. Possessing characteristics of both the microwave and electro-optical regions, millimeter wave radar attempts to blend the advantages of both its mi-

TABLE I

MILLIMETER-WAVE RADAR SYSTEM TRADEOFF CONSIDERATIONS

Advantages	Limitations
Physically small equipment	Component cost high
Low atmospheric loss[a]	Component reliability/availability low
High resolution	Short range (10–20 km)
Angular	Weather propagation[b]
Doppler	
Imaging quality	
Classification	
Small beamwidths	
High accuracy	
Reduced ECM vulnerability	
Low multipath/clutter	
High antenna gain	
Large bandwidth	
High range resolution	
Spread spectrum ECCM	
Doppler processing	
ECM	

[a] Compared to IR and visual wavelengths.

[b] Compared to microwave frequencies.

crowave and EO counterparts in such a manner to reduce or minimize their limitations or disadvantages.

1. *Frequencies of Operation*

Only a cursory examination of the well-known curves for atmospheric absorption or atmospheric attenuation versus frequency, such as that shown in Fig. 2, is necessary to identify the preferred radar-operating frequencies in the millimeter band. As previously indicated, the millimeter region is nominally considered to be that frequency band between 30 and 300 GHz. Within this region, four so-called "propagation window" frequencies exist where the atmospheric absorption is at a minimum. These occur at approximately 35, 95, 140, and 220 GHz and define the nominal radar operating regions in the millimeter band.

Applications requiring covertness or relatively short-range operation have occasionally resulted in radar or communications systems operating on or near one of the oxygen or water vapor absorption bands where molecule and electromagnetic interaction resonances produce regions of abnormally large absorption. Two of the more widely used absorption bands are near 22.5 and 60 GHz.

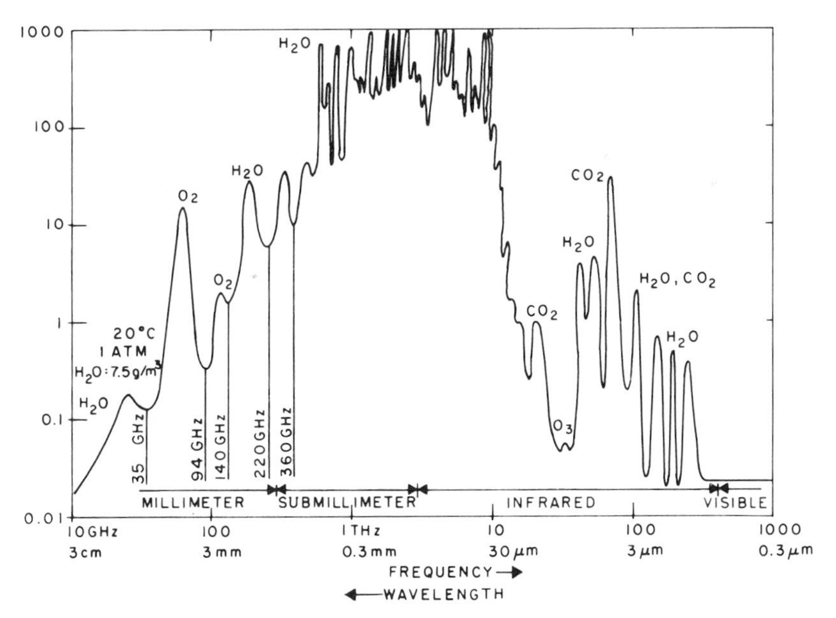

FIG. 2 Clear weather atmospheric attenuation of electromagnetic radiation as a function of frequency. (From Preissner, 1978.)

2. *Resolution*

Because of their extremely high frequencies and correspondingly short wavelengths, millimeter wavelength radars produce resolution extremes in both the spatial and frequency domains. High resolution and the resulting system advantages of high resolution is one of the primary "drivers" behind the increasing use of millimeter waves in operational radar systems.

a. *Spatial Resolution Properties.* In many applications of radar, mobility requirements, physical volume, and weight limitations severely restrict the radar's size, especially the radar antenna. However, in many of these applications, coupled with the size restrictions is a requirement for high spatial resolution—perhaps in both the range and angular dimensions. Thus, conflicting performance requirements requiring engineering tradeoffs are produced—a narrow antenna beamwidth with a small antenna aperture.

The angular resolution characteristics of a radar are determined by the antenna beamwidth. Classical electromagnetic diffraction theory is used to establish the relationship between the antenna aperture physical size and its illumination function and the far-field pattern of the antenna. The half-power points (or 3 dB points) of the antenna radiation pattern are normally specified as the limits of the antenna beamwidth for the purpose of defining resolution; two identical targets are, therefore, resolved in angle if they are separated by more than the antenna angular beamwidth. As discussed in Skolnik (1970), "the antenna beamwidth in a given plane of the aperture varies as the linear dimension across the aperture. The beamwidth is proportional to the ratio of the radiated wavelength, λ, to the aperture dimension, l, with the constant of proportionality, k, depending on the distribution of electromagnetic energy across the aperture." Therefore, for a diffraction limited antenna,

$$\theta = k\lambda/l, \tag{1}$$

where θ is the width of the antenna radiation pattern between the half-power or 3 dB points in the plane corresponding to the antenna dimension l.

The constant of proportionality k can vary from 0.88 for uniform aperture illumination and 13.2 dB sidelobe levels to approximately 1.47 for a highly tapered illumination function and relatively low sidelobe levels of perhaps 30 dB below the main beam peak. A value of $4/\pi$ is sometimes used for k to represent an intermediate, average, or typical pattern with sidelobes near -25 dB below the pattern peak. Accordingly, the angular resolution properties of many radars, microwave or millimeter, can be expressed as

$$\theta = 4/\pi(\lambda/l). \tag{2}$$

Antenna beamwidths (resolution) at several frequencies of interest from X band (10 GHz) through 220 GHz are given in Table II as a function of aperture size.

Following Woodward (1955) and Barton (1976), radar range delay resolution is directly proportional to the reciprocal of the bandwidth of the received signal, which is usually defined as the radar receiver bandwidth or the narrowest bandwidth in the complete receiving system. For a pulsed radar and a normal matched filter receiver, the pulse width, $\tau \cong 1/B$, is usually taken as the measure of the range delay resolution of the radar, where B is the receiver bandwidth. Therefore, in a matched filter radar, two targets could be theoretically resolved in range delay, if they were separated by more than the radar pulse width in time or, equivalently, by more than $c\tau/2$ in range, where c is the speed of light (3×10^8 m/sec). Radar range resolution is independent of transmitted frequency to the extent that the received signal bandwidth is independent of frequency. Thus a radar's basic resolution limit in the range dimension R_r is

$$R_r = c\tau/2. \tag{3}$$

b. *Frequency Properties.* In addition to providing estimates of target characteristics in the three spherical coordinates of position, radar has the ability to resolve targets in a fourth dimension—radial velocity. Through the so-called Doppler frequency shift, the radial velocity of a moving target can be directly estimated by a radar, independently of the other basic measurements.

All moving targets, when illuminated by an electromagnetic signal of frequency f_0 will reflect a signal shifted in frequency by the Doppler fre-

TABLE II

ANTENNA BEAMWIDTH (IN DEGREES) AS A FUNCTION OF RADAR FREQUENCY

Aperture size (m)	Frequency (GHz)				
	10	35	95	140	220
0.01	—	62.5	23.10	15.6	9.9
0.1	21.9	6.3	2.3	1.6	1.0
1.0	2.2	0.63	0.23	0.16	0.1

quency shift f_d. The amount of frequency shift is given by the Doppler equation

$$f_d = -2\gamma_r/\lambda \tag{4}$$

and

$$\lambda = c/f_o, \tag{5}$$

where γ_r is the time derivative of range, or radial velocity, of the target taken to be positive for receding or outbound targets, c the velocity of light, and λ the wavelength of the transmission. Rearranging Eq. (4) allows the target's radial velocity to be expressed in terms of the radar measured target characteristic, the Doppler frequency:

$$\gamma_r = -f_d\lambda/2. \tag{6}$$

According to Eq. (4), the amount of Doppler shift produced by a target is inversely proportional to the radar wavelength. Thus, for the same target radial velocity, a millimeter radar will result in Doppler shifts approximately ten times higher than microwave frequencies, i.e., *X* band compared to 95 GHz. Some examples of Doppler frequency shifts for a one mile per hour (MPH) radial velocity target are given in the following tabulation:

Radar frequency (GHz)	Doppler shift (Hz/MPH)
10	30
35	104
95	283
140	418
220	657

An airborne target approaching a 95-GHz radar at a radial velocity of 600 MPH would produce a Doppler frequency shift of 169.8 kHz; the same target approaching an *X*-band radar would produce a frequency shift of only 18 kHz. Hence the millimeter radar has the potential for greater target velocity resolution, assuming equal receiver signal processing bandwidths. However, the ability of a radar to measure and resolve differences in Doppler frequency shifts and, therefore, target radial velocity depends, in a rather complicated manner, on detailed characteristics of the radar transmitter, receiver, and signal processor, the techniques used to maintain phase coherency between the transmitted signal and reference signals used in the detection process, the transmitted spectrum, and the duration of the transmission. Barton and Ward (1969) discuss radar frequency resolution in considerably more detail.

3. *Bandwidth*

Overall millimeter-radar equipment limits on the total available system bandwidth range from the instantaneous bandwidth of the radar receiver that is normally matched to the transmitted pulse length (i.e., approximately 10 MHz for a 100 nsec pulse) to the mechanical (or perhaps electronic) tuning range of the transmitter source that might be 1–2 GHz for a typical pulsed extended interaction oscillator (EIO) operating in the millimeter-wavelength region. Certain millimeter sources have bandwidth characteristics that require a receiver bandwidth larger than the reciprocal of the pulse length. Intermediate to these limits are tuning ranges of some of the devices currently used as local oscillators. For example, tuning ranges for relatively low-power IMPATT and klystron sources suitable for local oscillators vary between perhaps 100 and 500 MHz.

Passive radar components such as circulators and rotary joints will also limit the available radar bandwidth. Percentage bandwidths of 5–10% are typical for these components (Strom, 1973). Finally, the atmospheric windows centered around the nominal radar center frequencies will tend to limit frequency excursions around these center frequencies to approximately 20% before the atmospheric attenuation becomes prohibitively large (Whicker and Webb, 1978).

B. Radar Range Equation

Millimeter-wave radar performance, like its microwave counterparts, is governed by, and can be predicted from, the basic radar range, beacon, and jamming equations. In the case of microwave radars, the free-space attenuation term in these equations can usually be neglected, whereas, for millimeter radar, it may be the most important factor limiting radar performance.

The performance of a radar system is governed by the complex interrelationships of several system parameters. Analytically, radar performance and tradeoff analysis is normally investigated by employing the radar range equation—a mathematical expression that can relate the radar maximum range performance to the complete set of system parameters. One form of the radar range equation is given below.

$$R_{\mathrm{m}}^{4} = P_{\mathrm{t}}G^{2}\lambda^{2}\sigma 10^{-0.2\alpha R}/(4\pi)^{3}(KT_{0}B)F_{\mathrm{n}}(\mathrm{S/N})_{1}L_{\mathrm{s}}, \qquad (7)$$

where R_{m} is the maximum radar range corresponding to minimum $(\mathrm{S/N})_{1}$, P_{t} the peak radar transmitted power, λ the radar wavelength, B the receiver bandwidth $\approx 1/\tau$ (for matched receiver), τ the pulse width (integration time for a cw radar), G the antenna gain, σ the target radar cross section, K the Boltzmann's constant (1.38×10^{-23} joule/deg), T_0 the standard reference temperature (290°K), F_n the receiver noise figure, α the atmospheric attenuation coefficient (dB/km), $(S/N)_1$ the minimum equivalent receiver output single-pulse signal-to-noise ratio required for the radar's function (detection or tracking), and L_s the system losses.

This form of the radar range equation incorporating a single-pulse signal-to-noise ratio does not measure the overall effectiveness of the radar since the effects of integration are not included. Integration can significantly enhance the performance of a radar for both detection and tracking. However, the single-pulse S/N represents a quantity that is widely used in the literature to compare and predict the performance of radars. Integration or signal processing gain can usually be included by adding a multiplicative factor in the numerator of Eq. (7).

As an example of the use of the radar range equation for a simple, first-cut calculation of radar performance, consider the set of postulated radar parameters shown in Table III. To avoid an iterative solution for determining maximum range, since the propagation loss is range dependent, it is necessary to define a constant total loss factor L_T, independent of range, which includes both system and atmospheric losses. For this calculation, assume $L_T = 6$ dB. Also, assume no integration or signal processing gain; furthermore, a circular antenna aperture having a gain of 37 dB is assumed.

Using these parameters, the maximum expected radar range is plotted in Fig. 3 as a function of target RCS. Indications of the general range of cross section of several classes of targets are also included at the top of this figure. The band of anticipated performance indicates the spread resulting from system compromises, atmospheric loss, range dependence, and possible field degradation that might be reasonably expected. These performance curves are for clear weather and represent, for the parame-

TABLE III

POSTULATED RADAR PARAMETERS FOR RADAR RANGE CALCULATION

Parameter	Value assumed
Wavelength λ	3 mm
Transmitter power P_t	4 kW
Antenna gain G	37 dB
Pulse length τ	50 nsec
Bandwidth B	20 MHz
Noise figure F_n	10 dB
Losses L_s	6 dB
Signal-to-noise ratio $(S/N)_1$	13 dB

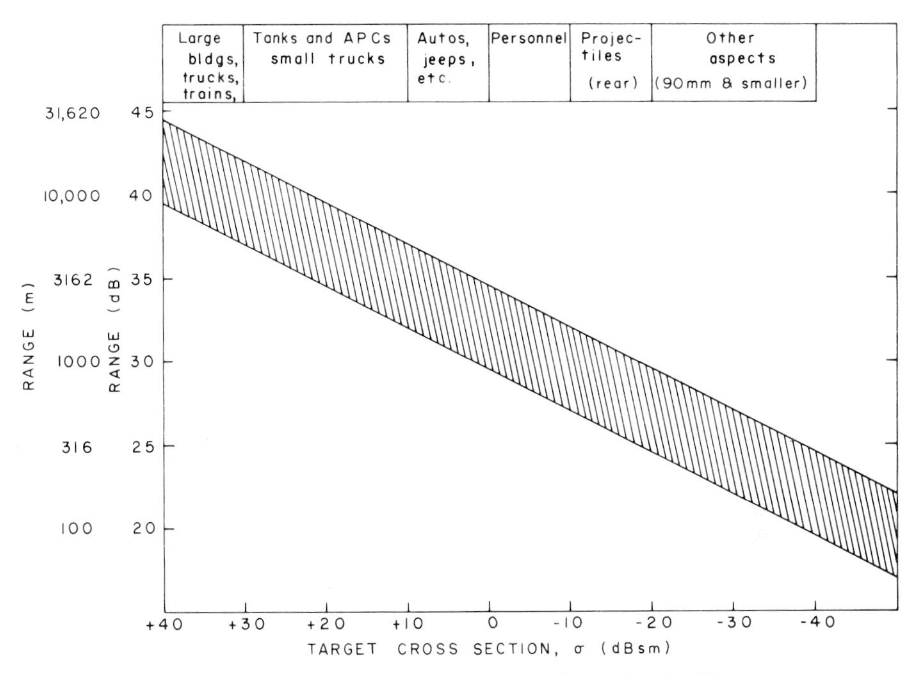

FIG. 3 Projected maximum range performance of postulated millimeter radar.

ters chosen, an upper bound on performance, since such factors as clutter have not been included. The range performance achieved in this example is strongly influenced by the assumed constraints on the antenna aperture size; choice of a larger aperture would result in a corresponding increase in range.

C. Components and Technology

Significant progress has recently been made in developing components and the basic technology to support radar programs in the millimeter-wavelength region. Whether this has been the result of, or is itself one of the driving factors behind, the increased utilization of this part of the spectrum is not entirely clear (the classic chicken and egg problem).

The performance characteristics of most of the critical millimeter components tend to decrease with increasing frequency. A wide range of components is currently available from commercial vendors up to, and including, the 95-GHz window. Above 95 GHz—at 140 GHz and, most especially, 220 GHz—generally available components consist primarily of laboratory devices lacking the basic structural and electrical integrity required for military and commercial use in a harsh field environment.

The foregoing comments notwithstanding, significantly improved technology at millimeter wavelengths exists in the mixer, source, and passive circuit arenas. For example, one of the primary performance-determining technologies in any radar development program is the amount of power available from the radar transmitter and the type of source available. This is particularly true for millimeter-wave systems, since they are invariably power limited. Summarized in Table IV are typical power levels for four devices that are extensively used as power sources in millimeter radar—the magnetron, EIO, gyrotron, and solid-state (Gunn and IMPATT) sources (Reedy and Eaves, 1979). All of these devices have improved power outputs, and/or operate at higher frequencies, and, in some cases, have longer lifetimes than previous versions of the device. Traveling wave tubes (TWT) operating at 94 GHz with a peak power output of approximately 1000 W, an instantaneous bandwidth of 4%, and a small signal gain of 60 dB are also becoming available, producing a capability for fully coherent master oscillator, power amplifier (MOPA) transmitters at 95 GHz with sizable peak output powers (Arnold, 1980).

Low noise mixers operating at millimeter wavelengths might be expected to exhibit typical double sideband noise figures of 5, 8, 9, and perhaps 15 dB for operation at 35, 95, 140, and 220 GHz, respectively (Seashore *et al.*, 1979). However, both transmitter peak powers and receiver noise figures are continually being improved as new devices, techniques, and technology become available.

Passive circuit components such as circulators, isolators, phase shifters, switches, duplexers, couplers, and waveguide are sometimes overlooked when the technology to support a millimeter-radar development program is considered, even though these components are absolutely necessary to the fabrication of a radar and, in fact, in many in-

TABLE IV

MILLIMETER-WAVE RADAR TYPICAL TRANSMITTER OUTPUT POWER (PEAK)

	Frequency (GHz)			
Power source	35	95	140	220
Magnetron	50–100 kW	1–6 kW	—	—
Extended interaction oscillator	2–3 kW	1–2 kW	200 W	60 W
Gyrotron[a]	200 kW	20 kW	9 kW	1 kW
Solid-state (Gunn, IMPATT)	18 W	10 W	2.5 W	0.4 W

[a] Higher power levels have been reported in the literature for this device.

stances may be the performance limiters. For example, circulators producing isolations in excess of 30 dB above 95 GHz are difficult to fabricate with current technology. In addition, insertion loss is relatively high for millimeter circulators, thus reducing available radar output power.

Two technology areas requiring considerable effort before the higher millimeter frequencies can be truly expected to be widely used in both military and civilian applications are test equipment and standards—measurements, calibration, and mechanical. For millimeter components to have the necessary level of system commonality and interchangeability, and in order to provide a measure of system and component performance comparison, a complete spectrum of test equipment and system of measurements standards must exist.

A more detailed treatment of the technology base required and available to support millimeter-wave radar will be discussed later in this chapter.

D. Applications for Millimeter Radar

The primary performance characteristics of millimeter radar (i.e., physically small, high resolution, large bandwidths, and relatively short range) dictate, to a large extent, potential applications. Some general categories of applications include the following:

(1) *Surveillance and target acquisition:* Millimeter radar would be especially effective against low-altitude airborne targets and ground targets where it is necessary to reduce the clutter return. Millimeter radar might also be effective for an exoatmospheric satellite target acquisition radar where space attenuation was minimal or ground intercept undesirable, for terrain following and obstacle avoidance radar, and for target acquisition and surveillance radars intended for RPV, helicopter, or aircraft platforms. Millimeter radar might also act as the target acquisition sensor in a hybrid or combined sensor configuration where a millimeter radar would acquire targets for hand over to, or as a cueing aid for, a higher resolution sensor such as a laser designator, optical device, etc.

(2) *Fire control and tracking:* High resolution allows millimeter radar to support short-range, closed loop fire control systems against low-altitude aircraft, missiles, ground targets, etc., and to provide the tracking radar for missile guidance in beam rider, command-guided, or semiactive modes.

(3) *Seekers and terminal missile guidance:* Their small physical size makes millimeter wave radars ideally suited for application as active seekers for terminal homing projectiles or missiles and for antiballistic missile terminal homing seekers.

(4) *Instrumentation and measurement:* High spatial resolution and correspondingly high accuracy give millimeter radar unique advantages in applications requiring extremely high target location accuracy, detailed target and clutter signature investigation, and for acquisition of high resolution meteorological data at relatively short ranges.

The potential applications summarized above certainly are not exhaustive. Additional applications are covered in Richard (1976) and Strom (1973) and a more detailed treatment of certain of these applications will be given later in this chapter.

II. Propagation Effects

No attempt will be made in this section to review and present data that completely describe millimeter-wave propagation effects, since an extensive summary of propagation information is presented elsewhere in this volume. Only that information required to make this chapter reasonably self-contained or, in some manner, unique to radar applications will be summarized here.

A. Attenuation and Reflectivity

Figure 4 illustrates, in block diagram form, the major factors affecting the signal received at any radar. Of particular interest are those attenuation and reflection or scattering influences present in the propagation path, which, for terrestrial radar, is normally the atmosphere. The primary attenuation-producing factors for millimeter radar are the molecular

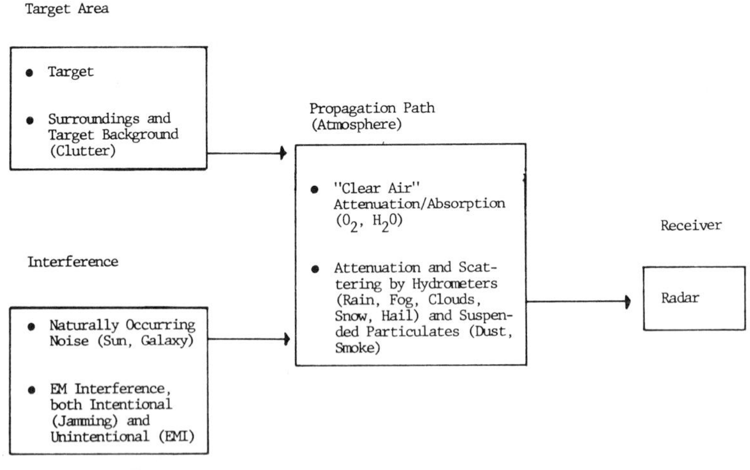

FIG. 4 Major factors affecting received radar signal.

absorption of water vapor and oxygen in a relatively clear atmosphere, absorption of condensed or suspended water in the form of droplets, in rain, fog, clouds, etc., and scattering from these water droplets. Backscatter of energy from the suspended water droplets in fog and clouds and from rain can severely limit radar performance, since the target return signal will be embedded in and corrupted by this noiselike reflected energy. Suspended particulate matter, such as dust particles and smoke, may also influence millimeter-wavelength propagation.

1. *Attenuation by Atmospheric Gases, Rain, and Fog*

Although the transmission of electromagnetic energy through the atmosphere is a very complex phenomenon, certain aspects of this phenomenon have been intensively and extensively studied and documented (Downs, 1976; Kulpa and Brown, 1979; McCartney, 1966). However, much of the basic propagation data necessary to predict the performance of millimeter-wavelength sensors accurately either does not exist or the

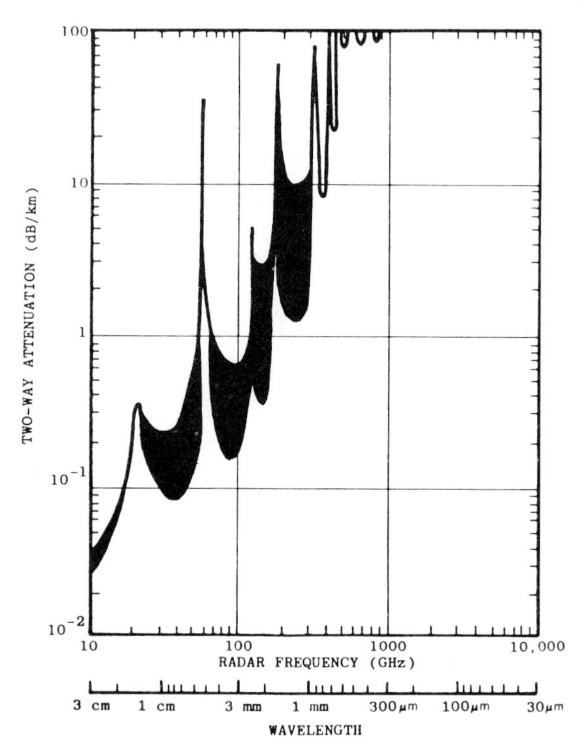

FIG. 5 Clear weather attenuation uncertainty for horizontal, sea level path. (From Strom, 1973.)

uncertainty associated with existing data is large enough to make accurate, analytical prediction of the sensor performance limits extremely difficult.

Figure 5 (Strom, 1973) illustrates this uncertainty by shading the window region two-way attenuation levels. Note that at 220 GHz the clear air attenuation varies between approximately 1.5 dB/km and, perhaps, 11 dB/km, depending on such parameters as atmospheric pressure, temperature, and water density or humidity. Large regions of uncertainty are present at all of the window frequencies. To reduce this level of uncertainty and to predict radar performance more accurately, Kulpa and Brown (1979) point out that a rather detailed specification of the conditions under which the sensor is expected to operate is required. For example, the geographic location of expected operation can be important since geographic as well as seasonal variations in humidity and temperature can affect clear air attenuation values by as much as 10–15 dB/km.

Nevertheless, it is still worthwhile to document propagation effects on millimeter waves as completely as possible so that equipment performance can be estimated as accurately as possible using available data. Attenuation produced by basically clear air atmospheric gases at 20°C, one atmosphere pressure, and 7.5 g/m³ water vapor is shown in Fig. 6

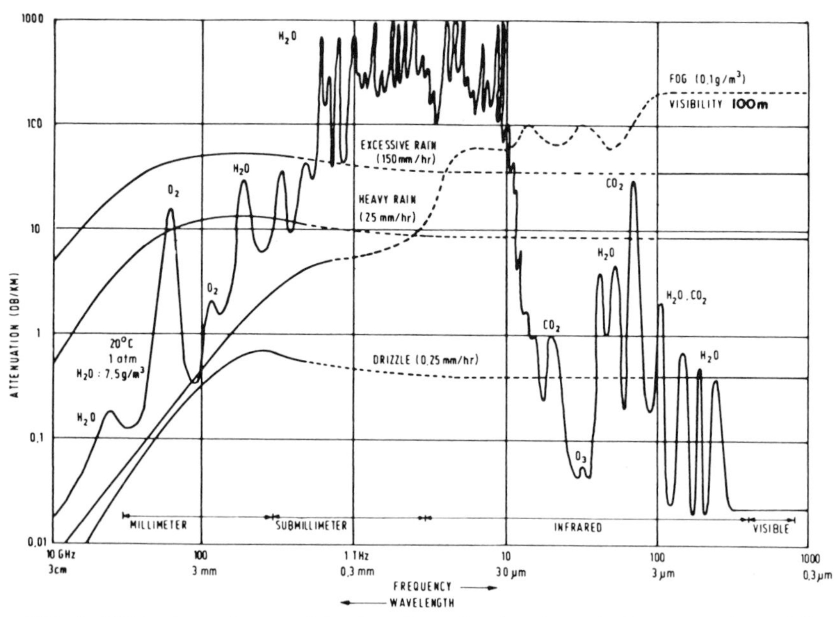

FIG. 6 Millimeter and near-millimeter attenuation by atmospheric gases, rain, and fog. (From Preissner, 1978.)

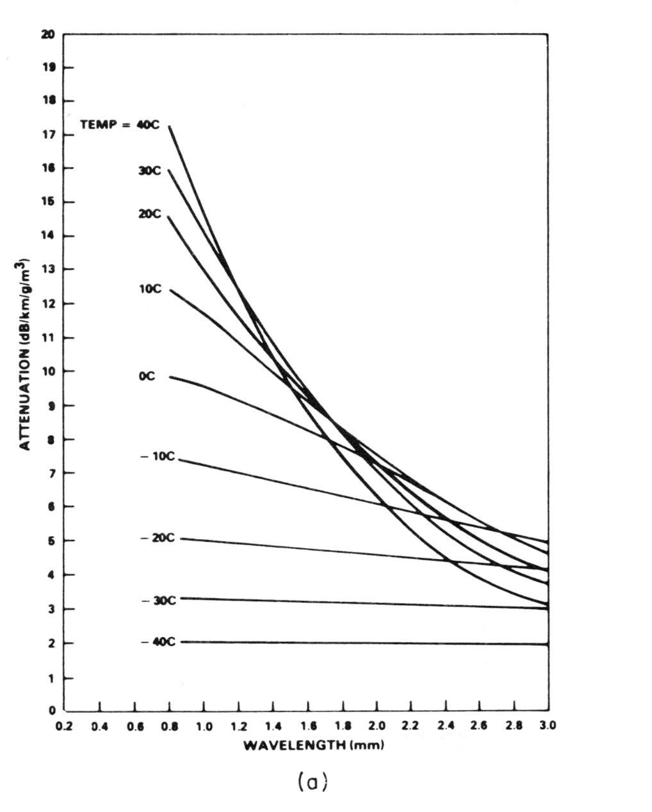

(a)

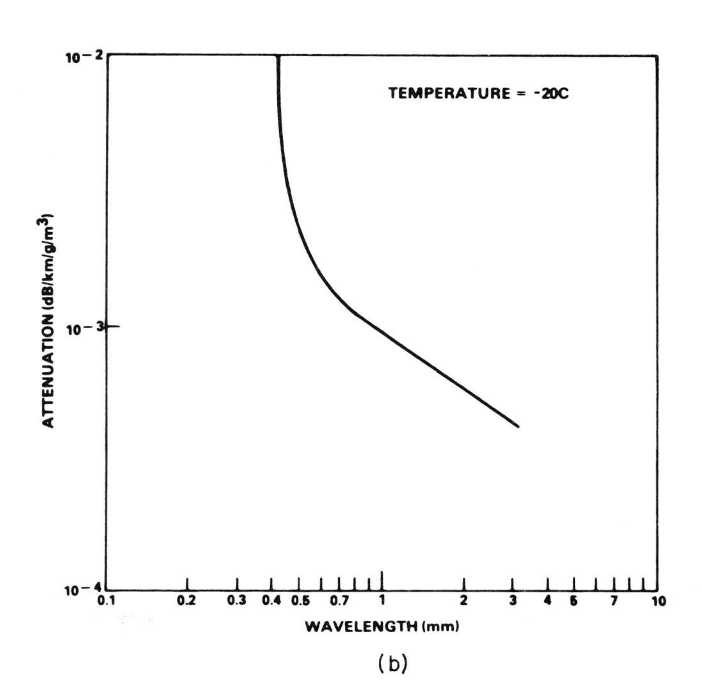

(b)

FIG. 7 Fog and cloud attenuation properties. (a) Cloud and fog attenuation, (b) ice cloud or fog attenuation. (From Kulpa and Brown, 1979.)

along with attenuation produced by certain hydrometers (rain and fog) (Preissner, 1978). Attenuation produced by rain increases as a function of frequency with maximums occurring in the 100–200-GHz region, after which it remains relatively constant. Some typical attenuation values for fog and water clouds and ice clouds as a function of frequency and temperature are shown in Fig. 7. Ice clouds generally produce negligible attenuation at these frequencies (Preissner, 1978). The interested reader is referred to other volumes of this series or to Kulpa and Brown (1979) for a more complete documentation and discussion of millimeter propagation in both clear air and in the presence of naturally occurring hydrometers such as rain, snow, and hail.

The uncertainty associated with the currently available propagation data base at millimeter waves cannot be overemphasized. For example, Table V (taken from Kulpa and Brown, 1979) summarizes the variation in total one-way attenuation at 220 GHz in a 100-m fog and 4-mm/hr rain using the best available data. Uncertainties of 10–15 dB in attenuation are present in the data, illustrating the difficulty in analyzing performance of millimeter equipment.

2. *Rain Backscatter*

Only a very limited number of theoretical and experimental investigations have attempted to quantify the radar backscattering properties of rain at millimeter wavelengths (Richard and Kammerer, 1975; Wilcox and Graziano, 1974; Currie *et al.*, 1975c; Strom, 1973). The basic source of experimental data relating rain radar backscatter characteristics to rain rates at the low millimeter wavelengths and X band is a series of measurements conducted at McCoy AFB, Florida, in August and September 1973, by the U.S. Army Ballistic Research Laboratories, Aberdeen, Maryland. The resulting data from these measurements are recorded and documented in Currie *et al.* (1975c) and Richard and Kammerer (1975). Reduction of the data from these experiments is shown in Fig. 8 where the mea-

TABLE V

UNCERTAINTY IN ATTENUATION AT 200 GHz[a]

Condition	One-way attenuation (dB/km)
Clear air	1.6–11.2
100-m fog	0.4– 4.7
Total	2.0–15.9
Clear air	1.6–11.2
4-mm/hr rain	1.0– 7.0
Total	2.6–18.2

[a] From Kulpa and Brown (1979).

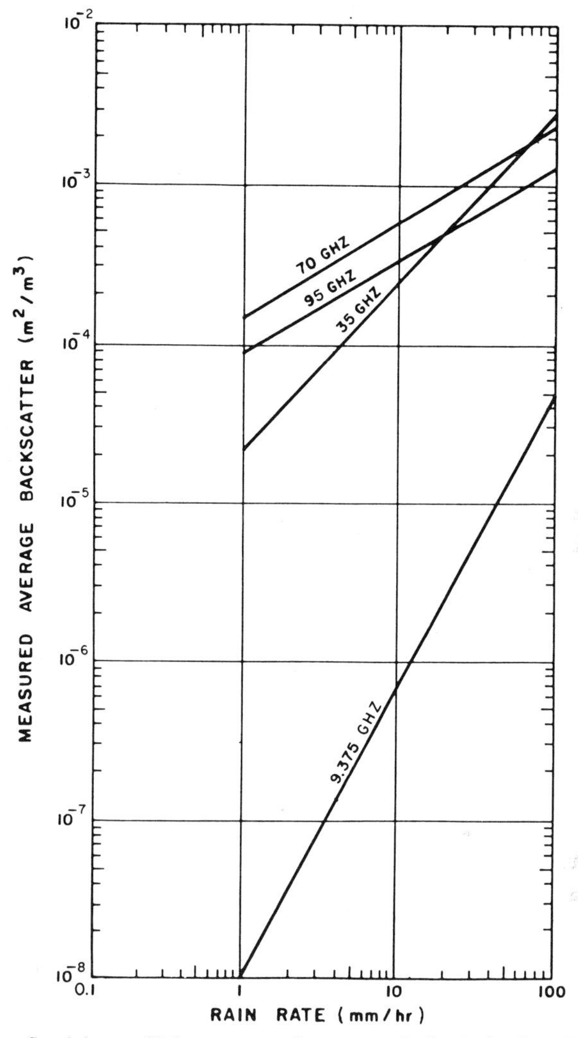

FIG. 8 Rain reflectivity coefficient versus rain rate, vertical polarization. (From Richard and Kammerer, 1975.)

sured average backscatter coefficient of rain is given as a function of rain rate for the operating frequencies of 9.375, 35, 70, and 95 GHz. One interesting property of rain reflectivity is apparent in Fig. 8: The rain backscatter coefficient is smaller at 95 GHz than at 70 GHz. Data shown in Fig. 8 were taken from Richard and Kammerer (1975) since their method of data reduction resulted in a larger and, thus, somewhat more conservative value for the backscatter coefficient, especially at the lower rain rates.

3. *Dust, Smoke, and Other Obscurants*

Even though one of the primary advantages of millimeter-wave sensors over their electrooptical counterparts is enhanced propagation characteristics through dust, smoke, and other battlefield and naturally occurring obscurants, very limited definitive data exist to specify attenuation and backscatter when propagation takes place under these conditions. Allen and Simonson (1970) and Downs (1976) presented some of the first comparative data for attenuation by smoke and dust. More recently, spurred on by the increased interest in, and usage of, the millimeter region, more investigators have examined the effects of smoke and dust on millimeter propagation (Morgan *et al.*, 1979; Martin, 1979).

All of the above references confirm that suspended dust and smoke in the atmosphere in amounts typical of a battlefield or naturally occurring environment produce a practically imperceptible amount of attenuation at frequencies of 140 GHz and below. Knox (1979) found that the only obscurant to affect transmissivity significantly during exercises conducted at Smoke Week II was dust and debris produced by exploding charges of C4 material that had been buried under several centimeters of dirt. Maximum attenuation measured during these trails was approximately 0.4 dB/km at 140 GHz.

Petito and Harris (1979) observed that only during periods when exploding artillery rounds hurled debris, principally soil and metallic fragments, into the air was there significant attenuation and backscatter from artillery bursts when observed with a 95-GHz radar.

4. *Turbulence Effects*

Refractive index inhomogeneities in the propagation path of a millimeter-wavelength signal causes certain propagation phase shifts to result, which in turn result in propagation scintillation effects, angle of arrival fluctuations, depolarization, and thermal blooming. Only scintillation and angle of arrival fluctuations are thought to be significant at millimeter wavelengths, however. Recent theoretical and experimental investigations of these effects at 94 and 140 GHz have shown that the scintillation effects are not likely to greatly affect millimeter-wave systems except at the extreme limits of performance, but that atmospheric turbulence effects may result in angle of arrival variations of perhaps 0.35 mrad, which is approximately the same level of accuracy required of many systems (McMillan *et al.*, 1979).

B. Multipath Effects

When an RF signal is incident on the surface of the earth, a portion of the incident electromagnetic energy is forward scattered; at the target, the

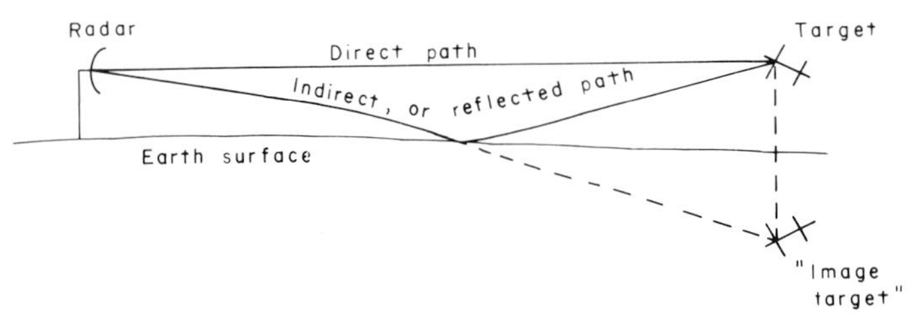

FIG. 9 Multipath geometry for a target being observed by a radar, showing both the direct and indirect (reflected) signal paths.

energy that reaches the target by the direct path from the transmitter to the target and by reflection from the surface of the earth combines vectorally and can add either in or out of phase. Effects produced by these reflected signals, collectively termed "multipath," can produce fluctuations in signal strength from a target, or in the case of a tracking system, can introduce considerable tracking errors. A simplified geometry for a multipath situation is shown in Fig. 9, showing both the direct and indirect path. The magnitude of these multipath-related effects is related to the amount of energy incident on the surface, the reflection coefficient of the surface, the amount of energy that reaches the target by the direct path, and the relative phase of the direct and indirect components.

While the analysis of a general multipath situation may be quite complex, considerable insight into the process may be gained by a simplified analysis. In such a simplified analysis, the surface is considered to be a randomly rough surface having known dielectric properties. The voltage reflection coefficient for a smooth surface of the same dielectric material can be calculated, and this modified by a factor to compensate for roughness of the surface in order to obtain an approximate description of its forward scattering properties. The forward scattered reflection coefficient of a smooth uniform dielectric surface ρ_0 can be calculated directly from the Fresnel equations, and Fig. 10 shows the results of such calculations for a sea water surface at frequencies of 35, 95, and 140 GHz for both horizontal and vertical polarizations.

The reflection coefficients of the type shown in Fig. 10 are then modified by the specular scattering factor ρ_s, where

$$\rho_s^2 = \exp[-(4\pi\sigma_h \sin\gamma/\lambda)^2], \tag{8}$$

where σ_h is the rms deviation of the surface height, λ the wavelength, and γ the grazing angle. Note that a surface is considered rough when ρ_s^2 is less than 0.5 or $\rho_s \leq 0.7$; this is approximately the Rayleigh roughness criterion.

The reflection coefficient ρ is given by

$$\rho = \rho_0\rho_s. \tag{9}$$

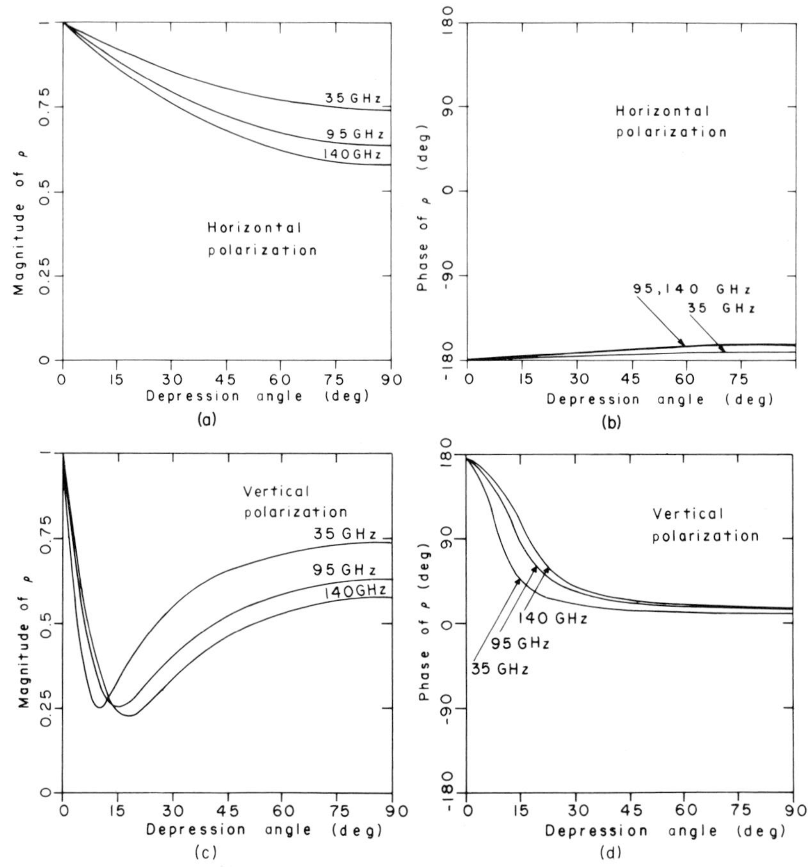

FIG. 10 Reflection coefficient (magnitude and phase) for smooth seawater at frequencies of 35, 95, and 140 GHz for vertical and horizontal polarizations.

In general, use of millimeter-wave radar may offer the advantage of a reduction in multipath effects, primarily because for a given aperture size, the antenna beamwidth decreases with increasing frequency, and such narrow beamwidths can be used to illuminate the target without directing as much energy toward the reflecting earth surface. Also, since roughness is dependent on σ_h/λ and since λ is small for millimeter-wave radars, a given surface appears rougher as the frquency increases, thus decreasing the specular scattering factor and thus decreasing ρ.

III. Clutter Characteristics

When a radar system illuminates the earth's surface (land or sea), a portion of the energy is forward scattered, giving rise to multipath effects described in Section II; in addition, a portion of the energy is reflected

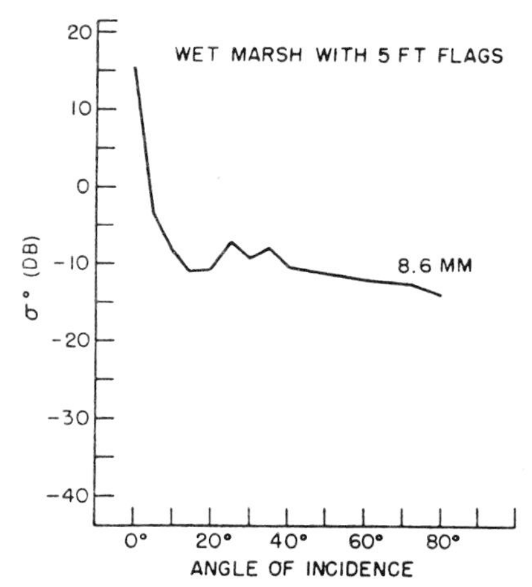

FIG. 11 σ° as a function of angle of incidence (note that 0° corresponds to looking straight down) for wet terrain covered with tall weeds or flags. (From Grant and Yaplee, 1957. Reprinted from *Proceedings of the IRE* **45,** 976–982. © 1957 IEEE.)

back toward the radar system. These unwanted signals are usually referred to as "clutter," and can seriously impact overall system performance for those situations where an appreciable amount of energy is backscattered toward the radar. Historically, such clutter has been described in terms of its radar cross section per unit area (σ°), a dimensionless quantity that, when multiplied by the resolution cell size of the radar system, gives the clutter radar cross section; particularly when high resolution radar systems are utilized, clutter no longer appears to be homogeneous with a uniform spatial distribution, and the concept of σ° should be used with some care. However, it is still a quite useful quantity, and one almost universally utilized to report the magnitude of clutter backscatter.

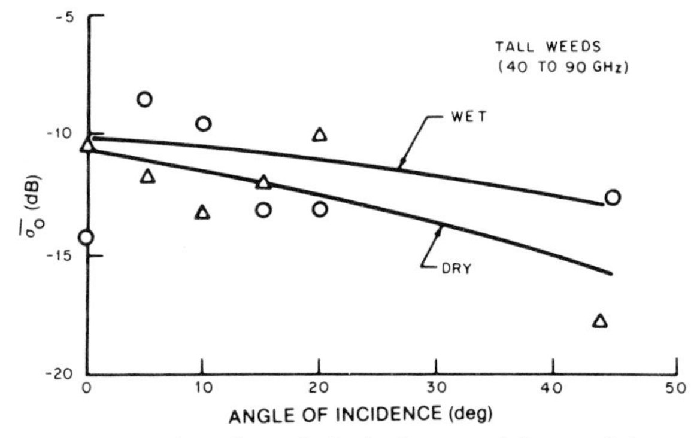

FIG. 12 Mean value of σ° for tall weeds for both wet and dry conditions as a function of incidence angle, representative of values over the 40–90-GHz frequency region; △, dry; ○, wet. (From Suits and Guenther, 1979.)

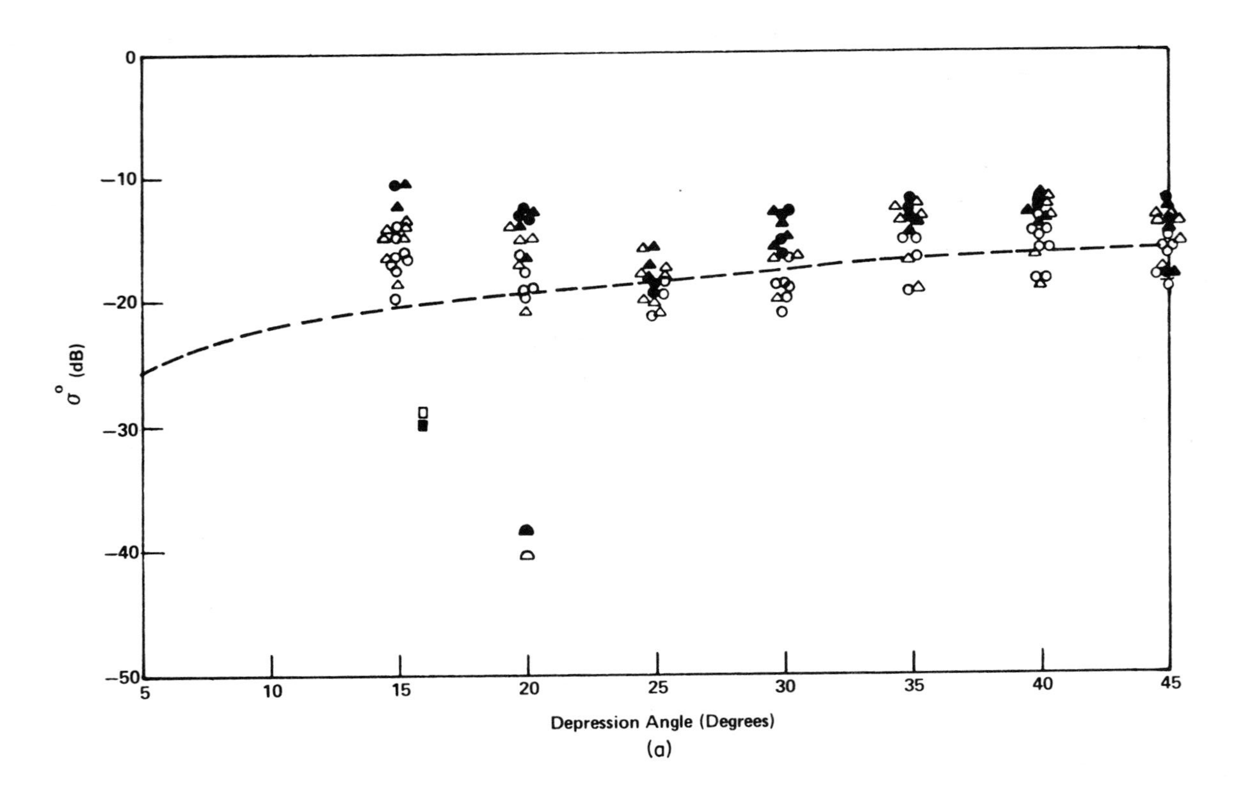
0
−10
−20
−30
−40
−50
σ^o (dB)
5
10
15
20
25
30
35
40
45
Depression Angle (Degrees)
(a)

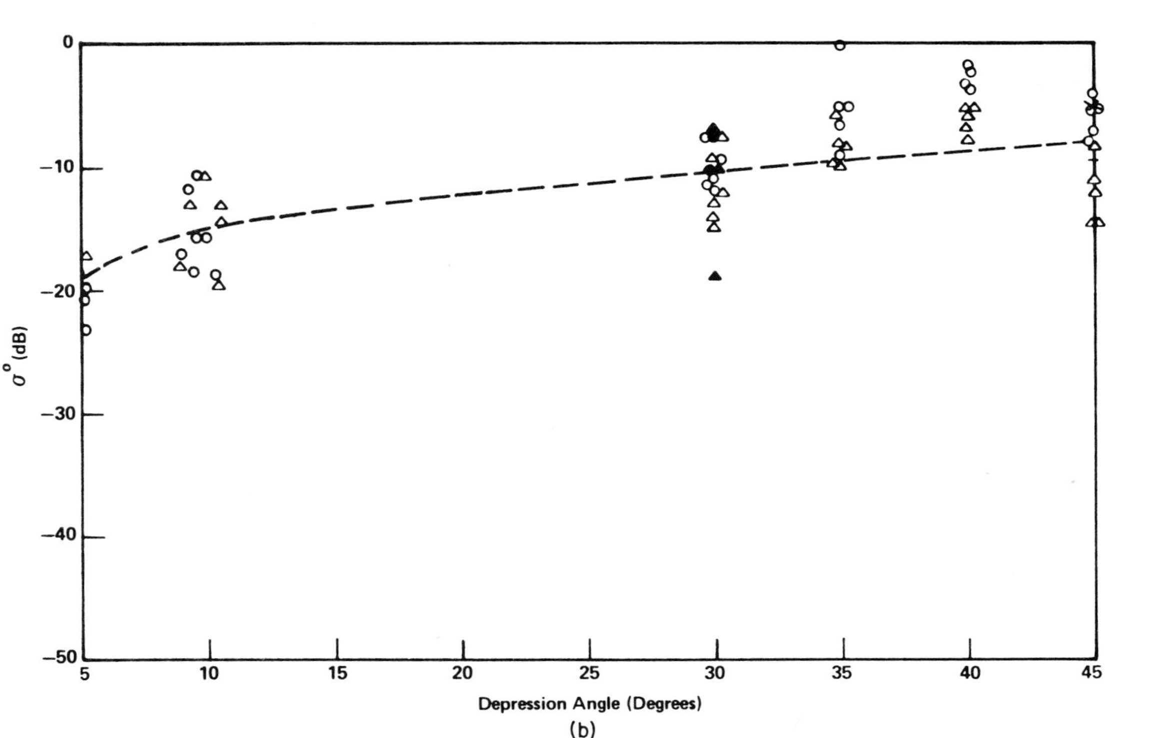

FIG. 13 Measured values of $\sigma°$ for deciduous trees as a function of depression angle for both wet and dry conditions at frequencies of (a) 35 GHz and (b) 95 GHz. (From Currie, 1979. Reprinted from *1979 International Symposium Digest—Antennas and Propagation* **2,** 504–507. © 1979 IEEE.) (a) ––, $\sigma° = -15 + 10 \log(\theta/25) - 8 \log(\lambda/0.32)$; ●, VV wet foliage; ▲, HH wet foliage; ○, VV dry foliage; △, HH dry foliage; ◠, VV lake; ◓, HH lake; ■, VV dry field; □, HH dry field. (b) ––, $\sigma° = -11 + 10 \log(\theta/25) - 8 \log(\lambda/0.32)$; ●, VV wet foliage; ▲, HH wet foliage; ○, VV dry foliage; △, HH dry foliage.

Another problem in characterizing clutter returns in a radar system is that the return is a complex function of a number of parameters, including frequency, polarization, depression angle, clutter type, and system resolution. Analysis and reporting of clutter data are further complicated by the fact that clutter returns are not constant, but fluctuate with time; thus, in addition to the average value, some measure of both the amplitude fluctuations and the temporal behavior (or spectral width) of the clutter must also be provided if a useful description of clutter is to be obtained.

A. Land Clutter

The properties of land clutter can vary widely. This variability is partially due to the fact that land clutter is present in a large number of forms, such as trees, fields, desert, snow, and cultural areas; even within such broad categories, different clutter regions can produce decided differences in reflectivity. In addition, local environmental conditions such as wind velocity, wind direction, and moisture content can affect clutter returns.

Experimental data describing the radar reflectivity from land at incidence angles above a few degrees at a frequency of 35 GHz were presented by Grant and Yaplee (1957); more recently, additional measurements at 35 and 95 GHz of land clutter and of snow-covered surfaces have been reported (e.g., Currie *et al.*, 1975a,b,c, 1976a,b; Hayes and Dyer, 1973; Hayes *et al.*, 1979; Currie, 1979; Dyer and Currie, 1978; Kosowsky, 1974; Suits and Guenther, 1979). Several different conventions to describe the angles involved have been used in reporting these data. The term depression angle or incident angle has been used where grazing incidence is 0° and vertical incidence is 90°. The term "angle of incidence" has also been used, for which 0° corresponds to vertical incidence and 90° to grazing. The convention of the original source will be used in reporting representative values in this chapter.

Figure 11 gives σ° as a function of angle of incidence for several nonhomogeneous terrain types at 8.6 mm. Fig. 12 is a plot of mean σ° as a function of depression angle for wet and dry weeds, representative of values from 40 to 90 GHz. Figure 13 is a more exhaustive set of experimental data describing σ° for deciduous trees at 35 and 95 GHz for both wet and dry conditions. In addition, 35-GHz data for a lake and dry field are also included as part of this data set. Figure 14 is another representation of σ° for a variety of terrain at 94 GHz, for depression angles from below 5° to above 20°. Figure 15 gives σ° over the range of depression angles from 40° to 90° for composite natural ground targets. Data of Fig. 15 are represented to be typical for the 40–90-GHz frequency range.

Information concerning the amplitude fluctuations of the radar return

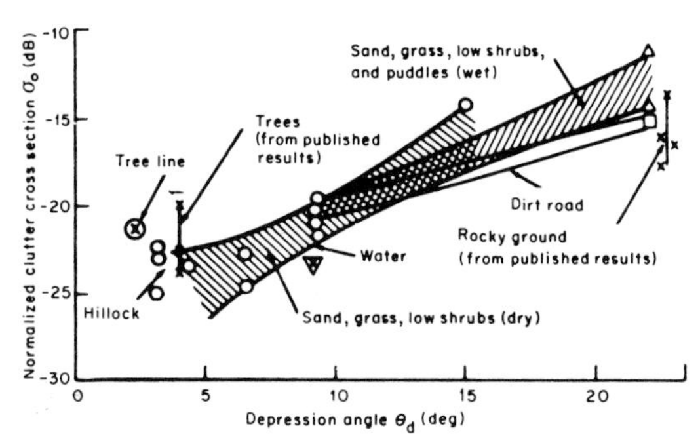

FIG. 14 σ° as a function of depression angle at 94 GHz. (From Suits and Guenther, 1979.)

from land clutter is important for probability of detection analyses and has not been widely reported. Currie *et al.* (1975a) have measured fluctuations from land clutter such as trees and reported that the distributions were decidely non-Rayleigh in shape, but could be approximated by a log normal distribution. Figure 16 shows histograms showing the probability of occurrence of various values of log normal standard deviation for the return from trees at 35 and 95 GHz. As can be seen, there is a consider-

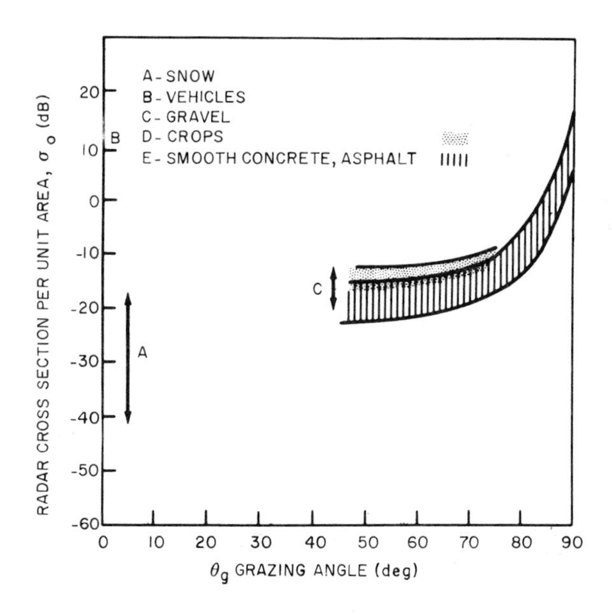

FIG. 15 σ° as a function of grazing or incidence angle for natural ground targets representative of values over the 40–90-GHz frequency range. (From Suits and Guenther, 1979.)

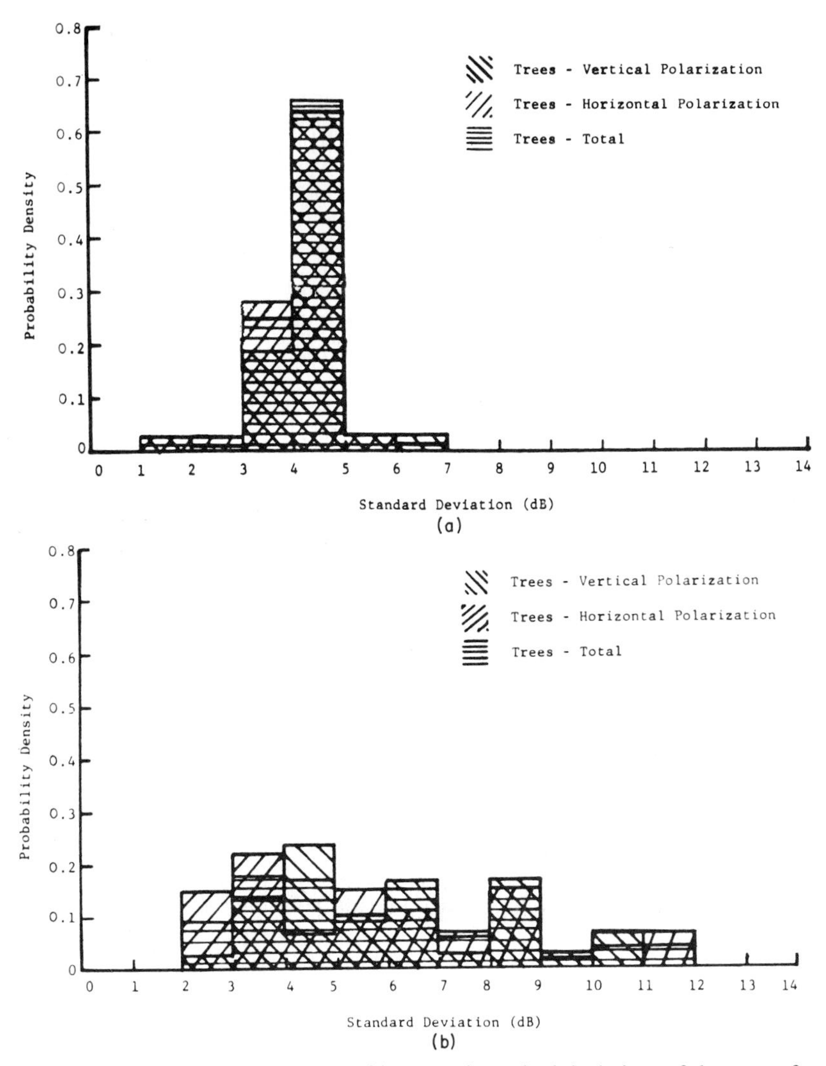

FIG. 16 Probability of occurrence of lognormal standard deviations of the return from trees for horizontal and vertical polarizations at frequencies of 35 and 95 GHz (Currie *et al.*, 1975a). (a) 35 GHz. (b) 95 GHz.

able spread in the nature of the fluctuations, but in general they have a range of amplitude fluctuations that is substantially wider than Rayleigh (which would correspond to a log normal standard deviation of from 2 to 3 dB).

The spectrum of the return is also important, and as far as is known, there have been no spectral investigations of the return from land clutter

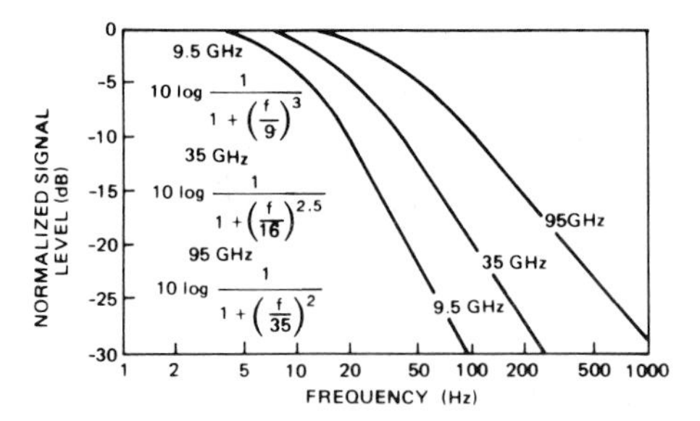

FIG. 17 Frequency spectra of the incoherent backscatter from trees at frequencies of 9.5, 35, and 95 GHz. Data have been normalized to the zero-frequency component. (From Trebits *et al.*, 1978.)

performed utilizing truly coherent systems. A limited amount of incoherent spectra have been collected and analyzed, and data for 35- and 95-GHz returns are presented in Fig. 17, along with an empirical fit to the spectral shape suggested by Fishbein *et al.* (1967). Figure 18 gives the same type of data for 70-GHz clutter. It has been suggested that the spectral shape is non-Gaussian, and Figs. 17 and 18 show spectral distributions that are substantially broader than Gaussian spectra. Decorrelation is intimately related to frequency spectra, and Fig. 19 gives decorrelation

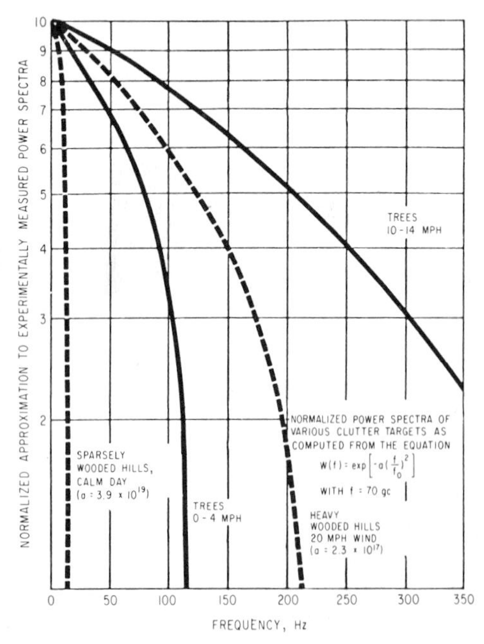

FIG. 18 Normalized clutter spectrum of the incoherent return from wooded ground at 70 GHz. (From Kosowsky, 1974.) Intermittent line, calculated; solid line, experimental.

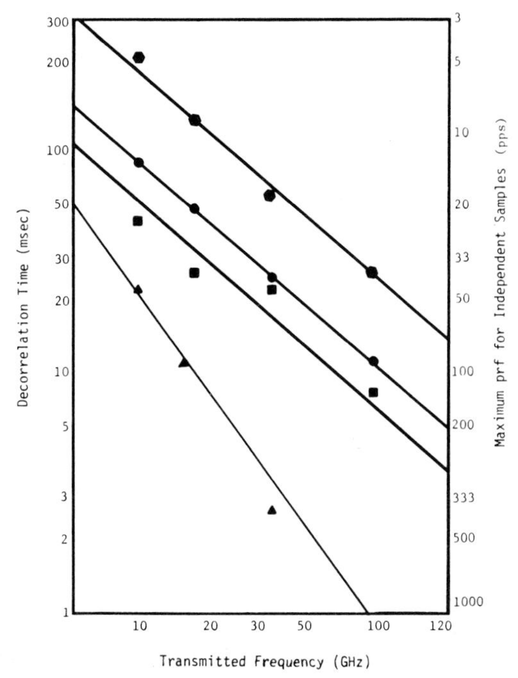

FIG. 19 Decorrelation time as a function of windspeed and frequency for the return from deciduous trees (Currie *et al.*, 1975a). Wind speed: ○, ~2 mph; ●, 3–5 mph; ■, 6–10 mph; ▲, 11–14 mph.

time as a function of wind speed for tree clutter over a range of millimeter frequencies.

The radar reflectivity of snow is a subject of some interest for a number of millimeter-wave system designers. While the reflectivity of snow is a function of a number of parameters, it appears to be primarily driven by the surface roughness, and by the free water content of the snow. The effect of free water is clearly shown by the variations in reflectivity during the course of a day; Fig. 20 is a plot of σ° as a function of time along with a plot of the percent of free water in the upper ten centimeters of the snow surface, clearly showing the connection between free water content and radar reflectivity. Recently, a series of measurements utilizing scatterometers at 35, 98, and 140 GHz have been carried out with a system that permits a rather wide range of incidence angles to be explored; unfortunately, the illuminated area is relatively small with this system, but there is substantial agreement between the results obtained and results from more realistic geometries. Figure 21 shows the parallel and cross-polarized backscatter coefficient as a function of incident angle for wet

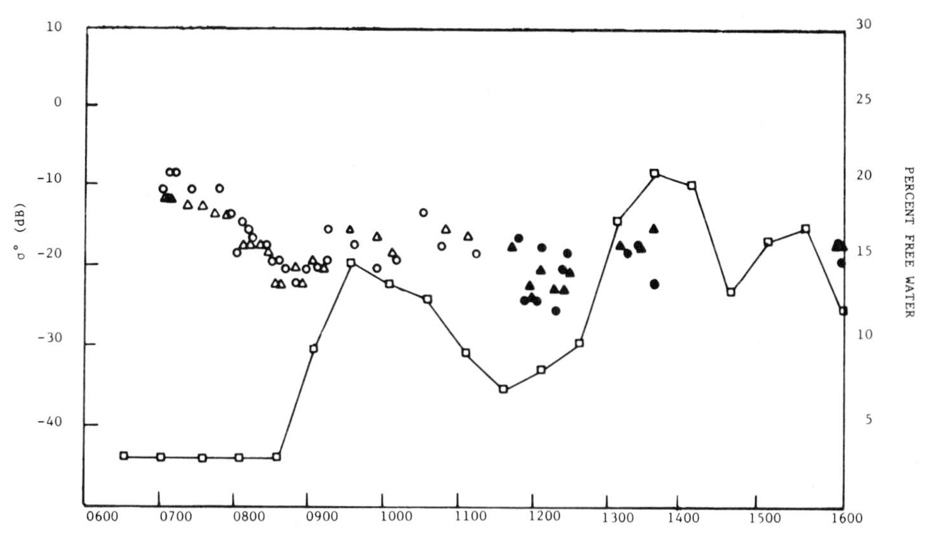

FIG. 20 σ° for snow-covered terrain at 35 GHz over an extended time period. The free-water content is also plotted, illustrating the correlation between free water content and snow reflectivity. (From Currie *et al.*, 1976a. Reprinted from *1976 Record AP-S International Symposium*, pp. 579–582. © 1976 IEEE.) April 13, 1976: △, HH 15° depression; ○, VV 15° depression; ▲, HH 13° depression; ●, VV 13° depression; □ % free water, upper 10 cm.

and dry snow at frequencies of 35, 98, and 140 GHz, measured using this scatterometer.

B. SEA CLUTTER

While one might expect the radar backscatter from the surface of the ocean to be a much more easily described process than that from land, here again a number of variables strongly impact the values of sea clutter return that are actually measured. Among these are depression angle, polarization, and sea state; there are a number of other dependencies that may be important, such as wind direction, surface currents, surface contaminants, and the presence of evaporation layers or ducts over the ocean's surface. One of the more dramatic dependencies of radar sea clutter return is its variation with depression angle, as shown in Fig. 22; here one can see the so-called plateau region, the interference region, and the high angle region. For many shore-based and airborne applications, depression angles in the plateau and interference regions are of primary interest.

Only fragmentary data are available that describe the radar sea clutter return at frequencies above 35 GHz (e.g., Tolbert *et al.*, 1957; Rivers, 1970; Long *et al.*, 1965; Long, 1975; Boring *et al.*, 1957; Wiltse *et al.*, 1957; Dyer *et al.*, 1974; Dyer and Currie, 1978; Trebits, *et al.*, 1979). Examina-

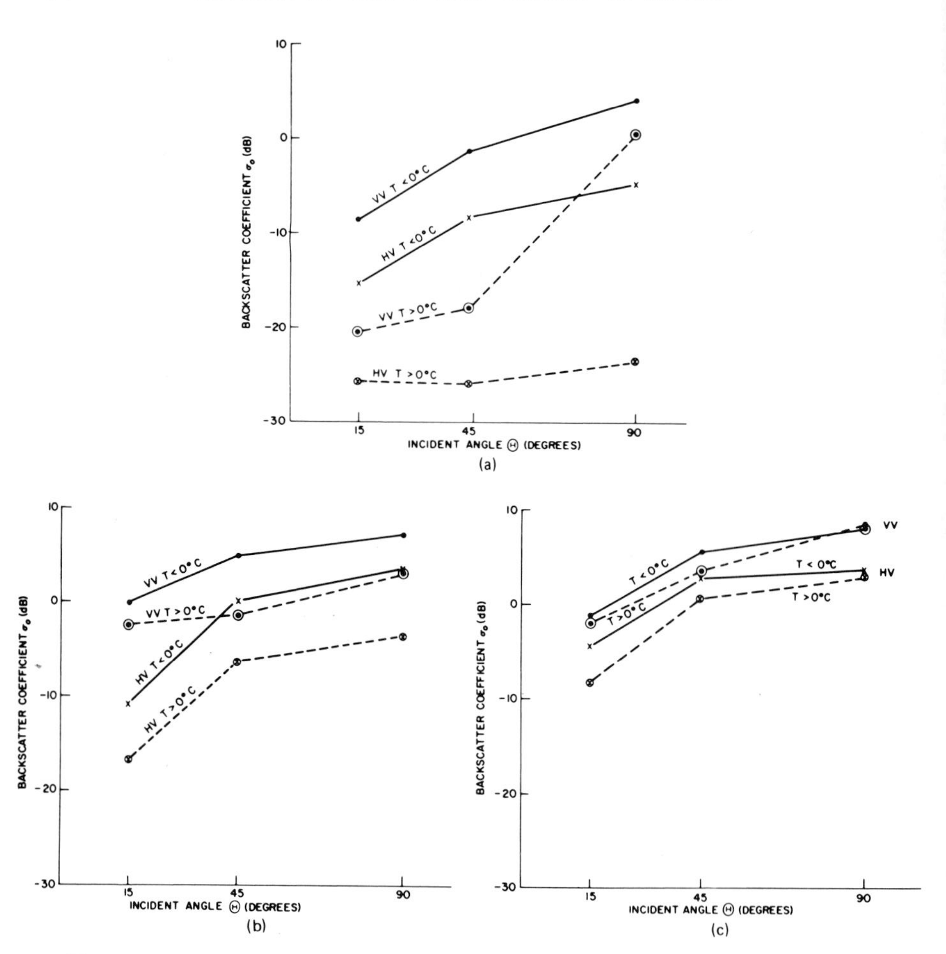

FIG. 21 σ° for wet and dry snow as a function of incident angle for wet ($T > 0°C$) and dry ($T < 0°C$) snow at (a) 35, (b) 98, and (c) 140 GHz. (From Hayes *et al.*, 1979. Reprinted from *1979 International Symposium Digest—Antennas and Propagation*, pp. 499–502. © 1979 IEEE.)

tion of a number of these returns indicates that the increases in radar sea clutter return with increasing frequency that had been observed at lower frequencies do not continue into the millimeter-wave region, particularly for vertically polarized returns. Figure 23 is a summary of a body of sea clutter data taken in the plateau region showing the less dramatic increase with frequency at millimeter wavelengths. Also note that the vertically polarized return is substantially smaller than the horizontally polarized signals, at 95 GHz, which is a distinct reversal of the behavior at lower frequencies, where the average value of HH polarized return is typically

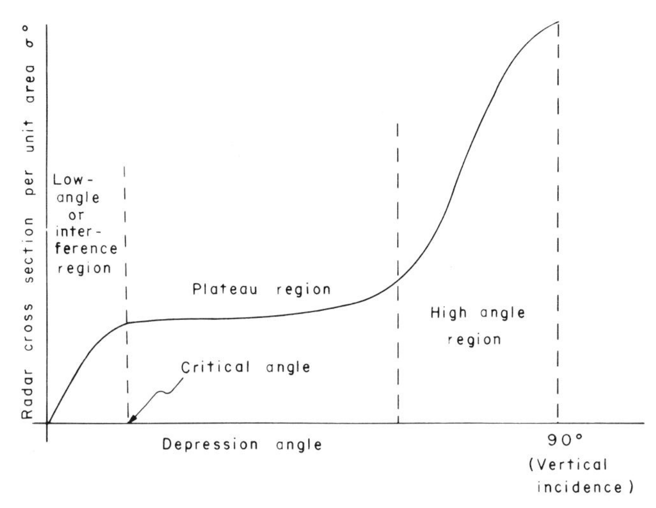

FIG. 22 General behavior of radar sea clutter as a function of incident angle, illustrating the three major angle regimes: the low angle or interference region, the plateau region, and the high angle region. The critical angle is the transition between the interference region and the plateau region.

smaller than VV polarized clutter. This is more graphically demonstrated in Fig. 24, which shows near-simultaneous measurements of σ°_{HH} and σ°_{VV} taken at 95 GHz; more recent measurements taken at the same frequency have verified that this behavior was not an isolated case. Figure 25 is a plot of σ° as a function of depression angle for "smooth" and "rippled" seawater over the angle regions from 45° to 90°, presented as representative values over the frequency range from 40 to 90 GHz.

At millimeter wavelengths, sea clutter returns have amplitude distributions that differ from Rayleigh, appearing to be approximately log normal, as shown in Fig. 26, which compares cumulative distributions for horizontally and vertically polarized returns with a Rayleigh distribution. Representative values of log normal standard deviation are 7 dB for horizontal and 4.7 dB for vertically polarized returns at 95 GHz (Rivers, 1970; Dyer *et al.*, 1974). The autocorrelation functions associated with millimeter-wave radar sea clutter evidence the same general properties as the sea clutter autocorrelation functions at lower frequencies, i.e., an initial rapid decorrelation, probably associated with the rearrangement of Doppler scatterers on the surface, followed by a more gradual decrease associated with rearrangement of the wave surface, which may be in turn followed by a periodic component associated with the passage of waves through the radar resolution cell.

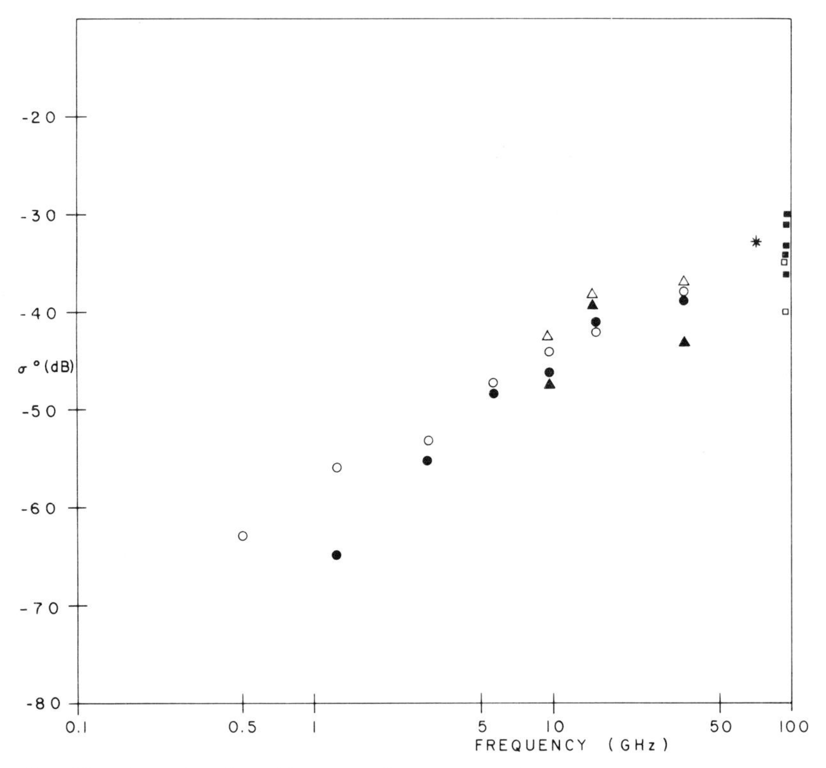

FIG. 23 σ° as a function of frequency for radar sea clutter returns in the plateau region. Note in particular that the low frequency trends are not extended into the millimeter region, particularly for vertical polarization. Grazing angle 1°–1.4°; sea state 2–3. ○, VV; ●, HH (Nathanson, 1969): 1°, averaged wind direction. △, VV; ▲, HH (Dyer *et al.*, 1974. Reprinted from *1974 International IEEE/AP-S Symposium and Digest*, pp. 319–322. © 1974 IEEE.): 1° upwind. □, VV; ■, HH (Rivers, 1970): 1.4° upwind. *,(Tolbert *et al.*, 1957).

IV. Technology Base

The configuration of many millimeter-wave radars is not dissimilar to that of radars operating at lower frequencies. However, there are a number of distinct difficulties encountered when implementing a radar at millimeter wavelengths, primarily associated with the decrease in efficiency of transmitter sources and local oscillators with increasing frequency, and the increased losses that are present in transmission lines at these shorter wavelengths.

A. RECEIVERS

While direct detection of a signal received by a millimeter-wave radar is a possibility, the lack of sensitivity of such an approach argues against its use in many applications. If low-noise/high-gain amplifiers were directly

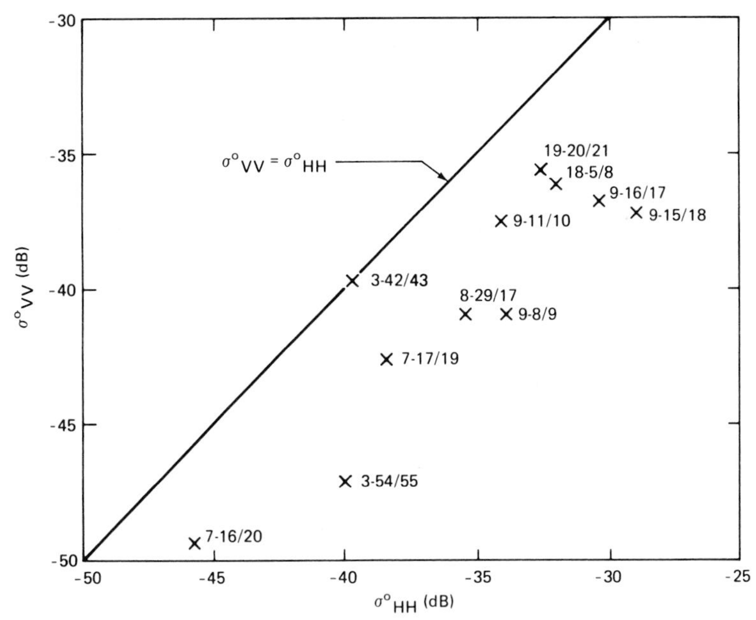

FIG. 24 Scatter diagram of σ° for vertical and horizontal polarizations at 95 GHz. (From Rivers, 1970.)

available at these frequencies, some of these disadvantages could be circumvented; since this is not the case, it is conventional practice to utilize a low-noise mixer to down-convert the received signal to frequencies that are more easily handled. Schottky diodes are almost universally utilized at millimeter frequencies for such mixers. The two common mixer arrangements are to utilize the diode as a conventional mixer, or to utilize harmonic mixing, where a local oscillator at a submultiple of the desired frequency is applied to the diode, and the nonlinear action of the mixer diode generates sufficient signal for mixing at the desired frequency.

A cross-section view of a Schottky barrier diode is given as Fig. 27, and a drawing of such a diode inserted into a waveguide structure is shown as Fig. 28. Figure 29 gives a representation of double sideband noise figure as a function of frequency for such diodes.

As outlined above, at higher operating frequencies the generation of suitable local oscillator signal may be a difficult problem, and harmonic mixing may be utilized to drive the mixer. Figure 30 shows such a harmonic mixing arrangement, and Fig. 31 gives the conversion loss as a function of harmonic number of such operation. It has been reported that there are some difficulties when employing harmonic mixing in radar receivers that operate with relatively short pulses.

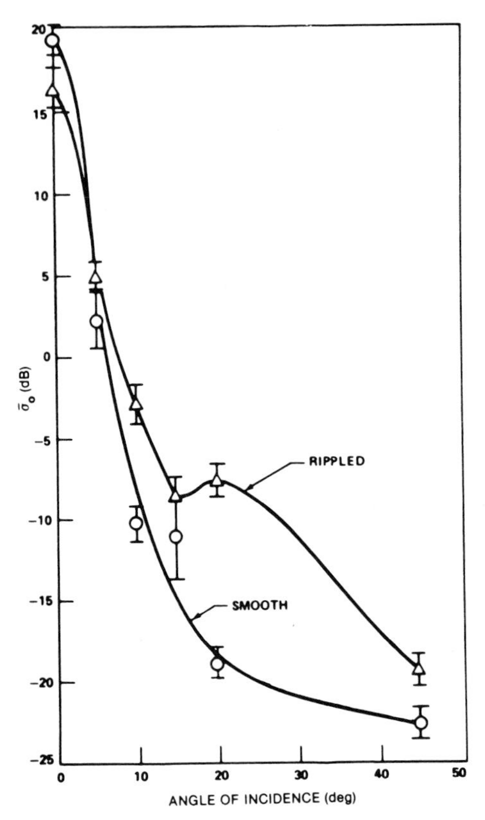

FIG. 25 Mean value of σ° for smooth (○) and for rippled (△) water as a function of angle of incidence, representative of returns over the 40–90-GHz frequency range. (From Suits and Guenther, 1979.)

B. TRANSMITTERS

Transmitters for millimeter-wave radars may be either solid-state or thermionic devices. The solid-state devices are primarily the IMPATT and the Gunn devices. Thermionic sources include magnetrons, traveling wave tubes (TWTs), klystrons, extended interaction oscillators (EIOs), backward wave oscillators (BWOs), and gyrotrons. Figure 32 shows several high-power transmitting devices, including EIOs, magnetrons, and an IMPATT diode oscillator.

Most of the high-power, pulsed, solid-state sources at millimeter wavelengths are IMPATT diodes, and they may be used as amplifiers or oscillators. A cross-sectional view of a millimeter-wave IMPATT diode package is given in Fig. 33. This package may be mounted in the wave-

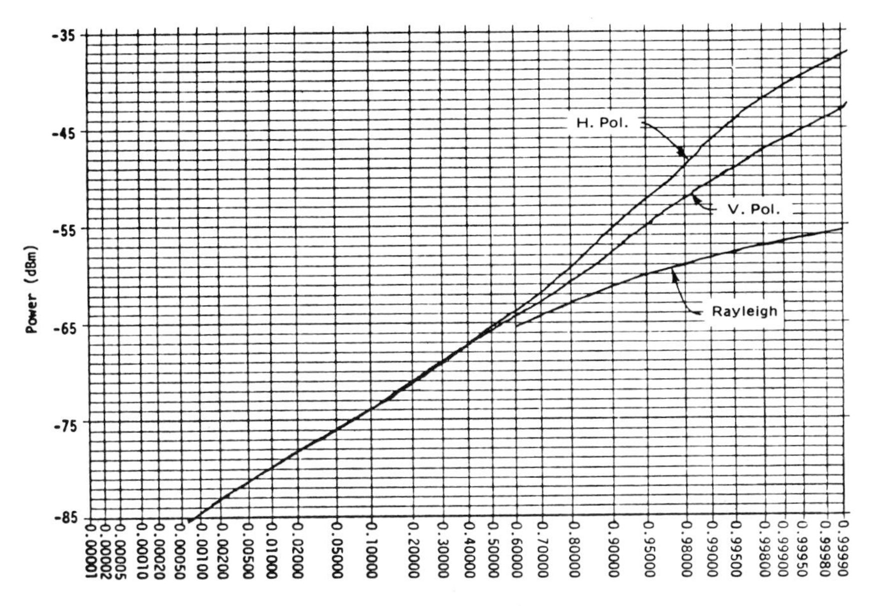

FIG. 26 Representative cumulative probability distributions for sea clutter for horizontal and for vertical polarizations from sea clutter. The cumulative probability for a Rayleigh distribution is included for comparison. (From Dyer *et al.*, 1974. Reprinted from *1974 International IEEE/AP-S Symposium and Digest*, pp. 319–322. © IEEE.)

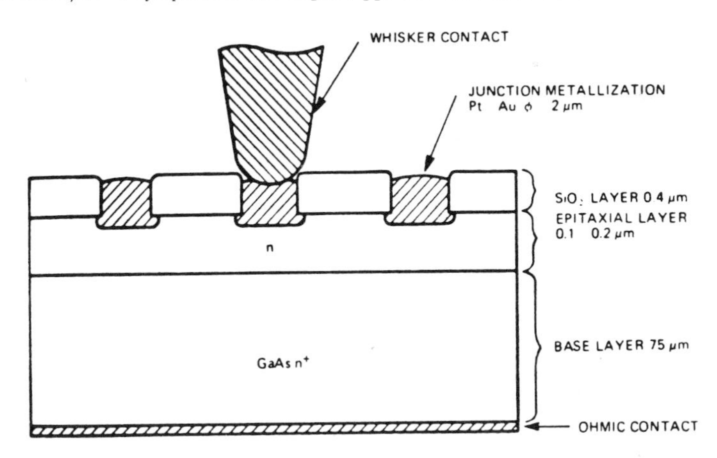

FIG. 27 Cross-section view of a Schottky barrier mixer diode. Shown are the array of "moat etched" junctions on GaAs substrate with the whisker contacting one of the chips. (From Schneider, 1979.)

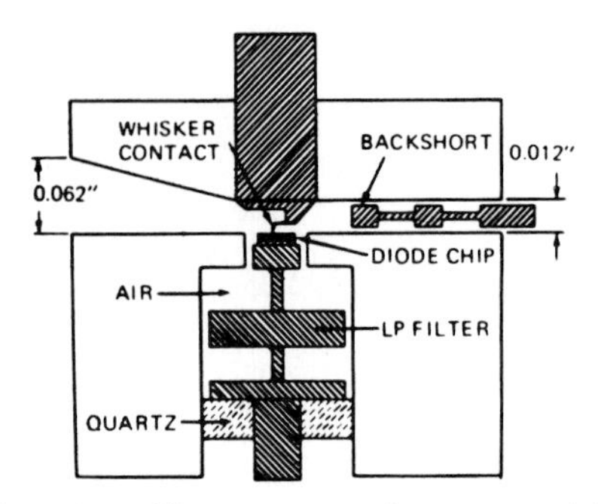

FIG. 28 Cross section of a millimeter-wave mixer mounted in a waveguide section. Shown is the taper to reduced height waveguide, sliding backshort, IF low-pass filter, and the diode mounted in the reduced height waveguide section. (From Schneider, 1979.)

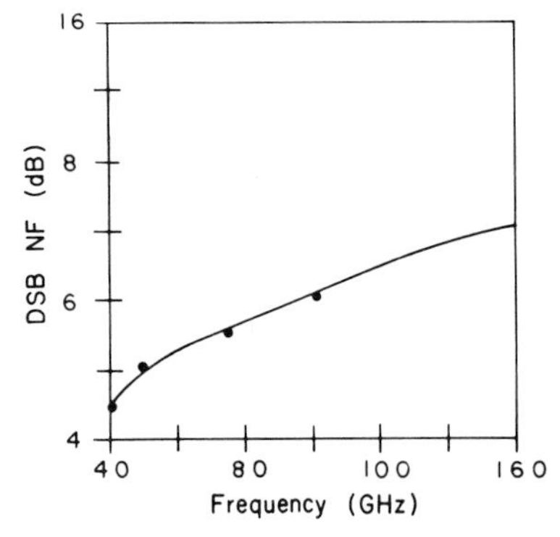

FIG. 29 Double sideband noise figure as a function of operating frequency. (Data from Hughes Millimeter-Wave Receiver Products Catalog, May 1979.)

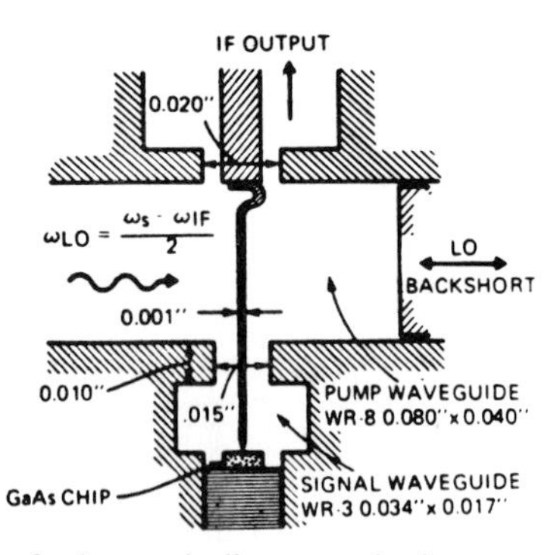

FIG. 30 Cross section of a harmonically pumped mixer operating at 230 GHz. (From Schneider, 1979.)

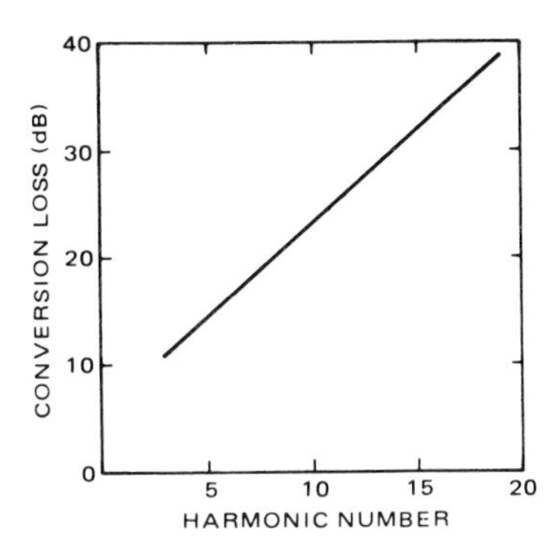

FIG. 31 Conversion loss for millimeter-wave harmonic mixers employing silicon Schottky diodes. The harmonic number indicates the multiple of the pump frequency. (From Kramer, 1979.)

guide in two different manners as shown in Fig. 34, providing tuning of the mount by both the waveguide and the coaxial section so that both the real and imaginary parts of the device may be matched. Figure 35 gives output power, frequency, and efficiency for a 94-GHz IMPATT oscillator as a function of diode or bias current. It should be noted that both current

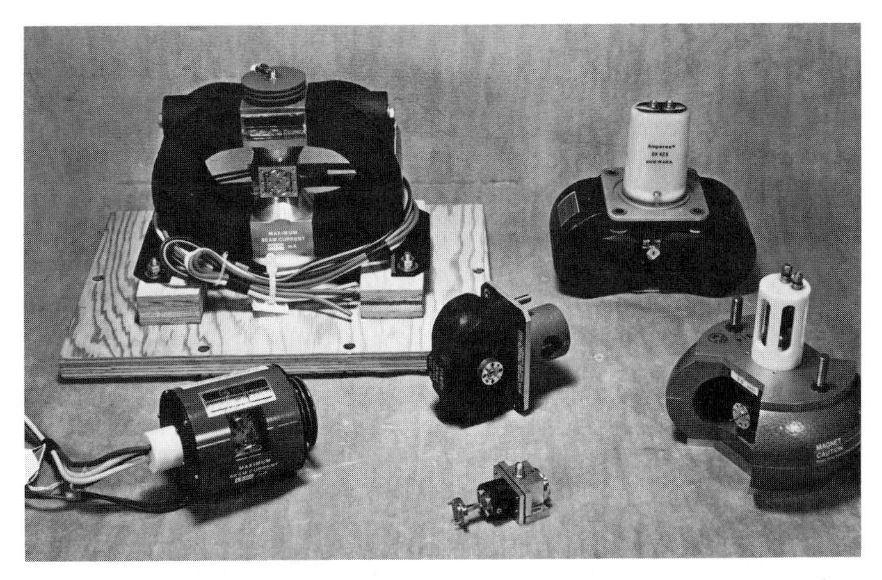

FIG. 32 Photograph of several commonly used high-power millimeter-wave transmitter sources. In the middle is a 5-W Gunn oscillator, surrounded by (from left to right) an EIO with a samarium–cobalt magnet, and EIO with an Alnico magnet, a 500-W 70-GHz magnetron, a 6-kW 95-GHz magnetron, and a 1-kW 95-GHz magnetron.

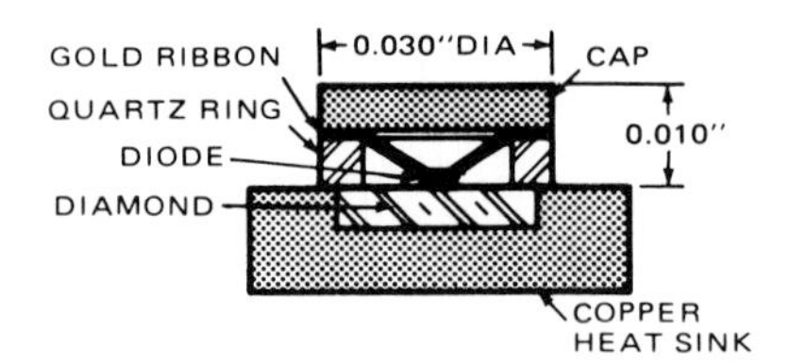

FIG. 33 Cross-section view of a millimeter-wave IMPATT diode package. (From Kuno and Fong, 1979.)

changes and thermal heating during the pulse may affect the frequency of oscillation of the pulse (Bellmare and Chudobiak, 1979); for that reason, control of pulse shape as shown in Fig. 36 is sometimes recommended to provide a degree of control of the transmitted pulse spectrum. However, even with such compensation, the output of IMPATT devices sometimes tends to be broadband with considerable phase noise, leading to efforts to injection lock such diodes; typically, injection locking with about 13 dB gain considerably improves the output spectral shape (Chang, *et al.*, 1979). Figure 37 gives power output versus frequency for both cw and pulsed IMPATT oscillators, and the outputs of several amplifiers may be combined to increase the total output power (Quine *et al.*, 1978). It may not be unreasonable to expect to see hundreds of watts produced by pulsed, combined IMPATT diodes in the next few years (Kuno and Fong, 1979).

At the present time, 5–10 W are available from pulsed IMPATTS at 95 GHz (Chang *et al.*, 1979), and laboratory development models have evidenced 3 W at 140 GHz (Ngan and Nakaji, 1979), with devices at 225 GHz under development (Kuno and Fong, 1978); both IMPATT and Gunn oscillators are suitable as cw sources and may be quite useful where a few milliwatts of output power are desirable for use as a FM-cw or a biphase coded transmitter.

Magnetrons may be used as sources for higher-power, pulsed trans-

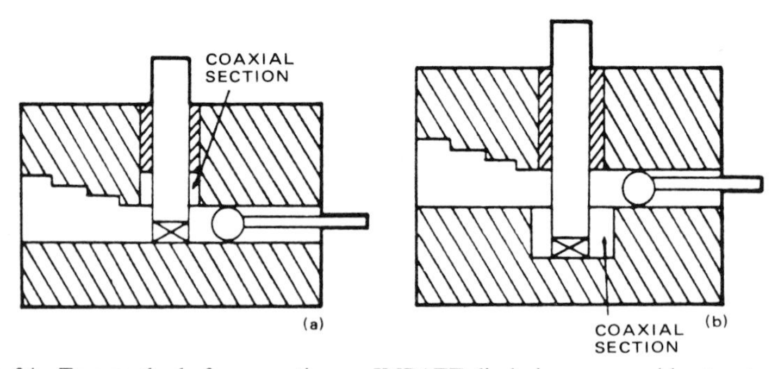

FIG. 34 Two methods for mounting an IMPATT diode in a waveguide structure, with provision for matching included. (From Kuno and Fong, 1979.)

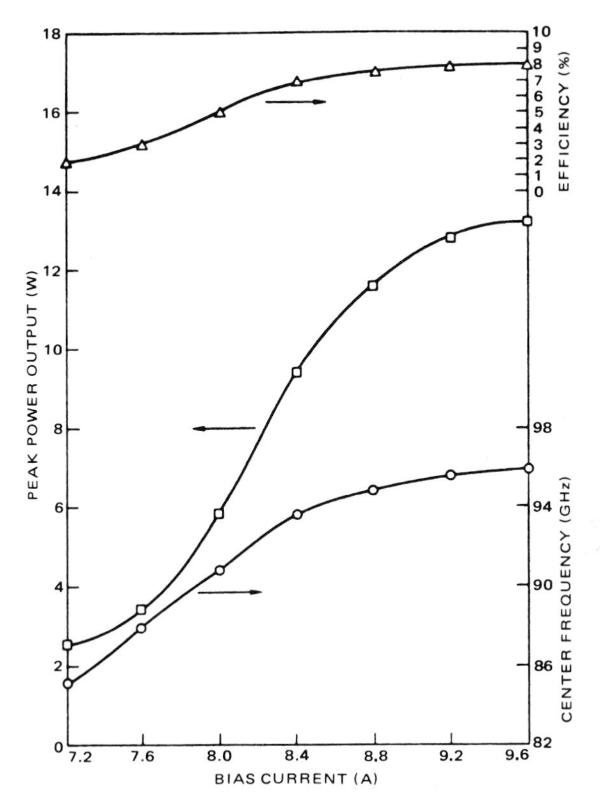

FIG. 35 Output power, frequency, and efficiency for a 94-GHz IMPATT oscillator as a function of current. (From Chang *et al.*, 1979. Reprinted from *1979 IEEE MTT-S International Microwave Symposium Digest,* IEEE Catalog No. 79CH1439-9MTT, pp. 71–72. © 1979 IEEE.)

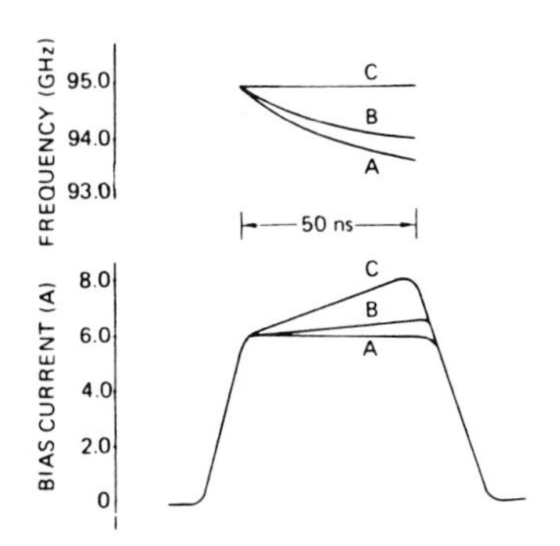

FIG. 36 IMPATT device frequency as a function of pulse shape. (From Kramer, 1979.)

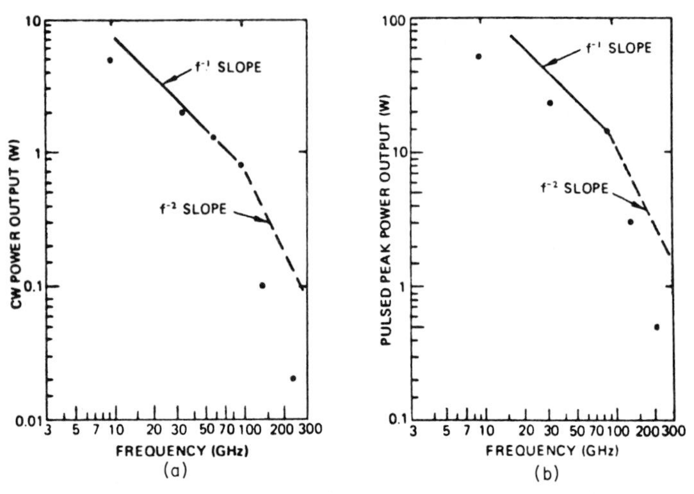

FIG. 37 Power output as a function of frequency for IMPATT oscillators. (From Kuno and Fong, 1979.)

mitters. At millimeter-wave frequencies, most magnetrons are of the "rising sun" type, although experimental inverted coaxial devices have been fabricated. Peak powers available range from 125 kW at 35 GHz, 10 kW at 70 GHz, 1–6 kW at 95 GHz, to 1 kW at 140 GHz. Typical duty cycles are 0.0002–0.0005, and efficiencies are of the order of 10%.

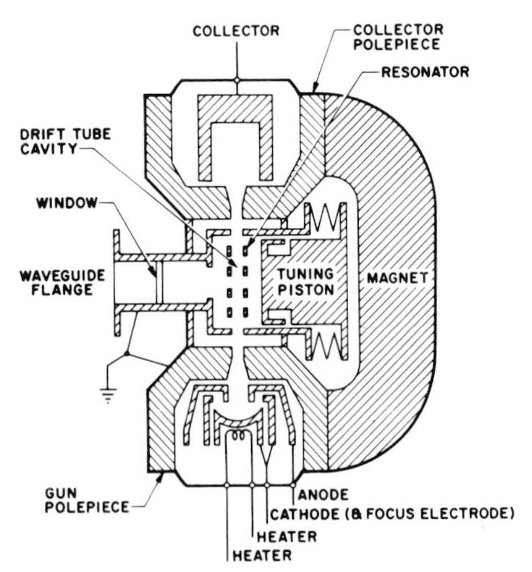

FIG. 38 Cross-section view of a millimeter-wave extended interaction oscillator. Note the removal of the cathode from the rf interaction region, permitting increased cathode area. (Courtesy Varian Associates of Canada.)

One of the problems with millimeter-wave magnetrons is the small cathode sizes that must be used; this results in high cathode current densities that may produce short tube life. In order to circumvent this problem, the extended interaction oscillator (EIO) was developed. Figure 38 shows a cross-sectional view of an EIO, illustrating the separation of the cathode from the interaction region, which permits a substantial increase in cathode area, and Table VI summarizes the characteristics of some pulsed millimeter-wave EIOs. The extreme voltage sensitivity of the EIO and its large value of stray capacity often necessitate special techniques for modulation of the tube (Ewell *et al.*, 1979, 1980; Ewell, 1981). Plans are under way for development of gridded EIOs and extended interaction amplifiers, but none are available at this time.

There has been little activity in millimeter-wave TWTs until recently, but Hughes has operated an experimental 1-kW peak, 250-W average power tube at 95 GHz (Henry, 1974), and a 60-GHz, 5-W tube for communications purposes has been developed (Pranter, 1978).

Backward wave oscillators (BWOs) are another source of cw power in the millimeter-wave region, and operation at frequencies as high as 1300 GHz has been achieved. However, more representative results are 10 mW over the 325–390 GHz band, or 5 W at 280 GHz (Epsztein, 1978).

In recent years, a new class of microwave and millimeter-wave oscillators and amplifiers has been developed. These devices, called gyrotrons, or electron cyclotron masers, show promise of providing peak powers at millimeter-wave frequencies that are considerably higher than obtainable using previous techniques (see e.g., Granatstein *et al.*, 1975; Hirshfield and Granatstein, 1977; Flyagin *et al.*, 1977; Zapevalov *et al.*, 1977; Ahn, 1978; Jory *et al.*, 1977, 1978).

Gyrotron devices typically utilize a relativisitic electron beam and convert constant electron energies to microwave energies in an intense electromagnetic field. Initial results typically involved operation at megavolt levels, and the use of superconducting magnets was necessary to obtain the extremely high magnetic fields required; however, it is possible to

TABLE VI

CHARACTERISTICS OF SOME PULSED MILLIMETER EIOs

Tube type	Mech. tuning range (GHz)	Power output (peak, W)	Beam voltage WRT cathode (kV)	Anode voltage WRT cathode (kV)	Electronic tuning range (MHz)
VKV 2443	92.7–96.0	950–1770	21	12	300–360
VKT 2419	139.7–140.3	270	20	7.6	370
VKY 2429	225.5	70	21.3	8.2	400

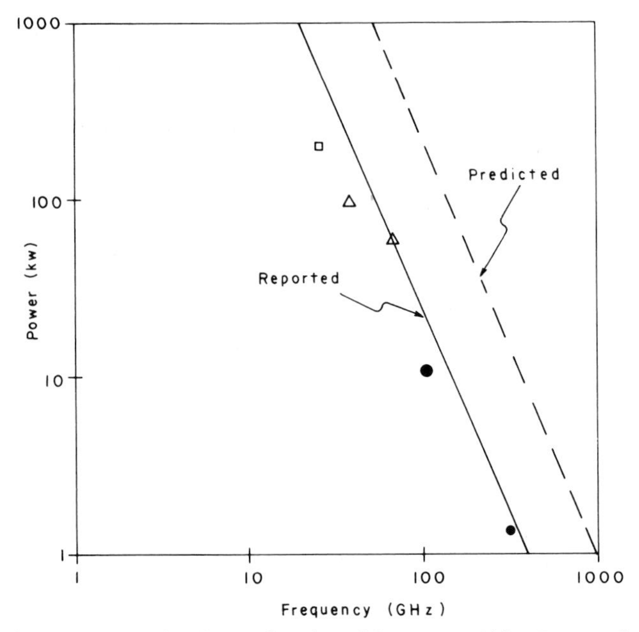

FIG. 39 Gyrotron power levels as a function of frequency. (After Jory *et al.*, 1977. Reprinted from *Proc. 1977 Int. Electron Devices Conf.*, pp. 234–237. © 1977 IEEE.) □, Varian millisecond pulsed; △, Soviet millisecond pulsed; ●, published Soviet cw data.

build devices having much more modest voltage and magnetic field requirements, although they operate at somewhat lower power levels.

Figure 39 is a presentation of the achievable and predicted gyrotron power levels over a range of frequencies, while Tables VII and VIII present additional details on a number of Soviet devices. It should be mentioned that relativistic devices are capable of substantially higher peak powers, but due to the large size and weight that are associated with such devices, it appears unlikely that they are suitable for many radar ap-

TABLE VII

PEAK POWER LEVELS FROM CYCLOTRON MASERS DRIVEN BY INTENSE RELATIVISTIC ELECTRON BEAMS

Wavelength (cm)	Peak microwave power (MW)	Accelerating voltage (MV)	Diode current (kA)
4	900	3.3	80
2	350	2.6	40
0.8	8	0.6	15
0.4	2	0.6	15

TABLE VIII

REPORTED GYROTRON OPERATING CONDITIONS AND OUTPUT PARAMETERS

Model number	Mode of oscillation	Wavelength (mm)	CW or pulsed	Harmonic number	B-field (kG)	Beam volts (kV)	Beam amps	Output power (kW)	Measured eff. (%)	Theoretical eff. (%)
1	TE_{021}	2.78	CW	1	40.5	27	1.4	12	31	36
2	TE_{031}	1.91	CW	2	28.9	18	1.4	2.4	9.5	15
	TE_{231}	1.95	Pulsed	2	28.5	26	1.8	7	15	20
3	TE_{231}	0.92	CW	2	60.6	27	0.9	1.5	6.2	5

plications. Thus, the data presented in Fig. 39 and Table VIII appear to be representative of present and projected gyrotron capability.

Up to this point, discussions have centered on the utilization of gyrotrons as oscillators; however, it is possible to configure gyrotrons as amplifiers as well, by providing appropriate input and output couplings in order to bunch initially the electron beams and to extract energy from the resulting beam. To date, most activities have concentrated on the implementation of a gyro-klystron as an amplifier in the lower portion of the millimeter-wave spectrum, and 200 kW have been achieved at 28 GHz.

C. Antennas

At millimeter-wave frequencies, conventional front fed reflector antennas as shown in Fig. 40 may be utilized; however, at millimeter wavelengths, waveguide losses associated with the length of the waveguide run to the feed located at the focus may become excessive. In such cases, the use of Cassegrain antennas, such as shown in Fig. 41, may provide a significant advantage.

Fig. 40 Front-fed paraboloidal dish millimeter-wave (35 GHz) antenna.

FIG. 41 Cassegrain 95-GHz antenna. Such systems permit short waveguide runs and may minimize aperture blockage.

There are a number of situations where it is not desirable to support a microwave feed or subreflector in the aperture, and for such a situation a dielectric lens antenna may be attractive. Such lenses are usually "zoned" or excess material removed, as shown in Fig. 42, in order to reduce weight while still maintaining its desirable focusing properties.

Generation of a rapidly scanning beam for target acquisition or surveillance is a challenging problem and at the higher microwave frequencies may become quite difficult. Due to the fact that phased array techniques are not readily applicable in the higher frequency portion of the millimeter-wave region, it is often attractive to consider the use of electromechanical scanning antennas for such applications. One useful approach is the geodesic Luneburg lens, which has the property of focusing a point source at one edge to a uniform phase front on the diametrically opposite edge, as shown in Fig. 43; the actual lens itself is a parallel plate metal lens. Figure 44 shows a cross-sectional drawing of a geodesic Luneberg lens fabricated for 70-GHz operation. This antenna utilizes a parabolic reflector to provide beam shaping in the vertical direction, and scans in the azimuth direction at up to 70 scans per second over a 30° scan sector; beamwidth is $0.55° \times 3.5°$ for this system and is "folded" to reduce its height (Goodman and Dyer, 1968). Such antennas are relatively low-loss devices and have antenna patterns that are not a sensitive function of frequency (Johnson, 1963).

FIG. 42 Zoned dielectric millimeter focusing lens.

Geodesic lenses are not the only type of parallel plate antenna that can be utilized. A pillbox-type antenna has been fabricated for operation at 95 GHz, and is shown in Fig. 45 (Bodnar *et al.*, 1973). This system generated an antenna beam 0.11° × 1.5°. This type of antenna may be used to scan the beam and to track targets in a track-while-scan mode.

It is sometimes desirable to utilize a conventional null tracking system, and conventional monopulse or conical scan approaches may be used at millimeter wavelengths. Conical scan operation may be achieved by rotating either the feed or a tilted subreflector in a Cassegrain system. Monopulse operation can be achieved using a Cassegrain antenna fed with a monopulse feed horn or by utilizing a lens fed by a monopulse feed arrangement. Figure 46 shows a four-horn monopulse feed designed for operation at 95 GHz and used to feed a stepped dielectric lens.

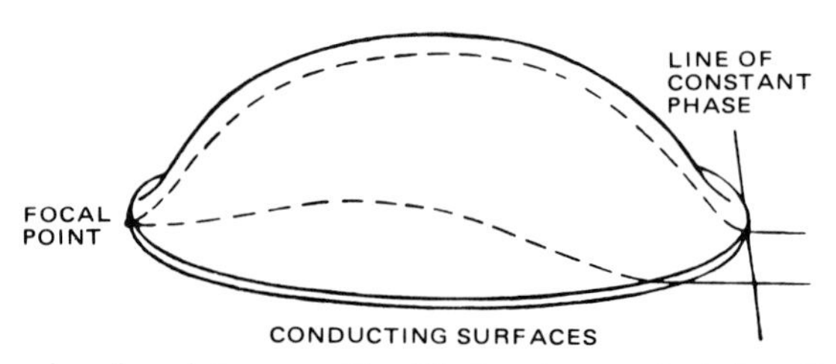

FIG. 43 Focusing of a point on one side of the lens to a constant phase front on the opposite side by means of a parallel metal surface geodesic Luneburg lens. (From Johnson, 1962.)

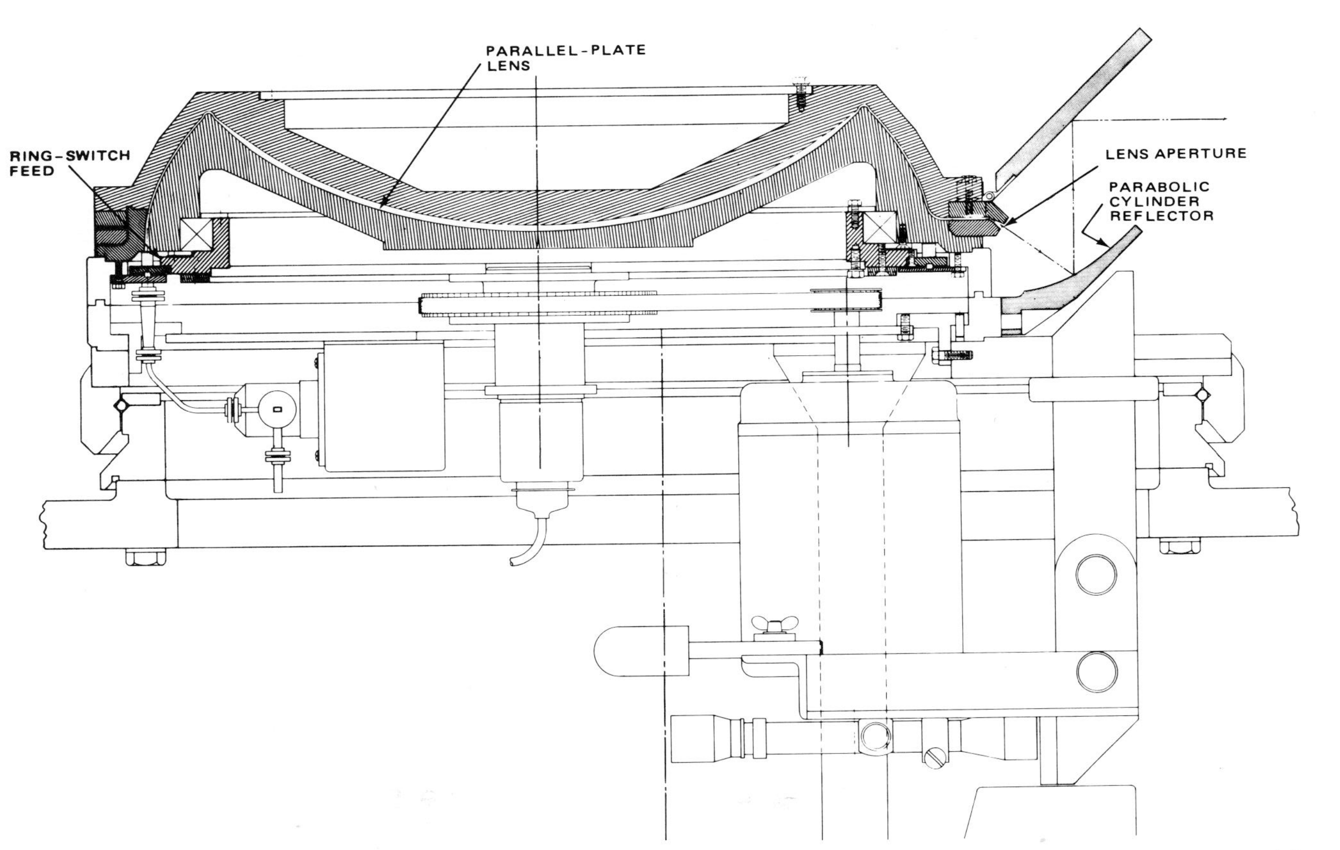

FIG. 44 Cross-section view of a 70-GHz Luneburg lens. The lens is folded to reduce height and uses a ring switch to scan rapidly the feed horn. (From Goodman and Dyer, 1968.)

FIG. 45 A 95-GHz pillbox-type antenna, producing a 0.11° × 1.5° beam. (From Bodnar *et al.*, 1973.)

D. Passive Components

A complete spectrum of the basic components required to support radar system development at 35 GHz is currently available. This includes such components as waveguide, connectors, transitions, attenuators, hybrids, filters (preselectors), circulators, couplers, isolators, and duplexers. Some advanced component development has also taken place at higher frequencies extending up to 95 GHz. Table IX presents some of the more important performance characteristics for several passive electro-formed components in the 35- and 60-GHz regions (Raue *et al.*, 1974; Davis, 1976).

At frequencies above 35 GHz, however, considerable work still needs to be done in developing low-loss, wide bandwidth, and high isolation components that are lightweight, low power, and complement the small size advantages of millimeter waves. This is especially true at 140 and 220 GHz. Recognizing this fact, the Aeronautical System Division of the Air Force System Command is undertaking a program to advance millimeter componentry in the 100–300-GHz region. In the passive component area, this program will include efforts to develop bandpass filters with perform-

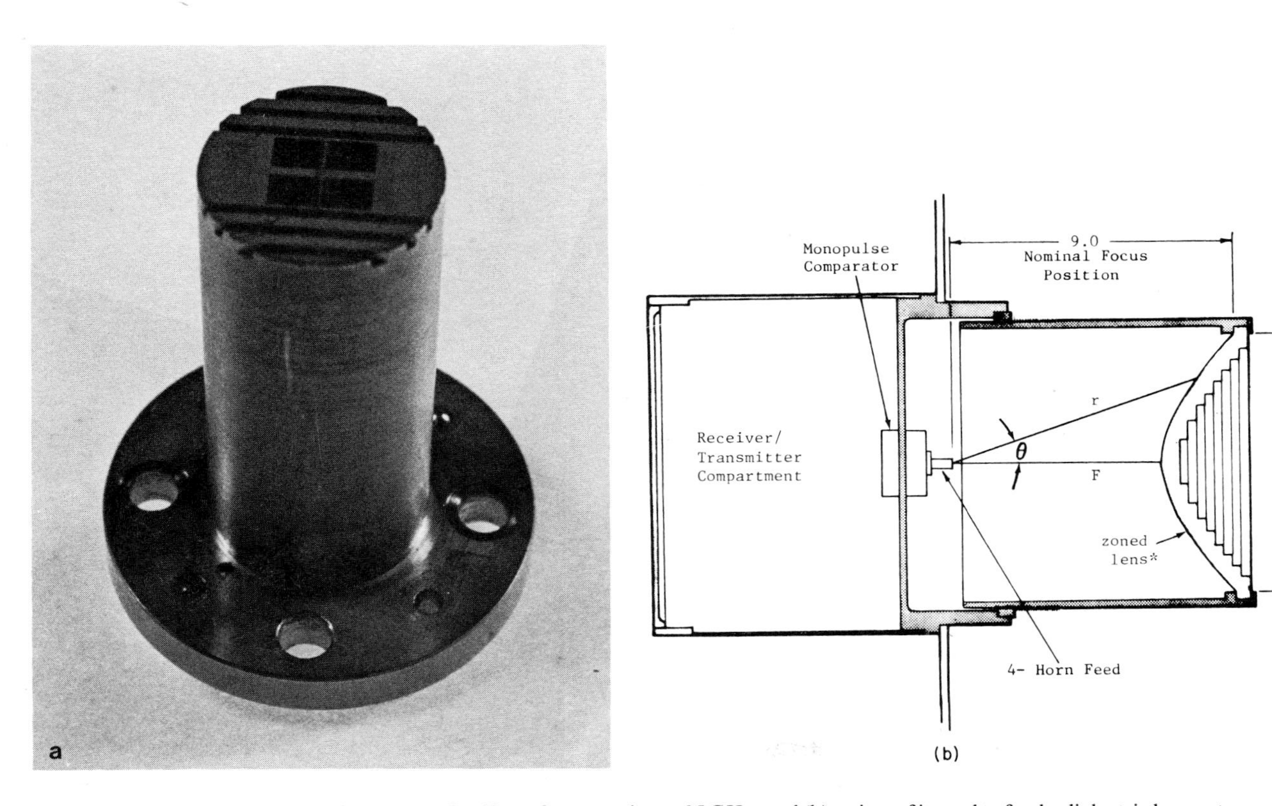

FIG. 46 (a) Four-horn monopulse antenna feed horn for operation at 95 GHz, and (b) a view of it used to feed a dielectric lens antenna. Each step equals 0.28041 in. (From Kozakoff and Britt, 1980. Reprinted from *1980 IEEE SOUTHEASCON*, pp. 65–68. © 1980 IEEE.)

TABLE IX

PERFORMANCE CHARACTERISTICS OF SEVERAL PASSIVE COMPONENTS[a]

Component	Center frequency (GHz)	Bandwidth	Insertion loss (dB)	Isolation (dB)	VSWR
3 dB short slot hybrid	30–37	4 GHz	0.2	25	1.1
	55–63	8 GHz	0.5	20	1.2
Fixed tuned bandpass filters	30–38	400 MHz	0.2	—	1.1
	55–65	200 MHz	0.3	—	1.2
High-pass filters	30–38	—	0.12	30[b]	1.2
	55–60	—	0.5	50[b]	1.25
Circulators	32–36	8 GHz	0.2[c]	20	1.1
	55–65	2 GHz	0.3	20	1.2
WG-coaxial transition	35	10 GHz	0.2	—	1.2
Passive power divider	37	6 GHz	0.2	20	1.2

[a] From Davis (1976).
[b] 1.0 GHz below design pass frequency.
[c] 0.1 dB over 5 GHz.

ance goals that include a 10-GHz bandwidth at frequencies centered at 140 and 240 GHz, 20-dB attenuation at 2 GHz from band edge, and 2–3 dB of midband insertion loss. Performance goals for the coupler development are ±1 dB coupling flatness across a full waveguide bandwidth, 1–3-dB insertion loss, >20-dB directivity, and coupling of 3, 6, or 10 dB at both 140 and 240 GHz.

In addition to the components listed above, the Air Force also specifies that ferrite materials be investigated to develop and demonstrate a junction circulator and Faraday rotation isolator. The circulators are to be operated at 140 and 240 GHz with 4- and 2-GHz bandwidths, respectively, 1–2-dB insertion loss, and 20-dB isolation. The isolator is to be designed for 20-GHz bandwidth, 1–2-dB insertion loss, and 30-dB isolation at the same frequencies.

Standard waveguide techniques and—to a limited extent—photolithographic, integrated circuits (microstrip) are being used to fabricate many of the passive millimeter components. However, because dimensions are extremely small and tolerance must be held within a fraction of a mil, photolithographic integrated circuits (ICs) for millimeter applications represent a difficult design and fabrication process not lending itself to low-cost, high-volume yields.

Above perhaps 100 GHz, where photolithographic techniques become difficult and inefficient, quasi-optical structures (dielectric waveguide, slabs, gratings, lenses, etc.) seem to perform well and offer advantages in construction because of their larger, dimensional tolerances. Consider-

TABLE X

PERFORMANCE CHARACTERISTICS FOR EXPERIMENTAL MILLIMETER PHASE SHIFTERS[a]

	Center frequency (GHz)	
Characteristic	34.5	75.5
Bandwidth (GHz)	34–35	74–77
Insertion loss (dB)	1.5	3.5–4.0
VSWR	<1.25	<1.25
Switching	3 bit of flux transfer	3 bit of flux transfer
Length (inches)	3	4.33

[a] From Babbitt *et al.* (1978).

able research is currently under way to develop quasi-optical techniques for frequencies above 100 GHz (Davis, 1976; Knox, 1976).

Only a limited amount of work has been reported in the literature on developing phase control elements at millimeter wavelengths, particularly above 35 GHz. Such devices are required for lightweight phase or electronically scanned antennas and for other switching and control functions in a practical millimeter radar. Experimental, nonreciprocal, toroidal phase shifters operating at center frequencies of 34.5 and 75.5 GHz have been developed with the characteristics shown in Table X (Babbitt *et al.*, 1978). However, tolerance control and machining requirements make fabrication of these devices with highly repeatable characteristics extremely difficult and expensive.

To circumvent the manufacturing and tolerance problems associated with the toroidal, nonreciprocal phase shifters, a new, and unique, arc plasma spray technique was developed by the U.S. Army Electronic Technology Laboratory. In this process, a toroid is formed by depositing a ferrite powder around a dielectric insert, thus forming a dielectrically loaded toroidal element. Using this technique, experimental phase shifters at 35, 65, and 95 GHz have been developed. Typical characteristics for these devices, as reported in Babbitt *et al.* (1978) are shown in the following tabulation:

	35 GHz	65 GHz	95 GHz
APS toroid length	3.43 cm	0.8 cm	0.64 cm
Insertion loss	0.7 dB	0.9 dB	1.5 dB
Bandwidth	0.6 GHz	1.0 GHz	1.0 GHz
Differential phase shift	422^0	87^0	38^0

An application of the arc plasma sprayed ferrite toroids as phase shifter elements in both a 35 and 95-GHz electronic scan line source antenna for a millimeter radar is discussed by Borowick *et al.* (1980).

V. Millimeter-Wave Radar Example

It is sometimes instructive to demonstrate certain basic concepts and limitations associated with a system design problem through a simple example. Some of the techniques required to determine approximate performance limits for a millimeter radar will be emphasized by analyzing the projected performance of two conceptual designs for a millimeter-wave, short-range, target acquisition system in this section.

The operational problem to be addressed is that of detecting a military target, such as a tank, jeep, or some other military ground vehicle, for subsequent engagement with a weapons system—perhaps a closed-loop, fire-control weapons system. Certain aspects of the problem are fixed, or are defined by constraints external to the radar, i.e., the antenna aperture size, available power, atmospheric constraints, etc. Antenna aperture size was chosen to be representative of a relatively small missile. A larger aperture would result in longer operational ranges. The fixed parameters for the two frequencies to be examined are given in the tabulation below. Many of these parameters have been either specified earlier in this chapter or previously calculated.

	Frequency (GHz)	
	94	140
Wavelength	3.2 mm	2.2 mm
Attenuation α (approximate)		
clear air	0.4 dB/km	1.5 dB/km
rain	3.0	3.2
Receiver noise figure, F_n	8.0 dB	10.0 dB
System losses L_s	4.0 dB	6.0 dB
Peak power P_t	10.0 W	2.5 W
(solid state source)		
Antenna aperture diameter	10 cm	10 cm
Beamwidth	2.3^0	1.6^0
Antenna gain G	36.9 dB	40.1 dB
Target cross section σ	30 m^2	30 m^2
(tank size target)		
Pulse width	50 nsec	50 nsec
(matched bandwidth B)	(20 MHz)	(20 MHz)

To reflect a more practical design situation and to include signal processing considerations in the problem, assume that the algebraic sum of signal processing gain (both coherent and incoherent integration of received pulses) and loss (mixer and detector losses, integration inefficiency, for example) for both radars totals 14 dB.

Using Eq. (7) in slightly rearranged form, the received S/N ratio as a function of range to the target can be calculated for both radars:

$$(\mathrm{S/N})_{\mathrm{R}} = \frac{P_t G^2 \lambda^2 \sigma 10^{-0.2\alpha R}}{R^4 (4\pi)^3 (KT_0 B) F_n L_s}. \tag{10}$$

Collecting numerical values for each parameter given above and expressing Eq. (10) in terms of dB, the following relationships apply:

$$94\ \text{GHz:}\ 10 \log(\mathrm{S/N})_{\mathrm{R}} = 134.7\ \text{dB} - 40 \log R - 2\alpha R_{\mathrm{km}}, \tag{11}$$

$$140\ \text{GHz:}\ 10 \log(\mathrm{S/N})_{\mathrm{R}} = 127.85 - 40 \log R - 2\alpha R_{\mathrm{km}}. \tag{12}$$

The 14 dB of assumed signal processing gain has not been included in these expressions.

The anticipated performance of these two conceptual designs can be graphically presented by solving Eqs. (11) and (12), adding the signal pro-

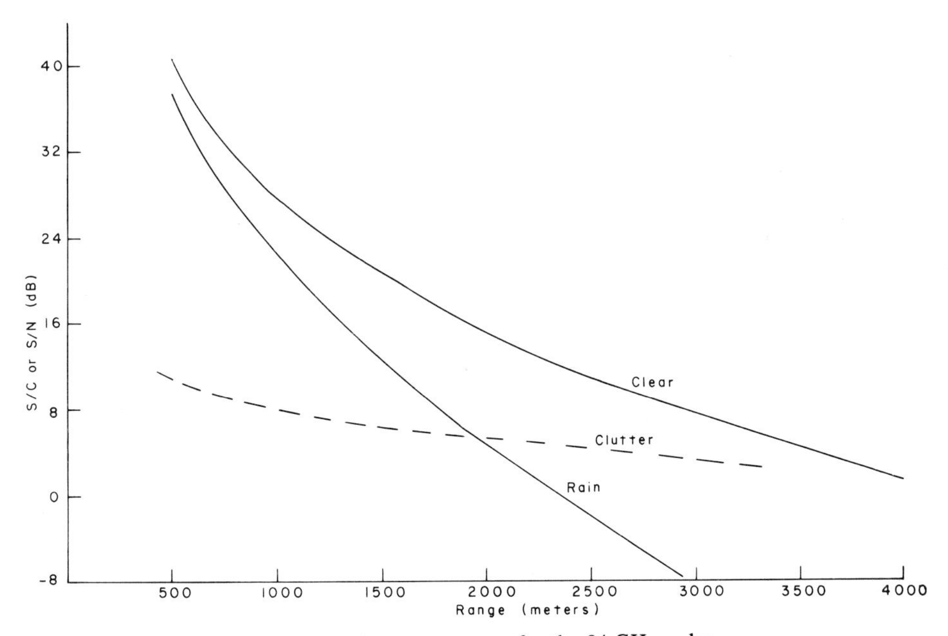

FIG. 47 Performance curves for the 94-GHz radar.

cessing gain and plotting the results. The resulting performance curves are shown in Fig. 47 for the 94-GHz radar, and in Fig. 48 for the 140-GHz radar.

Assume that a 12-dB S/N ratio is required to achieve a specified probability of detection P_d at a fixed probability of false alarm P_{fa} [Skolnik (1970) specifies a P_d of 0.7 at a P_{fa} of 10^{-6} for an integrated S/N ratio of 12 dB. From Figs. 47 and 48, the following performance limits can be determined for the two radars:

	Maximum range (at 12 dB S/N)	
	Clear	Rain
94 GHz	2350 m	1530 m
140 GHz	1400 m	1150 m

Thus the 94-GHz radar has a range advantage over the 140-GHz system of approximately 950 m in the clear and 380 m in the rain, under the conditions specified in this problem.

One additional aspect of this example should be examined: the effect of ground clutter in the same resolution cell masking the target and degrading detection probability. Using conventions established in preceding sections of this chapter, assume the following radar cross section per unit area values σ° to be typical for the depression angles of interest in this example:

	Frequency	
	94 GHz	140 GHz
σ^0	−15 dB	−12 dB

The values given above are to be considered as estimates of typical values for a general class of land clutter (grass, low trees, partially clear ground, etc.), not exact values since only a very limited data base for land clutter reflectivity currently exists at these frequencies.

The signal-to-clutter ratio S/C is the detection determining factor in situations where target detection is limited by clutter and is given by the ratio of the effective radar cross sections of the target and the clutter, σ and σ_c, i.e.,

$$S/C = \sigma/\sigma_c. \tag{13}$$

The effective clutter cross section σ_c is given by the product of the clutter cross section per unit area and the area of the land surface illuminated by

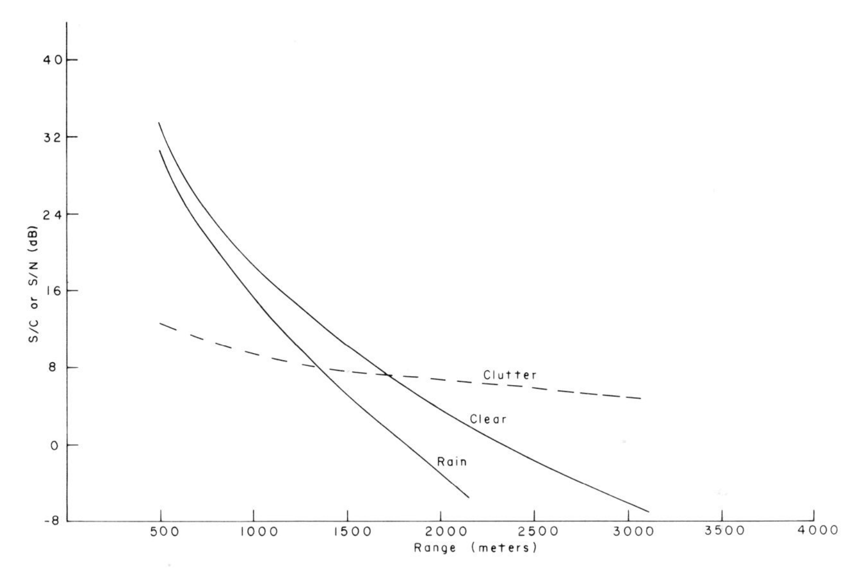

FIG. 48 Performance curves for the 140-GHz radar.

the radar A. For a pulse length limited situation at relatively low grazing angles ψ (Skolnik, 1970).

$$A = R\theta c\tau/2 \sec \psi, \tag{14}$$

where R is the range to target/clutter, θ the horizontal beamwidth, c the velocity of light (3×10^8 m/sec), and τ the pulse length. Therefore

$$\sigma_c = \sigma^\circ R\theta c\tau/2 \sec \psi. \tag{15}$$

Although land clutter can resemble thermal receiver noise under some conditions, in general, returns from land clutter have a much slower fluctuation rate and thus remain correlated for a much longer time. Because of this, receiver integration is much less effective for land clutter than for receiver noise. However, as Fig. 19 illustrates, millimeter-wave radar realizes some advantages in this regard since, typically, clutter decorrelation times t_c decrease with increasing frequency. Assume the following values are typical for the example under consideration:

	Frequency	
	94 GHz	140 GHz
t_c	10 msec	5 msec

Furthermore, assume that each radar has an integration time of 20 msec. Under these conditions, the 95-GHz radar will realize a 3-dB integration gain in clutter while the 140-GHz radar realizes a 6-dB gain. These values compare with the 14-dB integration gain achieved in a thermal noise-limited situation.

The S/C versus range for the two radars is also shown in Figs. 47 and 48. Both radars are "clutter limited" out to the range where the clutter S/C line crosses the rain limited (by attention only) S/N after which the radar is "noise limited." Moving target indication (MTI) techniques could be used to reduce the "fixed" clutter returns and increase the "visibility" of the target at the near-in ranges where the radars are clutter limited.

VI. Millimeter-Wave Radar Applications

In an effort to summarize millimeter-wave radar technology, potential applications of millimeter radar will now be reviewed in the four generic application areas previously defined: surveillance and target acquisition, instrumentation and measurement, seekers and terminal missile guidance, and fire control and tracking.

A. Surveillance and Target Acquisition

There are a number of applications where millimeter-wave radars operating at short-to-moderate ranges are quite attractive for surveillance and target acquisition. The advantages of millimeter waves for such applications include small size and weight coupled with high resolution in both azimuth and range, providing excellent resolution of the area under surveillance. Often, for such applications, a rapid scan over an extended angular sector is desirable; the unique capabilities of an electromechanically scanned geodesic lens have resulted in several prototype systems being fabricated for operation at millimeter wavelengths that incorporate such a scanning concept. The basic operation of the scanning geodesic lens was covered in the earlier section on antennas; characteristics of two 70-GHz radar systems, which employ such antennas, are summarized in Table XI.

The resolution achievable with a short transmitted pulse and narrow azimuthal beamwidth permits a considerable amount of detail to be displayed to the operators. In addition, the rapid scan rates achievable (up to 70 scans/sec) permit a virtually flicker-free display to be realized without the necessity for digital or analog storage techniques. Geodesic antennas may also rapidly stop the scan for "searchlighting" on a given target in order to investigate the Doppler signatures that might be associated with

TABLE XI

CHARACTERISTICS OF TWO 70-GHz SURVEILLANCE RADARS

Frequency	70 GHz	70 GHz
Pulse width	50 nsec	20–45 nsec
PRF	10 kHz	5–25 kHz
Az. beamwidth	0.2°	0.55°
Elevation beamwidth	0.3° (shaped)	3.5°
Peak power	15 kW	500 W
Scan sector	30°	45°
Scan rate	Up to 40 scans/sec	Up to 70 scans/sec

the specific targets of interest within the surveillance area. The fact that the system rapidly scans an extended area also permits the utilization of track-while-scan techniques, thus permitting simultaneous accurate tracking of a number of targets simultaneously while maintaining area surveillance. Figure 49 is a photograph of the antenna and transmitter–receiver portion of one such surveillance radar system, and Fig. 50 shows a B-scope presentation generated by such a system. The degree of detail achievable has permitted navigation of vehicles using only such radar-generated information without use of optical information.

FIG. 49 Antenna and transmitter–receiver of a 70-GHz rapid scan system used for battlefield surveillance. (From Long and Allen, 1960.)

FIG. 50 B-scope (bearing versus range) display generated by a millimeter surveillance radar system, illustrating the degree of detail obtainable. Range cursor is at 360 m, total display range is 500 m. (From Alexander *et al.*, 1976.)

B. INSTRUMENTATION AND MEASUREMENT

In order to perform meaningful systems analyses at millimeter wavelengths, it is necessary to have available signature and cross-section data from targets and clutter of interest. In order to acquire such data, a number of different instrumentation and/or measurement radar systems have been built. An example of a relatively high-power system utilized for such purposes is a 95-GHz system, employing a pulsed magnetron or EIO oscillator that has been used for data collection from a number of different target and clutter types (Currie *et al.*, 1978). A list of principal radar characteristics is given as Table XII, a block diagram of the system as Fig. 51, and a photograph of the complete radar in Fig. 52. Note, in particular, the incorporation of a variable attenuator in the receiver front end, permitting calibration over the system's dynamic range utilizing a single fixed external corner reflector, and the capability for interfacing a number of different data acquisition systems with the receiver output, depending on the specific data requirements.

There are a number of applications where more sophistication in signal parameters is required than provided above. An example of such a system is a lower power, coherent, frequency and polarization-agile 35-GHz solid-state radar (Currie *et al.*, 1978), shown in block diagram form as Fig. 53. The characteristics of such a system are summarized in Table XII.

The two earlier experimental systems are configured such that they can

TABLE XII

OPERATING PARAMETERS OF PULSED 35- AND 95-GHz INSTRUMENTATION RADARS

Transmitter	35 GHz		95 GHz	
rf frequency	35.24 ± 0.40 GHz (fixed or agile)		95 GHz (nominal)	
Peak power	200 mW (10 nsec to CW) 3 W (10–500 nsec)		1kW	
Pulse width	10 nsec to cw		20 nsec	
Pulse repetition frequency	variable to 40 kHz		0–4000 Hz	
Antennas				
Type	18″ Dish	5″ Horn/lens	12″ Cassegrain	3″ Horn/lens
Beamwidth	1.25°	4.5°	0.7°	3°
Gain	43 dB	31 dB	47 dB	35 dB
Sidelobes	−20 dB	−20 dB	−20 dB	−20 dB
Polarization:	Dual polarized		Dual polarized	
Receiver:	H,V		H,V	
Transmit:	H,V,RC,LC (agile or fixed)		H,V,RC,LC (agile or fixed)	
Receiver				
Type	Coherent, integrated mixers/preamps			
	Linear	Logarithmic	Logarithmic	
Bandwidths	500 MHz	160 MHz	100 MHz	
Dynamic range	30 dB	60 dB	70 dB	
Sensitivity	−80 dBm	−85 dBm	−82 dBm	
Detection	Square law amplitude Coherent (phase) Pseudocoherent (phase)		Amplitude and Pseudocoherent (phase)	
IF	240 MHz		160 MHz	

be operated simultaneously with radars in other frequency bands, in many cases permitting truly simultaneous multifrequency pulse-by-pulse measurements to be made from targets and clutter. Another example of a system that provides multiple-frequency band measurement capability is the RATSCAT radar cross-section measurement facility at Holloman AFB. This facility has the capability for making millimeter-wave reflectivity measurements at 35, 52, 70, and 94 GHz utilizing antennas that yield measurement beamwidths of the order of one degree.

One of the uses of a millimeter reflectivity range such as RATSCAT is for frequency-scaled measurements on model targets, i.e., predicting the

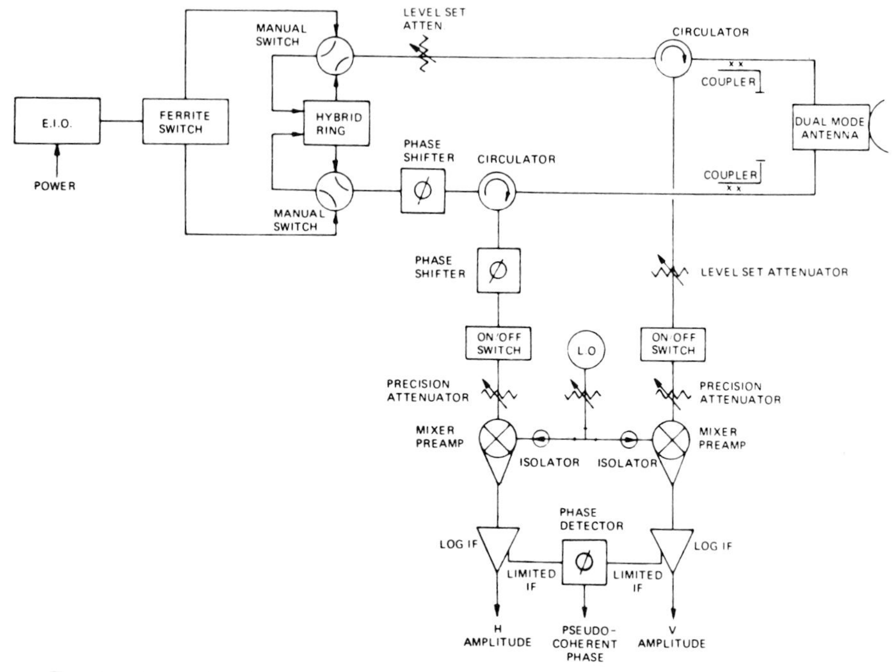

FIG. 51 Simplified block diagram of a 95-GHz instrumentation radar system. (From Currie *et al.*, 1978.)

low frequency radar cross-section behavior of a target by viewing a scaled-down version of the object at a higher operating frequency. An example of a facility dedicated to such measurements is the U.K. National Radio Modeling Facility, run by EMI Electronics Ltd. in England (Cram and Woolcock, 1978). This facility has radars operating as high as 890 GHz, consisting of over thirty different specialized systems. Both coherent and incoherent systems are utilized, and both amplitude and angle tracking measurements have been performed at this facility. A plan view of the modeling facilities is given as Fig. 54, showing provisions for measurements of radar cross section of suspended targets and of reflections from artificially generated water waves.

Often, the data required must be gathered from actual targets in the field and clutter in a natural environment rather than from scaled models. In the case of tracking data, this often means that true tracking systems must be implemented in order to acquire the desired information. One such system is a 70-GHz monopulse system, consisting of an 18-in. paraboloidal antenna having a 0.6°-beamwidth, a 10-kW transmitter operating at 70 GHz, logarithmic receiver, and 100-nsec pulse length (Kosowsky *et*

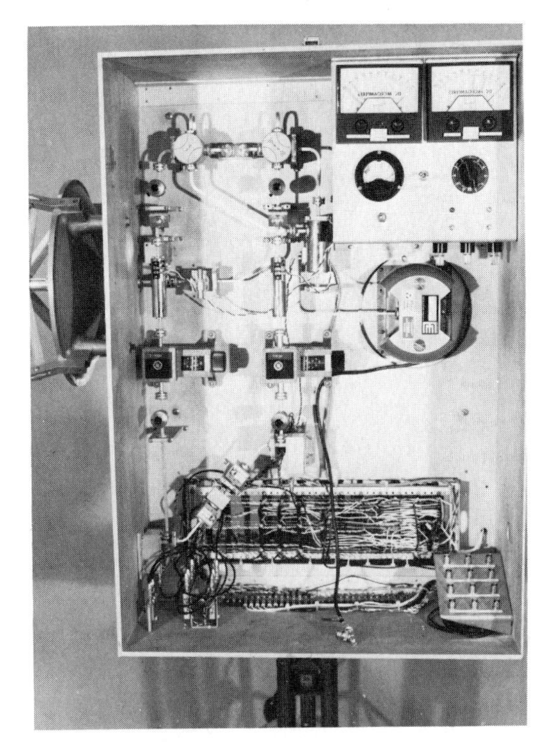

FIG. 52 Photograph of 95-GHz measurements radar showing the antenna and portions of the waveguide and receiver system. The modulator is separately housed on the other side to reduce interference.

al., 1976). This system has been demonstrated to yield milliradian accuracy and has been used to acquire data from a number of different targets and clutter, including the spectral data presented earlier in Fig. 18.

In many instances, it is desirable both to track a target with a high degree of precision as well as to maintain surveillance over an extended area; in order to accomplish this, track-while-scan systems, rather than the null tracking type of system described earlier, may be quite attractive. One such experimental system utilizes the geodesic lens described in Section IV.C, operating at 70 GHz, and this system has been utilized extensively in track-while-scan investigations in both the azimuth and elevation planes.

C. GUIDANCE AND SEEKERS

The size and weight advantages inherent in millimeter-wave sensors make them ideal for applications involving missile seekers and terminal

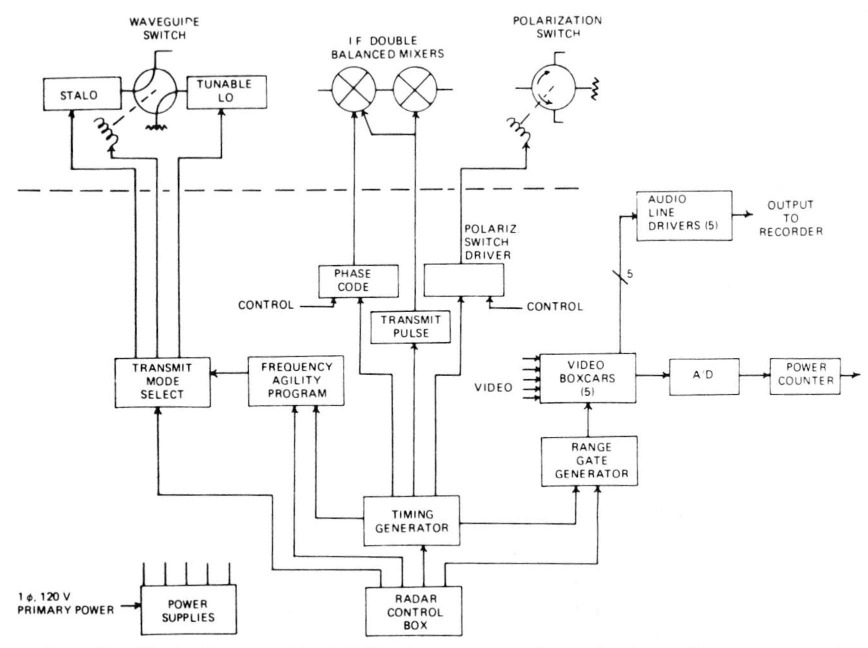

FIG. 53 Block diagram of a 35-GHz frequency and polarization agile measurements radar. (From Currie *et al.*, 1978.)

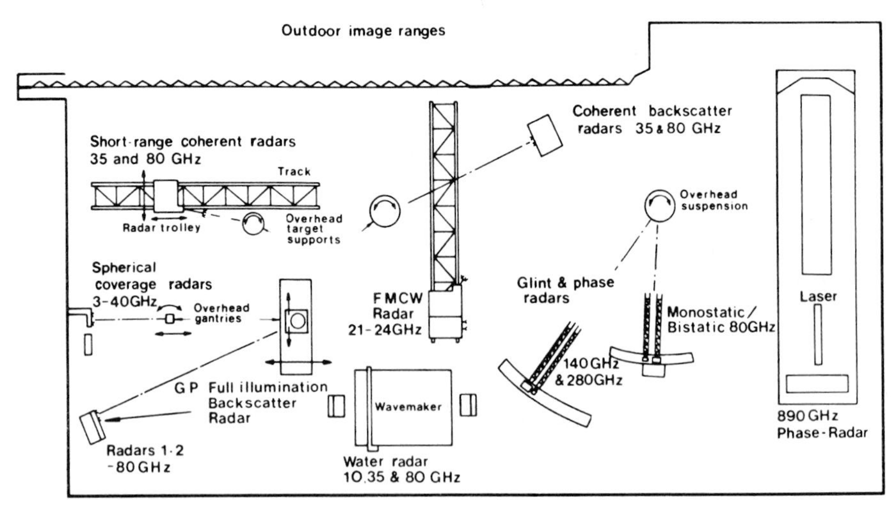

FIG. 54 Plan view of the U.K. National Radio Modeling Facility. (From Cram and Woolcock, 1978.)

guidance. In fact, this appears to be the most active research area currently incorporating millimeter-wave sensors (both passive and active). Passive, fully active, semiactive, beamrider and dual mode sensors and guidance techniques are all being investigated and evaluated for millimeter applications (Strom, 1973; Werner *et al.*, 1976; Bryant *et al.*, 1979; Brown *et al.*, 1977; Green and King, 1976; Bernues, *et al.*, 1979).

Shown in Fig. 55 is a 94-GHz missile seeker for air-to-ground applications incorporating either an active monostatic pulsed radar mode or a passive and/or bistatic cw radar mode. The antenna is a 10-in. Cassegrain with a conical scan feed arrangement. The radar is all solid state, utilizing an IMPATT transmitter (Bernues *et al.*, 1979). Dual mode operation allows the seeker to circumvent active radar "aim point wander" and target-induced "glint" track errors in the terminal engagement phase by switching to the passive mode.

Figure 56 is a conceptual drawing of a millimeter beamrider guidance concept, a technique receiving considerable attention for application at millimeter wavelengths (Green and King, 1976; Shackelford and Gallagher, 1977). In the most simplified terms, a millimeter beamrider guidance system can be defined as a system for guiding missiles to a target that incorporates a narrow pencil beam directed at the target along which a missile flies. Receiving and guidance equipment in the missile determines

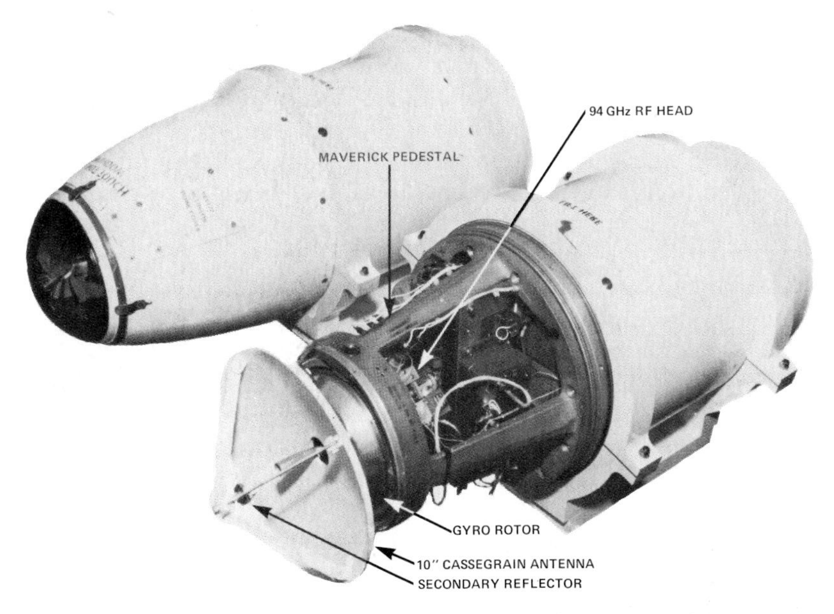

FIG. 55 A 94-GHz missile seeker. (From Bernues *et al.*, 1979.)

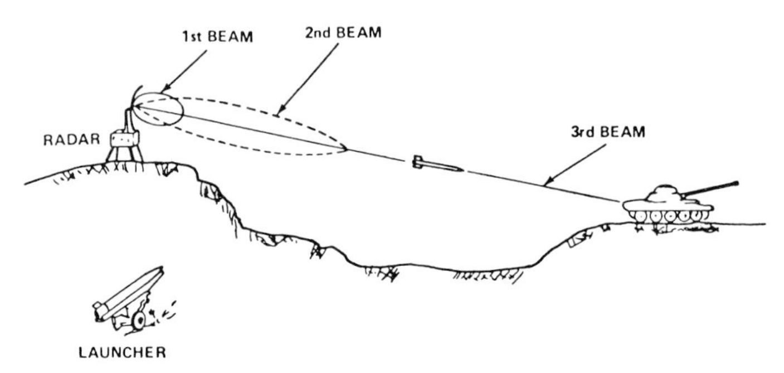

FIG. 56 Millimeter beamrider guidance concept. (From Green and King, 1976.)

the missile's position with respect to the beam center and develops commands to control the missile's aerodynamic control surfaces in such a manner as to keep the missile trajectory in the middle of the beam. Hardware is currently under development to demonstrate this technique.

A more complete and extensive discussion of millimeter guidance technology and applications is given in Chapter 3 of this volume.

D. Fire Control and Tracking

Short-range, closed-loop, fire control applications, such as that illustrated in Fig. 57, where extremely high spatial accuracy, small size and weight, and high mobility are the prime system requirements, are well suited to millimeter-wavelength radar. Such systems might serve as point defense antiaircraft weapons, ship defense systems, or for close-in battlefield air defense applications. Radars supporting these applications should provide some inherent target acquisition capability, although they might also be cued from a lower frequency, lower resolution target acquisition system, a false target discrimination capability, and, at least, a first-order target classification capability.

Multipath propagation and ground main beam clutter are classical problems in low angle tracking for air defense fire control systems—in particular, gunfire control systems. In this environment, millimeter radar can provide significant, potential advantages due to its narrow beamwidths and high resolution properties. The U.S. Army Armament Research and Development Command is currently developing an advanced fire control radar system to test the feasibility of millimeter-wave radar to direct and control antiaircraft gun systems with high precision against low-flying aircraft and helicopters. The radar system is a dual-band tracker, operating at K_u band and 94 GHz, utilizing a common monopulse antenna at both frequencies.

FIG. 57 An artist concept for a short-range, antiaircraft weapons system incorporating a millimeter-wave radar directed, rapid fire cannon mounted on an M113 armored personnel carrier.

Millimeter-radar techniques are currently under investigation for such diverse applications as a target acquisition (detection and tracking) and engagement system (limited fire control) having capabilities commensurate with current main battle tank fire control performance (Balcerak *et al.*, 1977; Fawcette, 1978), a fire control radar operating near the absorption band frequencies for ship defense (Harris, 1974), and an advanced, modular gunfire control system for use on fast patrol boats (Sundaram, 1979).

The tank target acquisition and engagement system referred to in the preceding paragraph is shown conceptually in Fig. 58 (Backus, 1979). This radar is intended to demonstrate the feasibility of millimeter waves for tactical systems. Preliminary specifications are given below (Balcerak *et al.*, 1977; Fawcette, 1978).

FIG. 58 Surveillance and target acquisition radar for tank location and engagement (STARTLE) system. (From Backus, 1979.)

Frequency	94 GHz (3.2 mm)
Beamwidth	11 mrad
Average power	0.1–0.5 W
Antenna aperture	14 in.
Field of view	15° × 7.5° in wide mode
	5° × 2.5° in narrow tracking mode
Frame time	<2 sec
Target detection	3000 m with 100-m visibility
Target tracking	0.5 mrad accuracy at 2 km with 100-m visibility

The radar is intended to complement, and be integrated with, the infrared tank thermal sight such that both systems share the same display. Spread spectrum waveforms and both coherent MTI and area MTI techniques are being investigated for detection of moving targets. The radar incorporates a solid-state IMPATT transmitter power source.

One of the simplest and most unique applications of millimeter-wave radar to a closed-loop fire control problem is discussed in Strom (1973). In this reference, a design is developed for a millimeter-radar fire control system for the Army infantryman and his rifle. The capabilities and performance parameters of this radar were examined and compared with the human eye.

The millimeter-radar application examples discussed in this section were certainly not all-inclusive or complete, but were selected primarily to illustrate the variety of operational problems involving target acquisition, tracking, guidance, and instrumentation where millimeter radar has some inherent advantages and shows promise for providing improved performance over either higher- or lower-frequency radars.

REFERENCES

Ahn, S. (1978). *Proc. 1978 Int. Electron Devices Conf.* pp. 394–395.

Alexander, N. T., Craven, T. S., and Foster, W. S. (1976). Final Rep., Contract N00014-76-C-0860. Eng. Exp. Stn., Georgia Inst. of Technol., Atlanta.

Allen, G., and Simonson, B. (1970). Tech. Rep. EATR 4405. U.S.Army Edgewood Arsenal, Aberdeen, Maryland.

Arnold, K. (1980). R&D Tech. Rep. DELET-TR-78-3015-1. U.S. Army ERADCOM, Fort Monmouth, New Jersey.

Babbitt, R. W., Stern, R. A., Whicker, L. R., and Young, C. W. (1978). *Symp. Millimeter Submillimeter Wave Propag. and Circuits; AGARD Conf. Proc.* No. 245, pp. 34/1–34/16.

Backus, P. H. (1979). *Electron. Def.* **11** (3)24–29.

Balcerak, R., Ealy, W., Martino, J., and Hall, J. (1977). "Effective Utilization of Optics in Radar Systems," Vol. 128, pp. 172–184. SPIE, Redondo Beach, California.

Barton, D. K. (1976). "Radar System Analysis." Artech House, Durham, Massachusetts.

Barton, D. K., and Ward, H. R. (1969). "Handbook of Radar Measurements." Prentice-Hall, Englewood Cliffs, New Jersey.

Bellemare, Y., and Chudobiak, W. (1979). *Proc. IEEE* **67,** 1667–1669.

Bernues, F. J., Ying, R. S., and Kaswen, M. (1979). *Microwave Syst. News* **9**(5), 79–86.

Bodnar, D. G., Moore, R. A., Cofer, J. W., Goodman, R. M., and Stapleton, L. A. (1973). Final Tech. Rep., Phases I and II, APL/JHU Subcontract 372113. Georgia Inst. of Technol., Atlanta.

Boring, J. G., Flynt, E. R., Long, M. W., and Winderquist, V. R. (1957). Final Tech. Rep., Contract NObsr-49063. Georgia Inst. of Technol., Atlanta.

Borowick, J., Stern, R. A., and Babbitt, R. W. (1980). *Annu. Tri-Serv. Radar Symp., 26th, West Point, N.Y.*

Brown, J. L., Dasarathy, V. V., Graf, E. R., and Weathers, G. D. (1977). Tech. Rep. TE-CR-77-9. M&S Computing, Huntsville, Alabama.

Bryant, J. L., Miller, J. M., Temme, D. H., and Vote, E. W. (1977). *Proc. DARPA/Tri-Serv. Millimeter Wave Conf., 6th, Eglin AFB, Fla.* 160–169.

Chang, K., Sun, C., English, D. L., and Nakaji, E. M. (1979). *1979 IEEE MTT-S Int. Microwave Symp. Dig.* IEEE Catalog No. 79CH1439-9MTT, pp. 71–72.

Cram, L. A., and Woolcock, S. C. (1978). *Symp. Millimeter Submillimeter Wave Propag. Circuits; AGARD Conf. Proc.* No. 245, Sect. 6.

Currie, N. C. (1979). *1979 Int. Symp. Dig.—Antennas Propag.* **2,** 504–507.

Currie, N. C., Dyer, F. B., and Hayes, R. D. (1975a). Tech. Rep. No. 3, Contract No. DAAA25-73-C-0256. Georgia Inst. of Technol., Atlanta.

Currie, N. C., Martin, E. E., and Dyer, F. B. (1975b). Tech. Rep. No. 4, Contract No. DAAA25-73-C-0256. Georgia Inst. of Technol., Atlanta.

Currie, N. C., Dyer, F. B., and Hayes, R. D. (1975c). Tech. Rep. No. 2, Contract DAAA75-73-C-0756. Georgia Inst. of Technol., Atlanta.

Currie, N. C., Dyer, F. B., and Ewell, G. W. (1976a). *1976 Rec. AP-S Int. Symp.* pp. 579–582.

Currie, N. C., Dyer, F. B., and Martin, E. E. (1976b). *1976 Rec. AP-S Int. Symp.* pp. 575–578.

Currie, N. C., Scheer, J. A., and Holm, W. A. (1978). *Microwave J.* **21**(8), 35–44.

Davis, R. T. (1976). *Microwaves* **15**(3), 32–43.

Downs, A. R. (1976). Memo. Rep. No. 2710. U.S. Army Ballist. Res. Lab., Aberdeen, Maryland.

Dyer, F. B., and Currie, N. C. (1978). *Symp. Millimeter Submillimeter Wave Propag. Circuits; AGARD Conf. Proc.* No. 245, Sect. 2.

Dyer, F. B., Gary, M. J., and Ewell, G. W. (1974). *1974 Int. IEEE*/AP-S Symp. Dig. pp. 319–322.

Dyer, F. B., Reedy, E. K., Currie, N. C., Horst, M. M., and Scheer, J. A. (1977). Intern. Tech. Rep. 77-01, Georgia Inst. of Technol., Atlanta.

Epsztein, B. (1978). *Symp. Millimeter Submillimeter Wave Propag. Circuits; AGARD Conf. Proc.* No. 245, Chap. 36.

Ewell, G. W. (1981). "Radar Transmitters." McGraw-Hill, New York. To be published.

Ewell, G. W., Ladd, D. S., and Butterworth, J. C. (1979). *1979 MTT-S Int. Microwave Symp.* 450–452.

Ewell, G. W., Ladd, D. S., and Butterworth, J. C. (1980). *Microwave J.* **23**(8), 57–70.

Fawcette, J. (1978). *Microwave Syst. News* **9**(7), 23–24.

Fishbein, W., Graveline, S. W., and Rittenbach, O. E. (1967). Tech. Rep. ECOM-2808. U.S. Army Electron. Command, Fort Monmouth, New Jersey.

Flyagin, V. A., Gaponor, A. V., Petelin, M. I., and Yulpatov, V. K. (1977). *IEEE Trans. Microwave Theory Tech.* **25,** 514–521.

Goodman, R. M., Jr., and Dyer, F. B. (1968). Final Rep., Contract DA-49-186-AMC-275(A). Eng. Exp. Stn., Georgia Inst. of Technol., Atlanta.

Granatstein, V. L., Sprangle, P., Parker, R. K., and Herndon, M. (1975). *J. Appl. Phys.* **46,** 2021–2028.

Grant, C. R., and Yaplee, B. S. (1957). *Proc. IRE* **45,** 976–982.

Green, A. H., and King, F. (1976). Intern. Tech. Note RE-77-3. U.S. Army Missile Command, Redstone Arsenal, Alabama.

Harris, R. L. (1974). *Rep. ARPA/Tri-Serv. Millimeter Wave Workshop, Appl. Phys. Lab., Johns Hopkins Univ.* pp. 119–138.

Hayes, D. T., Lammers, U. H., Marr, R. A., and McNally, J. J. (1979). *1979 Int. Symp. Dig.—Antennas Propag.* pp. 499–502.

Hayes, R. D., and Dyer, F. B. (1973). Tech. Rep. No. 1, Contract No. DAAA25-73-C-0256. Georgia Inst. of Technol., Atlanta.

Henry, J. F. (1974). *Rep. ARPA/Tri-Serv. Millimeter Wave Workshop, Appl. Phys. Lab., Johns Hopkins Univ.* p. 149.

Hirshfield, J. L., and Granatstein, V. L. (1977). *IEEE Trans. Microwave Theory Tech.* **25,** 522–527.

Johnson, R. C. (1962). *Microwave J.* **5**(8), 76–85.

Johnson, R. C. (1963). *Microwave J.* **6**(7), 68–70.

Jory, H. R., Friedlander, F., Hegji, S. J., Shively, J. F., and Symons, R. S. (1977). *Proc. 1977 Int. Electron Devices Conf.* pp. 234–237.

Jory, H. R., Hegji, S., Shively, J., and Symons, R. (1978). *Microwave J.* **21**(8), 30–32.

Knox, J. E. (1979). *Proc. DARPA/Tri-Serv. Millimeter Wave Conf., 8th, Eglin AFB, Fla.* 127–133.

Knox, R. M. (1976). *Microwaves* **15**(3), 56–67.

Kosowsky, L. H. (1974). *Rep. ARPA/Tri-Serv. Millimeter Wave Workshop, Appl. Phys. Lab., Johns Hopkins Univ.* pp. 139–146.

Kosowsky, L. H., Koester, K. L., and Graziano, R. S. (1976). *New Devices, Tech., Syst. Radar; AGARD Conf. Proc.* No. 197, Sect. 35.

Kozakoff, D. J., and Britt, P. P. (1980). *Proc. IEEE SOUTHEASCON,* 65–68.

Kramer, N. B. (1979). *Microwave J.* **22**(8), 57–61.
Kulpa, S. M., and Brown, E. A. (1979). Rep. HDL-SR-79-8, Vol. 1. Harry Diamond Lab., Washington, D.C.
Kuno, H. J., and Fong, T. T. (1978). *Symp. Millimeter Submillimeter Wave Propag. Circuits; AGARD Conf. Proc.* No. 245, pp. 14/1, 14/2.
Kuno, H. J., and Fong, T. T. (1979). *Microwave J.* **22**(6), 47.
Long, M. W. (1975). "Radar Reflectivity of Land and Sea." Lexington Books, Lexington, Massachusetts.
Long, M. W., and Allen, G. E. (1960). Final Rep., Contract DA 36-039-SC-74870. Georgia Inst. of Technol., Atlanta.
Long, M. W., Weatherington, R. D., Edwards, J. L., and Abeling, A. B. (1965). Final Rep., Contract No. N62269-3019. Georgia Inst. of Technol., Atlanta.
McCartney, E. J. (1966). Rep. No. AB-1272-0057. Sperry Rand Corp., Great Neck, New York.
McMillan, R. W., Wiltse, J. C., and Snider, D. E. (1979). *EASCON '79 Rec.* pp. 42–47.
Martin, E. E. (1979). Final Tech. Rep., GIT Rep. A-2104. Georgia Inst. of Technol., Atlanta.
Morgan, R. K., Stattler, J. D., and Tanton, G. A. (1979). *Proc. DARPA/Tri-Serv. Millimeter Wave Conf., 8th, Eglin AFB, Fla.* 119–125.
Nathanson, F. E. (1969). "Radar Design Principles." McGraw-Hill, New York.
Ngan, Y. C., and Nakaji, E. M. (1979). *1979 IEEE MTT-S Int. Microwave Symp. Dig.* IEEE Catalog No. 79CH1439-9MTT, pp. 73–74.
Petito, F. C., and Harris, R. L. (1979). *Proc. DARPA/Tri-Serv. Millimeter Wave Conf., 8th, Eglin AFB, Fla.* 135–145.
Pranter, N. (1978). *Symp. Millimeter Submillimeter Wave Propag. Circuits; AGARD Conf. Proc.* No. 245, Chap. 37.
Preissner, J. (1978). *Symp. Millimeter Submillimeter Wave Propag. Circuits; AGARD Conf. Proc.* No. 245, pp. 48/1–48/13.
Quine, J., McMullen, J., and Khandelwal, D. (1978). *1978 IEEE Int. Microwave Symp. Dig.* IEEE Catalog No. 78CH1355-7, MTT, pp. 346–483.
Raue, J. E., Bayuk, F. J., Ohashi, A. I., and Yuam, L. T. (1974). *Proc. 1974 Millimeter Waves Tech. Conf. NELC/TD308 Vol. 2 Nav. Electron. Lab. Cent., San Diego, Calif.* pp. D5-1–10.
Reedy, E. K., and Eaves, J. L. (1979). *Mil. Electron. Expos., '79, Anaheim, California.*
Richard, V. W. (1976). Memo. Rep. No. 2631. U.S. Army Ballist. Res. Lab., Aberdeen, Maryland.
Richard. V. W., and Kammerer, J. E. (1975). Rep. No. 1838. U.S. Army Ballist. Res. Lab., Aberdeen, Maryland.
Rivers, W. (1970). Final Tech. Rep., Contract No. N62269-70-C-0489. Georgia Inst. of Technol., Atlanta.
Schneider, M. V. (1979). *Microwave J.* **22**(8), 78–83.
Seashore, C. R., Miley, J. E., and Kearns, B. A. (1979). *Microwave J.* **22**(8), 47–57.
Shackelford, R. G., and Gallagher, J. J. (1977). Tech. Rep. TE-CR-77-7. U.S. Army Missile Res. Dev. Command, Redstone Arsenal, Alabama.
Skolnik, M. I. (1970). "Radar Handbook." McGraw-Hill, New York.
Skolnik, M. I. (1978). *1978 Tri-Serv. Radar Symp. Rec.* 145–146.
Strom, L. D. (1973). Rep. No. 108. System Planning Corp., Washington, D.C.
Suits, G. H., and Guenther, B. D. (1979). *In* "Near-Millimeter Wave Technology Base Study," Vol. 1, Chap. 6. U.S. Army Materiel Dev. Readiness Command Def. Adv. Res. Projects Agency, Washington, D.C.

Sundaram, G. S. (1979). *Int. Def. Rev.* **2,** 271–277.

Tolbert, C. W., Britt, C. O., and Straiton, A. W. (1957). Rep. No. 95, Contract Nonr 375(01). Electr. Eng. Res. Lab., Univ. of Texas, Austin.

Trebits, R. N., Hayes, R. D., and Bomar, L. C. (1978). *Microwave J.* **21**(8), 49.

Trebits, R. N., Currie, N. C., and Dyer, F. B. (1979). *Conf. Record, EASCON* **2,** 261–264.

Werner, R., Belanger, B., DiDomizio, R., and Smith, P. (1976). Rep. 1267R0005. Norden Div., United Technologies Corp., Norwalk, Connecticut.

Whicker, L. R., and Webb, D. C. (1978). *Symp. Millimeter Submillimeter Wave Propag. Circuits; AGARD Conf. Proc.* No. 245, pp. 1/1–1/5.

Wilcox, F. P., and Graziano, R. S. (1974). *1974 Millimeter Waves Tech. Conf., Nav. Electron. Lab. Cent., San Diego, Calif.* **1,** B5-1–B5-21.

Wiltse, J. C., Schlesinger, S. P., and Johnson, C. M. (1957). *Proc. IRE* **45,** 220–228.

Woodward, P. M. (1955). "Probability and Information Theory with Applications to Radar." McGraw-Hill, New York.

Zapevalov, V. Y., Korablev, G. S., and Tsimring, S. Y. (1977). *Radio Eng. Electron. Phys. (USSR)* **22,** 86–94.

CHAPTER 3

Missile Guidance

Charles R. Seashore

Honeywell, Inc.
Millimeter-Wave Technology Center
Bloomington, Minnesota

I. INTRODUCTION 95
A. *Terminal Guidance Problem* 95
B. *Millimeter-Wave Advantages* 97
II. PROPAGATION AND TARGETS 98
A. *Atmospheric and Propagation Effects* 99
B. *Target and Background Characteristics* 102
III. RANGE EQUATIONS 112
A. *Conical Scan Passive* 113
B. *Conical Scan Active* 118
IV. MILLIMETER-WAVE SEEKER DESIGN 122
A. *Missile Guidance Options* 123
B. *Seeker Search and Track* 126
C. *Waveforms and Processing* 132
D. *Seeker Hardware* 145
V. COUNTERMEASURES 148
REFERENCES 150

I. Introduction

A. Terminal Guidance Problem

Several types of millimeter-wave terminal guidance systems are currently being developed for launch-and-leave targeting applications. These applications include air-to-surface and surface-to-surface missiles as well as free-fall and parachute-suspended munitions. The objective of this activity is to obtain precise fixes and improve guidance under adverse weather conditions, electronically and without manual assistance. It is part of an effort to make weapons more autonomous, improve kill probabilities, and take the man out of the loop. The missile terminal guidance scenario is shown in Fig. 1. A delivery vehicle places the missile in an acquisition basket such that it is generally aligned with the targeting area. The missile then typically operates in a fully autonomous mode to search, detect, identify, and track the target.

ISBN 0-12-147704-5

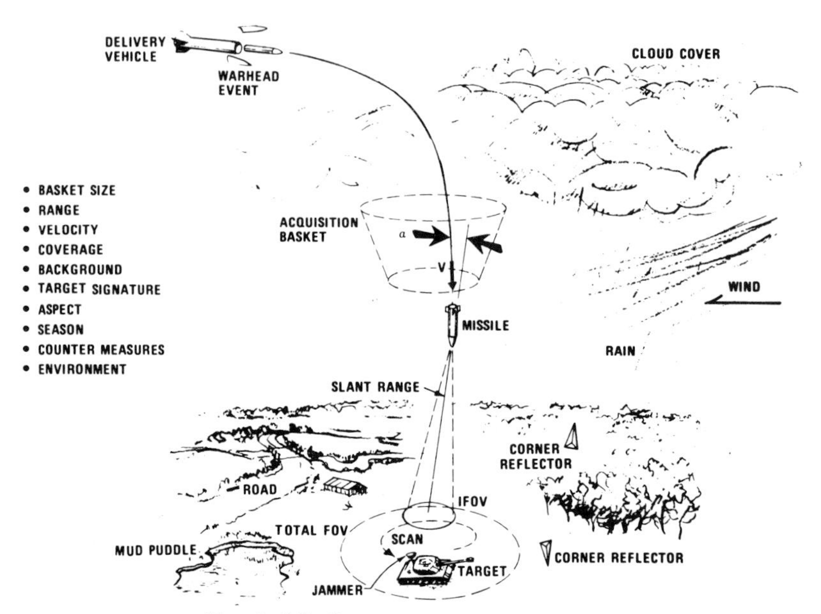

FIG. 1 Missile terminal guidance scenario.

Several factors illustrated in Fig. 1 influence the design and effectiveness of the missile in its role of delivering a munition on a terrain-based target array. These include

- acquisition basket size
- slant range to target
- terminal velocity
- ground coverage in search mode
- background clutter environment
- target radar and radiometric cross section
- target aspect angle relative to missile trajectory
- seasonal weather conditions
- battlefield environment including false targets
- deployed countermeasures

Due to the available techniques for microminiaturization of analog-digital components and state-of-the-art advances in millimeter-wave components, it has now become quite feasible to investigate radar and radiometry for small, air-to-surface terminally guided missiles. Until now, radar in particular has not been used for autonomously guiding small diameter air-to-surface missiles due to the need for high resolution in detecting targets embedded in a natural clutter environment.

A seeker operating at millimeter-wave frequencies can provide the desired high resolution for antenna diameters of six inches or less by reducing the size of the beam intercept pattern on the terrain and hence the amount of clutter competing with the target signal.

B. Millimeter-Wave Advantages

The millimeter-wave portion of the electromagnetic spectrum lies between the microwave and far infrared regions. Frequencies generally attributed to this portion of the spectrum are in the range 30–300 GHz corresponding to wavelengths of 10–1 mm. Extremely high angular resolution and heavily degraded adverse weather properties of optical systems are approached at the high end of this spectrum, and the all-weather propagation with broader resolution properties of radio waves is realized at the low frequency end of this spectrum. Millimeter waves have three unique characteristics that make them very useful for applications involving target tracking, terminal guidance, proximity fuzing, and secure information transmission. These characteristics include an interaction with natural atmospheric constituents and gases, a large effective rf bandwidth, and a narrow effective beamwidth with a small antenna diameter. Considering radar and radiometric sensors operating in missile guidance applications, the following characteristics are worth noting

1. *Small Size*

The shorter wavelengths compared to microwaves make it possible to reduce the size of rf componentry and so build smaller systems. This is particularly advantageous in missile-seeker applications where size and weight restrictions are quite significant.

2. *Wide Bandwidth*

At each millimeter transmission window, extremely large bandwidths are available. For example, at the four main windows 35, 94, 140, and 220 GHz, the available bandwidths are of the order of 16, 23, 26, and 70 GHz, respectively. This means that the equivalent of all the lower frequencies, including microwaves, can be accommodated in the bandwidth available at any one millimeter-wave window. Thus many individual frequencies can be used to provide increased immunity to interference from multiple users and achieve a high level of electromagnetic compatibility. It also makes jamming more difficult unless the exact frequency to be jammed is known. In radar, the range resolution can be increased, while in radiometry, higher detection sensitivities are possible.

3. *Narrow Beamwidth*

For a given antenna size, smaller radiated beamwidth can be achieved, providing higher resolution and hence improved accuracy when compared with microwaves. This is very important with seekers where precision target tracking is necessary for achieving more target detail and for discriminating better against small targets. Narrow beamwidth minimizes losses and noise due to sidelobe returns, a major problem in microwave radars, and, similarly, reduces errors due to multipath propagation. Jamming becomes more difficult and friendly interference is reduced. Small size apertures produce small beamwidths; for example, a 12-cm diameter antenna provides 1.8° beamwidth at 94 GHz compared to 18° at 9.4 GHz.

4. *Atmospheric Losses*

Atmospheric absorption and attenuation losses are relatively low in the transmission windows, compared to those encountered by laser or infrared transmissions in rain, fog, and smoke. Millimeter-wave sensors are thus more effective than electrooptic ones in adverse weather or battlefield smoke–dust conditions. The high attenuation encountered in the absorption bands limits the effective range, but can be used to advantage for covert or quiet radar applications.

Table I provides a very general comparison of radar and radiometric seeker capabilities for three regions of the electromagnetic spectrum. This illustrates that operation at millimeter wavelengths provides advantages over microwaves in tracking accuracy and over optical techniques in adverse weather performance and in volume search. Some system concepts are being considered wherein both millimeter-wave and optical sensors are combined in a common aperture for improved target detection performance in a countermeasure and false target battlefield scenario. This dual-sensor technique can offer exciting possibilities for a variety of missile seeker applications.

II. Propagation and Targets

Factors that are very important in millimeter-wave seeker design for missile guidance include atmospheric and propagation effects as well as target and background characteristics. In recent years, a number of measurement programs have been carried out that attempted to quantify the above factors in the frequency range 30–140 GHz. The material presented in this section is intended only as an introduction to the topic; the inter-

TABLE I

COMPARISON OF SEEKER OPERATION IN THREE SPECTRAL REGIONS

Radar/radiometer capability	Microwave[a] (3–30 GHz)	Millimeter wave[a] (30–300 GHz)	Optical[a] (0.4–14 μm)
Volume search	3	2	1
Classification/identification	1	2	3
Tracking accuracy	1	2	3
Adverse weather performance	3	2	1
Smoke performance	3	3	1
Covert capability	1	3	3
Day/night performance	3	3	2

[a] 3, good; 2, medium; 1, poor.

ested reader is invited to examine the work of Porter and Parker (1960), Altshuler *et al.*, (1968), Constantine and Deitz (1979), Kulpa and Brown (1979), Skolnik (1970), Richard (1976), Chen (1975), and Currie *et al.* (1975) for the details required for a comprehensive system analysis and design.

A. ATMOSPHERIC AND PROPAGATION EFFECTS

A significant consideration in developing millimeter-wave guidance systems is the effect of weather on the wave propagation. In the case of rain, the drops tend to be of the same diameter as the operating wavelength and can cause appreciable attenuation and backscatter. Although weather effects can be minimized by using high-gain antennas, circular polarization, frequency agility, and short pulse transmitters, they cannot be circumvented.

Attenuation of millimeter waves by atmospheric aerosols occurs both by scattering and absorption with the respective levels dependent on aerosol size, liquid-water content, dielectric constant, temperature, and humidity. The absorption of millimeter waves in the lower atmosphere is the result of both free molecules and suspended particles, such as water drops condensed into fog and rain. In clear air atmosphere, oxygen and water vapor are the substances that cause absorption. Figure 2 illustrates the attenuation of the atmosphere in decibel per kilometer over the wavelength region 3 cm–0.3 μm. As rain rates exceed 50 mm/hr, the attenua-

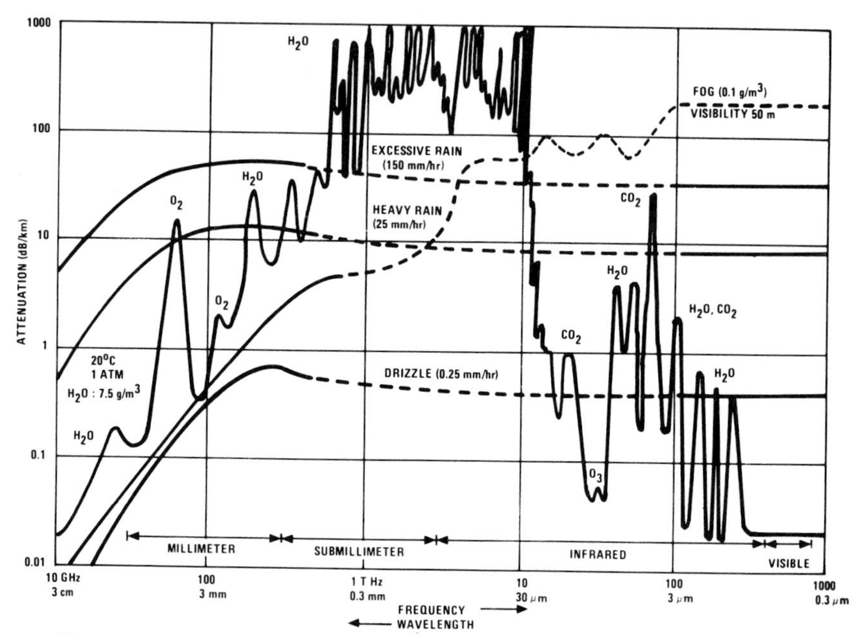

FIG. 2 Atmospheric attenuation characteristics for wavelengths 3 cm to 0.3 μm.

tion rates climb steadily with increase in frequency to approximately 100 GHz where they begin to level off. Table II lists one-way attenuation versus rainfall rate at the two window frequencies of 35 and 94 GHz. Millimeter-wave system design is typically carried out using a rain rate of 4 mm/hr.

The propagation of millimeter waves through fog results in power losses proportional to the liquid-water content of the aerosols. In addition, there is a significant temperature dependence. The attenuation coefficient is given in Table III for a liquid-water content of 0.1 g/m^3. This corresponds to an optical visibility of 120 m in a radiation fog and 300 m in advection fog. From Table III it is noted that the fog attenuation increases with frequency, hence the lowest frequency (35 GHz) should be selected for adequate fog performance. The backscatter coefficient of fog in the millimeter-wave region is more than two orders of magnitude smaller than that of rain and therefore has negligible effect on radar system performance.

Clouds also have to be considered as an attenuation mechanism for millimeter waves. The primary differences between attenuation effects by clouds and rain are produced by liquid-water content and extent of the respective absorption elements. Attenuation rates may be comparable, but total attenuation is usually much less in clouds since their range extent

TABLE II

ONE-WAY ATTENUATION VERSUS FREQUENCY AND RAIN RATE

Rain rate (mm/hr)	35 GHz $\alpha = 0.27R^{0.985}$ dB/km	94 GHz $\alpha = 1.6R^{0.64}$ dB/km
1	0.272	1.6
2	0.542	2.494
3	0.810	3.232
4	1.066	3.886
5	1.328	4.482
6	1.600	5.038
7	1.850	5.560
8	2.110	6.056
9	2.368	6.530
10	2.628	6.986
16	4.2000	9.436
20	5.200	10.884
25	6.528	12.556
50	12.824	19.564
150	35.0	50.0

is much smaller than for rain cells. The greatest amount of attenuation comes from the rain-bearing or cumulonimbus clouds. Typical root mean square attenuation values are 0.4 dB at 35 GHz and 2.07 dB at 94 GHz over a 1-km path in cumulonimbus clouds. Liquid-water content varies from 7.5 g/m^3 for cumulonimbus to 15 g/m^3 for cirrocumulus. Thus the total power loss from cloud attenuation is normally not large since the average cloud thickness is of the order of a few hundred meters.

The performance of a radar operating in rain is degraded by the absorption and scattering of energy by the individual raindrops. Radiometer performance is primarily influenced by the absorption effect of rain. The scattered energy reflected back to the radar is of interest since it contrib-

TABLE III

ATTENUATION COEFFICIENT OF FOG

Frequency (GHz)	Attenuation coefficient[a] (dB/km)	
	0°C	40°C
35	0.11	0.034
70	0.36	0.138
94	0.47	0.22

[a] Liquid water content is 0.1 g/m^3.

utes additional noise-to-receiver system. Forward scatter can become signigicant in the 70 and 94 GHz. Table IV shows the attenuation and backscatter coefficient for three operating frequencies. As can be seen, the backscatter cross section per unit volume in heavy rain is comparable at 70 and 94 GHz and is actually less than that for 35 GHz operation. Thus it can be concluded that a 70- or 94-GHz radar operating in rain would have less rain clutter than a 35-GHz system with the same size aperture. Trading off 70 and 94 GHz, the latter would have higher attenuation and lower backscatter at a rain rate of 16 mm/hr.

Table V provides a summary of the atmospheric and propagation characteristics for the two window frequencies where a majority of millimeter-wave seeker development is currently being carried out. For seekers of a 6-in. diameter or less, operating in an active radar mode, the frequency range of 90–100 GHz offers many attractive advantages.

B. Target and Background Characteristics

Target and background characteristics are key features in the design of a millimeter-wave seeker for terminal guidance applications. The target may take the form of a large man-made object such as a bridge, building, and ship, or smaller objects such as armored vehicles and trucks. Due to its delicate nature in determining system effectiveness, the majority of target signature data for explicit target types resides in the closed literature. The interested reader is referred to the work of Briscoe *et al.* (1972), Richer *et al.* (1975), Nelson (1976), and Beebe and Salzman (1976) for detailed active and passive signature data on various military targets. In this section, some general data are presented, largely derived from extensive

TABLE IV

Backscatter Coefficient and Rain Attenuation at Three Frequencies

	Rain rate[a] (1 mm/hr)		Rain rate (16 mm/hr)	
Frequency (GHz)	Attenuation (dB/km)	Backscatter (cm^2/m^3)	Attenuation (dB/km)	Backscatter (cm^2/m^3)
35	0.24	0.21	4.0	4.9
70	0.73	0.72	6.9	4.1
94	0.95	0.89	7.4	3.9

[a] Condition	Rain rate (mm/hr)
Light rain	1
Moderate rain	4
Heavy rain	16
Cloud burst	100

TABLE V

SUMMARY OF ATMOSPHERIC AND PROPAGATION EFFECTS AT TWO WINDOW FREQUENCIES

Parameter	One-way loss (dB/km)	
	35 GHz	94 GHz
Clean air attenuation	0.12	0.4
Rain attenuation (mm/hr)		
−0.25	0.07	0.17
−1.0	0.24	0.95
−4.0	1.0	3.0
−16.0	4.0	7.4
Cloud attenuation		
Rain	5.14	35.04
Dry	0.50	3.78
Fog attenuation (g/m^3)		
0.01 (light)	0.006	0.035
0.10 (thick)	0.06	0.35
1.0 (dense)	0.6	3.5
Snow (0°C)	0.007	0.0028

Honeywell in-house signature measurements carried out over the past six years utilizing the facilities shown in Figs. 3–6.

The side-by-side dual-frequency sensor concept is now generally accepted as the standard method of deriving comparative target-background signature data. These measurement sensors were configured to provide pulse radar operation with linear–circular polarization selection and a total power/Dicke-switched radiometer operation with linear–circular polarization selection. The sensors were mounted on a 95-ft boom, a 110-ft stationary tower, and a small helicopter during various measurement activities. Targets were placed on a buried turntable to achieve the desired aspect angle variations for reflectance measurements. A television camera is colocated with the two sensors to provide an image of ground truth for more effective data correlation and analysis. Figure 7 illustrates the data management system used for analysis of the target-background signature data. An IRIG-B compatible time code generator is used to synchronize all recorded outputs for accurate performance evaluation. A TV display including bore-sighted reticles, gimbal-controlled tracking gate (cursor), time display, and various ASC II codes to display mode and control functions are used for the pilot and instrumentation crew in the helicopter-mounted test bed. Although instrumentation presented in Fig. 7 is configured for a bistatic radar system, it is very effective also in the monostatic radar seeker evaluation process.

FIG. 3 35 GHz and 94 GHz side-by-side radar–radiometer.

The factors influencing the radar or radiometer input signal characteristic are illustrated in Fig. 8. The target is imbedded in a background scene, and the target signature is altered by countermeasures. Interfering electromagnetic sources include galactic elements as well as competing high-power search radars and deceptive jammers. The atmospheric and propagation characteristics can significantly alter the received signal strength. The target is characterized for a radar sensor by its radar cross section (RCS). The instantaneous radar signature of complex targets is influenced by such factors as contour shape, aspect angle, and the number of scattering centers. All of these characteristics are wavelength dependent and result in angle noise or target glint. In addition, the motion of these scattering centers causes interference patterns that impart a time variation to the radar signature, which is also directly related to the incident frequency. Thus as the frequency is increased, the glint can be expected both to change its probability density function and broaden its power spectral density. To minimize the effect of angular glint, most lower frequency systems utilize frequency agility to achieve rapid decorrelation. Although this technique is very effective, it increases the cost

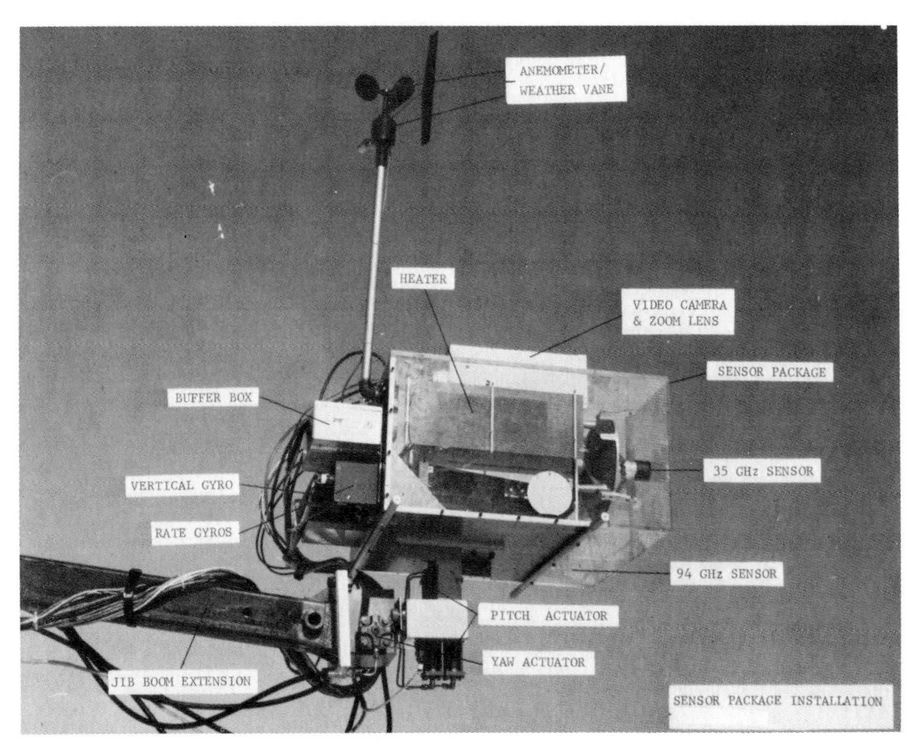

FIG. 4 35 and 94 GHZ sensors mounted on 95-ft boom.

and complexity of the radar system. As noted previously, because of the naturally occurring interferences due to target relative motion or change in aspect angle, rapid decorrelation can be realized by employing millimeter wavelengths at a constant frequency. Table VI shows 35-GHz RCS data statistics by depression angle for a tank target; Table VII supplies further detail on the variation of target RCS with aspect angle and depression angle at 35 GHz. Finally, Fig. 9 presents a statistical compilation of RCS at 35 GHz for numerous vehicular targets. Generally, larger targets such as trucks with numerous scattering centers produce the larger effective RCS for millimeter wave radar sensor with a pencil-beam antenna. For general system design, a target RCS in the range 25–125 m^2 may be selected.

The target is characterized for a radiometer sensor by its effective reflecting area and a series of radiometric temperature contours. These contours are illustrated in Fig. 10 for a tank target at 35 GHz. The target is thus described with a passive cross section by combining the radiometric temperature and physical reflecting area in a term such as 750 m^2 K for a typical tank target measured at 35 GHz. From signature data analysis, it

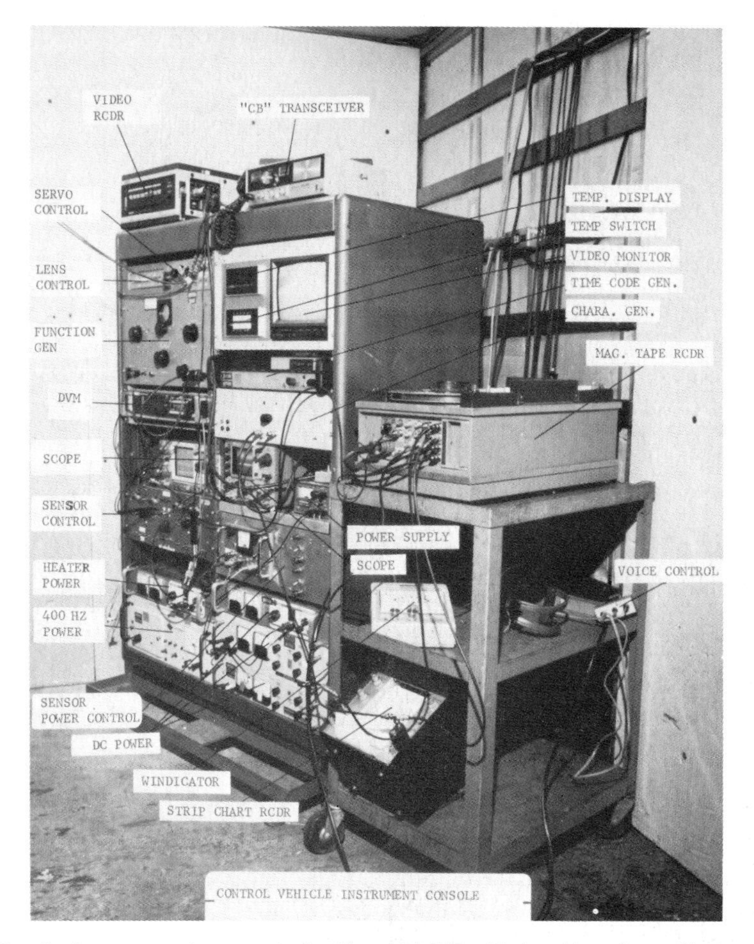

FIG. 5 Instrumentation console for 35 and 94 GHz side-by-side radar–radiometers.

has been noted that the active and passive target cross-section data are quite comparable for 35 and 94 GHz operation.

The background has been characterized for both radar and radiometer sensor operation. Table VIII summarizes measured terrain cross-section characteristics at 35 GHz. As the clutter cross-section level approaches 0 dB m^2/m^2, the sensor finds it very difficult to distinguish the target from the background without very sophisticated signal processing. There are two general coefficients used to describe clutter levels. They are γ, the background reflectivity, and σ_0, the normalized clutter radar cross section. They are related by

$$\gamma = \sigma_0/\sin\theta, \qquad (1)$$

FIG. 6 Boom truck and tower facilities used for target signature measurements.

TABLE VI

TANK TARGET RCS DATA STATISTICS BY RADAR DEPRESSION ANGLE AT 35 GHz

Depression angle	Number of points	Mean (m^2)	Standard deviation (m^2)	Median (m^2)	Minimum (m^2)	Maximum (m^2)
20°	95	68.7	44.6	55	15	239
30°	136	59.8	51.1	42	7	246
45°	98	90.8	52.6	76	29	232

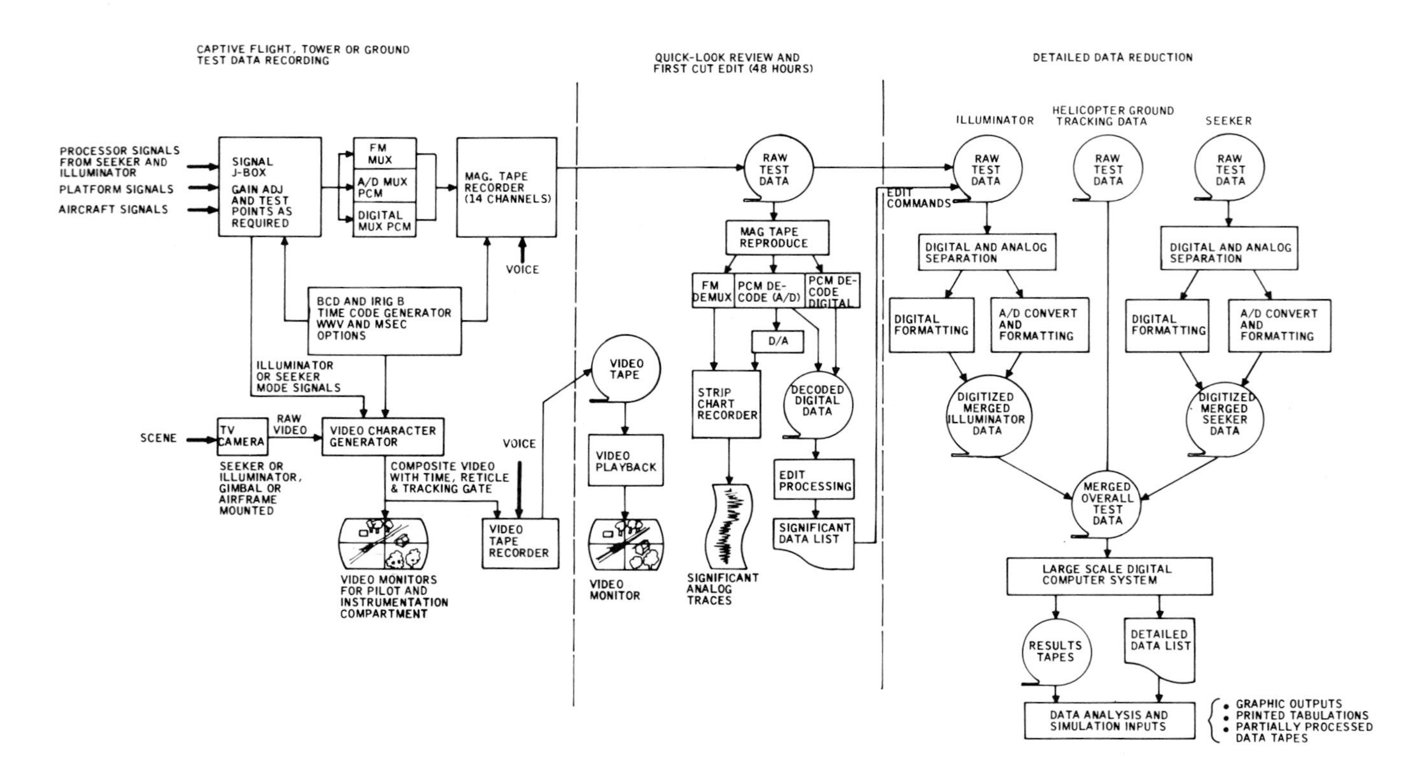

FIG. 7 Data management system for analysis of target/background signature data.

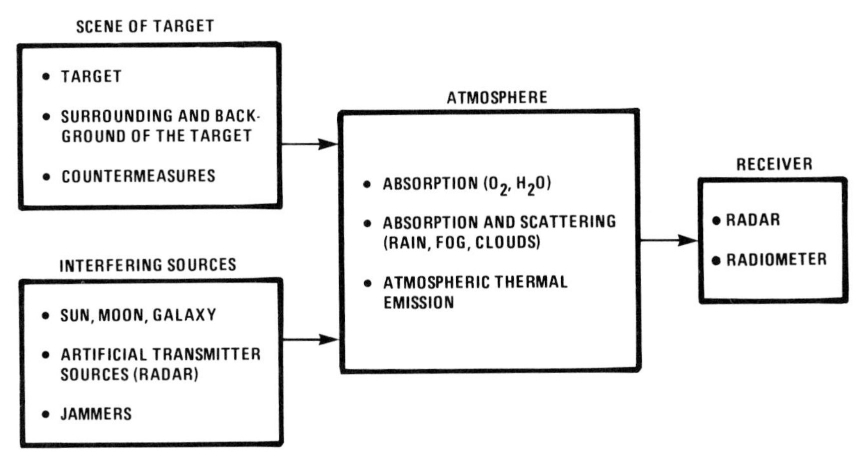

FIG. 8 Factors influencing receiver input signal characteristics.

where θ is the radar depression angle. These coefficients are related to a radar clutter cross section by

$$\sigma_c = \sigma_0 A_c, \tag{2}$$

where A_c is the clutter cell area. In the radiometer mode, the background is characterized by emissivities, since those objects with a high emissivity will not be good reflectors compared to a metallic target. Table IX provides measured emissivity data for a number of materials that form a

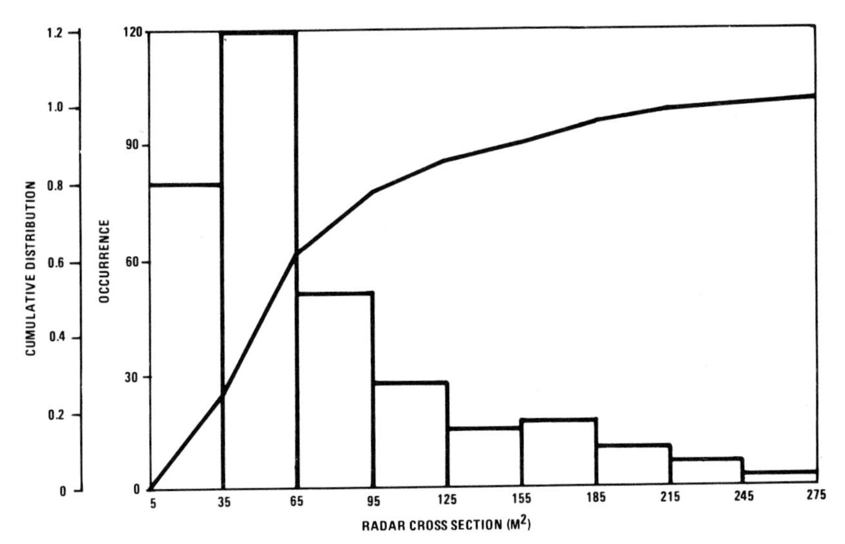

FIG. 9 Measured 35 GHz RCS distribution for numerous vehicular targets.

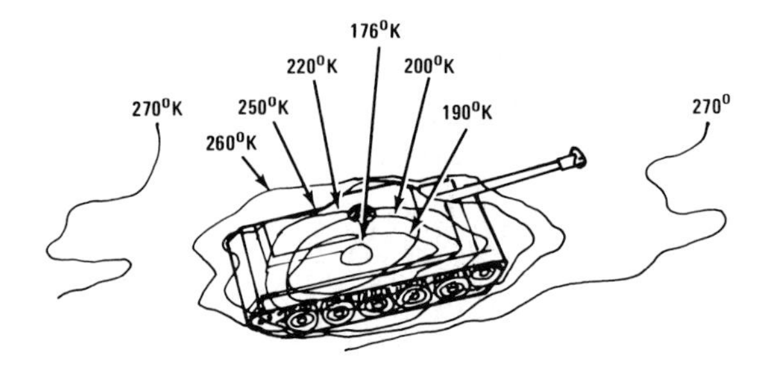

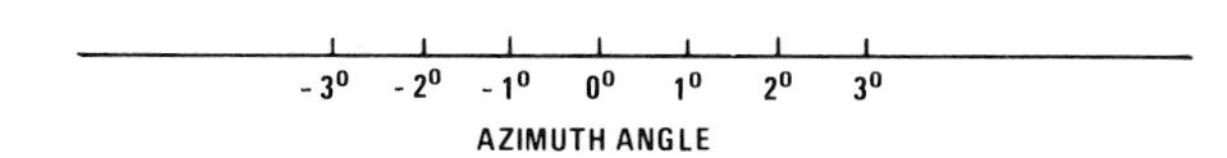

FIG. 10 Measured passive temperature for tank target at 35 GHz. Measurement conditions: antenna beamwidth, 20°; depression angle, 45°; antenna polarization, vertical; target range, 400 ft; radiometer frequency, 35 GHz.

background scenario. The radiometric temperature of the target is given by

$$T_{\mathrm{rt}} = \varepsilon_{\mathrm{t}} T_{\mathrm{t}} + \rho_{\mathrm{t}} T_{\mathrm{sky}}, \tag{3}$$

where ε_{t} is the target emissivity, ρ_{t} the target reflectivity, T_{t} the target thermodynamic temperature, and T_{sky} the radiometric sky temperature. Figure 11 shows the theoretical variation of effective radiometric sky tem-

TABLE VII

MEASURED TANK RCS VERSUS RADAR DEPRESSION ANGLE AND TARGET ASPECT ANGLE AT 35 GHz

DEPRESSION ANGLE	0°	90°	180°	270°
20°	45 m²	90 m²	90 m²	110 m²
30	25	80	500	200
40	75	90	360	115
50	95	110	180	125
60	130	95	220	80
70	50	105	140	90

0°
270°
90°
180°
ASPECT ANGLE DEFINITION

TABLE VIII

TERRAIN CROSS SECTION FOR 30° GRAZING ANGLE

Material	Cross section (dBm^2/m^2)
Grass	−22
Sand	−30
Trees	−18
Gravel	−24
Concrete	−33
Road	−15

perature versus observation angle for two conditions of relative humidity at 35 and 94 GHz. The sky is effectively warmer at 94 GHz; hence less contrast exists between target and background at this frequency. A rule-of-thumb for millimeter-wave sensors with antenna diameters of 6 in. or less is that radiometric operation is practical only up to about 40 GHz due to the sky-warming effect. This is further illustrated in Table X where degrading atmospheric conditions produce even more pronounced sky warming at 94 GHz. When the radiometric sensor is placed in an air-to-ground tactical mission involving dynamic search, detect, and track functions, the background exhibits a passive clutter characteristic. This is caused by the varying background emissivity and reflectivity as well as variable beamfill factor of discrete background features during antenna scan. Based on numerous measurements, it has been found that the clutter limited range in the passive mode is given by

$$R_c = R_n/1.78, \tag{4}$$

where R_n is the noise limited range.

TABLE IX

MATERIAL EMISSIVITIES AT NORMAL INCIDENCE

Material	Emissivity	Material	Emissivity
Sand	0.90	Heavy vegetation	0.93
Asphalt	0.83	Smooth rock	0.75
Concrete	0.76	Dry grass	0.91
Plowed ground	0.92	Dry snow (28–75 cm thickness)	0.88–0.76
Coarse gravel	0.84	Metal	0.0

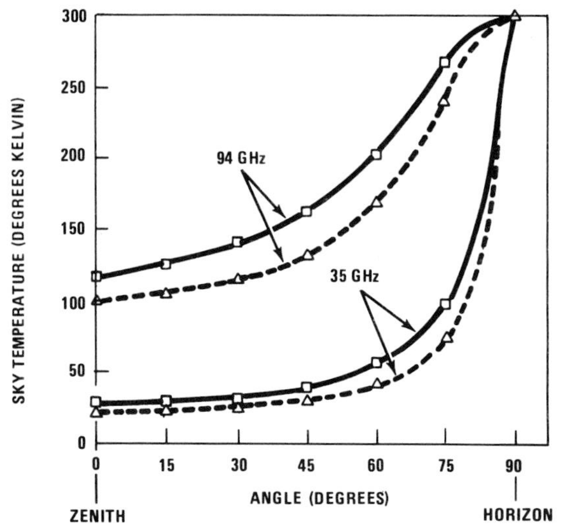

FIG. 11 Radiometric sky temperature at 35 and 94 GHz, theoretical. △, 50% relative humidity (RH); □, 75% RH.

III. Range Equations

There are typically three operating modes for autonomous, lock on after launch (LOAL) millimeter-wave seekers:

(a) passive acquisition and track to target impact;
(b) active acquisition and track to target impact; and
(c) active acquisition and initial track with passive terminal track to impact.

The latter active–passive concept has been investigated during recent years with extensive tower and captive flight testing. The pure passive mode is acquisition range limited due to a reduced target signal from poor beamfill for antenna diameters of 6 in. or less. The pure active mode is limited in terminal aimpoint accuracy due to glint and other effects gener-

TABLE X

RADIOMETRIC SKY TEMPERATURE VARIATIONS WITH ATMOSPHERIC CONDITIONS

Condition	35 GHz	94 GHz
Clear	34°K	60°K
Fog (0.32 g/m^3)	58	150
Rain (2 mm/hr)	77	143
Rain (4 mm/hr)	120	225

ally described by the terminology aimpoint wander. The active–passive design has been termed a contrast seeker since it derives its targeting signal from a backscatter and/or temperature contrast relative to the background clutter.

Figure 12 illustrates the development of range equations for active and passive operating modes. These relationships involve the hardware, propagation, target, and background characteristics. The following derivation is carried out for a conical scan seeker design since a majority of seeker development currently under way involves a mechanically scanned reflector or horn-lens antenna. As monopulse operation matures at millimeter-wave frquencies, the active range equation can be appropriately modified.

A. Conical Scan Passive

A simplified block diagram for the conical scan radiometer is shown in Fig. 13. A range equation for this configuration can be derived to include the signal-to-noise ratio (S/N) with appropriate effect for conical scan modulation.

The total power at the antenna terminals of the radiometer can be considered to be a combination of the signal power received by the antenna and the receiver noise power referred to the antenna terminals. Each of these power sources can be related to an equivalent noise temperature.

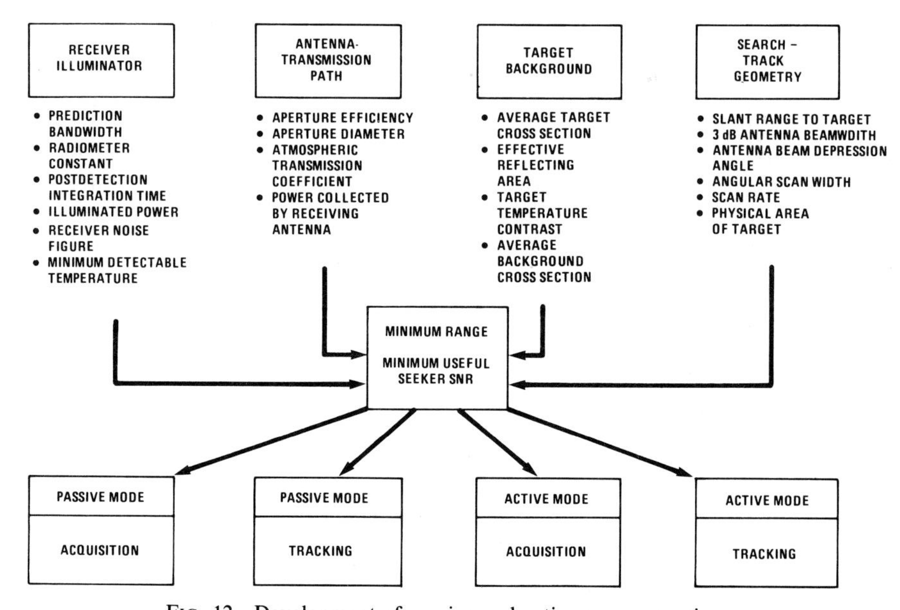

Fig. 12 Development of passive and active range equations.

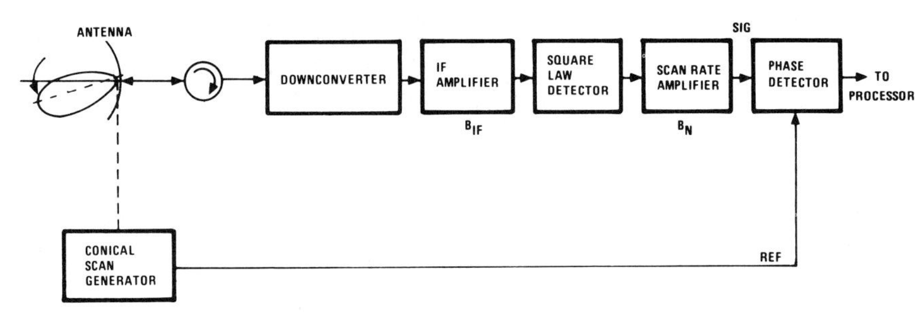

FIG. 13 Block diagram of passive conical scan radiometer.

The total temperature is called the radiometer operating temperature (T_{op}), which for a superheterodyne receiver is given by

$$T_{op} = 2(T_a + T_{eR}), \tag{5}$$

where T_a is the antenna temperature and $T_{eR} = (F - 1)\, T_0$ the effective temperature of the receiver. F is taken as the system noise figure, and T_0 is the standard noise temperature, typically 290°K. The factor 2 is due to the image response.

The statistical characteristics of the broadband power received by the antenna and the receiver noise are identical. Both are white or uniform power spectrum over the rf band. Taking T_{op} as the operating temperature of the radiometer, referred to the antenna terminals, the power density at the square-law detector input is

$$\text{IF Power density} = \tfrac{1}{2}kT_{op}G, \tag{6}$$

where k is Boltzmann's constant and G the overall gain of the rf, mixer, and IF components. For a constant operating temperature, the square-law detector produces both a dc and fluctuating ac power output. The input and output spectral densities of the square-law detector are shown in Fig. 14. A double-sided power spectrum is used, and the double-frequency detector output is assumed to be filtered.

In a total power radiometer, the signal power is taken to be the dc component of the output power while the noise power is the fluctuating part in the output bandwidth $2B_N$. The factor 2 is present because of the negative frequencies. Thus, in a total power radiometer, the power S/N at the video filter output is

$$P_S/P_N = \text{S/N} = K(\Delta T_{op})2B_{IF}2/KT_{op}2B_{IF}2B_N, \tag{7}$$

where the change in operating temperature

$$\Delta T_{op} = 2\Delta T_a \tag{8}$$

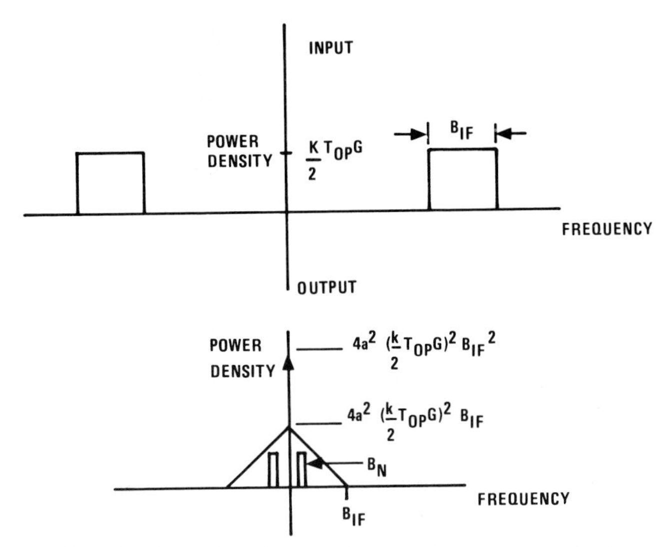

FIG. 14 Square-law detector input and output power density.

is caused by a change in the antenna temperature. This simplifies to

$$S/N = (2\Delta T_a/T_{op})^2(B_{IF}/2B_N) \text{ (total power radiometer)} \qquad (9)$$

for the total power radiometer configuration. Also the filter output S/N is proportional to the square of the antenna input S/N and related to both IF and video bandwidths.

The sensitivity of a number of switched radiometers has been described in great detail in the literature. Differences in sensitivity are due to the type of filtering used in the post-detection amplifier and in the type of phase detector or multiplier used.

Letting a constant K_R represent the particular type of radiometer processing, the S/N can be written as

$$S/N = [\Delta T_a/K_R(T_a + T_{eR})]^2(B_{IF}/2B_N). \qquad (10)$$

The minimum detectable rms temperature is found by setting S/N = 1. Therefore, ΔT_{min} is

$$\Delta T_{min} = K_R(T_a + T_{eR})/\sqrt{B_{IF}/2B_N}, \qquad (11)$$

the general expression for the sensitivity of a radiometer. Tables have been published that given values of K_R for a number of different types of radiometer and forms of radiometer signal processing. K_R is 1.0 for a total power radiometer and $2\sqrt{2}$ for a Dicke-switched radiometer with narrow band scan rate amplifier and phase detector.

A radiometer detects a target by observing a change in antenna temperature ΔT_a. Thus ΔT_a can be related to the target radiometric temperature contrast ΔT_T by

$$\Delta T_a = \eta_R(\Omega_T/\Omega_A)\Delta T_T, \tag{12}$$

where η_R is the antenna radiation efficiency, Ω_T the solid angle subtended by the target, and Ω_A the equivalent solid angle of the antenna (including effects of sidelobes). The ratio of the target solid angle to the antenna beam solid angle is referred to as the beam dilution (or fill) factor. The antenna solid angle may be given as

$$\Omega_A = \eta_R 4\lambda^2/\eta_A \pi D^2 = \Omega_M/\eta_B, \tag{13}$$

where η_A is the antenna aperture, η_B the beam efficiency, D the diameter of circular antenna, and Ω_M the main lobe solid angle. The solid angle subtended by a target of projected physical cross-section area A_T is related to the range R by

$$\Omega_T = A_T/R^2. \tag{14}$$

Thus the passive radiometer range equation can be expressed in the form

$$R = [\eta_A \pi D^2 A_T \Delta T_T \sqrt{B_{IF}/2B_N}/4\lambda^2(T_A + T_{eR})K_R \sqrt{S/N}]^{1/2}. \tag{15}$$

Substituting receiver noise figure into the equation gives the following:

$$R = \left[\underset{\text{(antenna)}}{\left(\frac{\eta_A \pi D^2}{4\lambda^2}\right)}\underset{\text{(target)}}{\left(\frac{A_T\,\Delta T_T}{1}\right)}\underset{\text{(radiometer)}}{\left(\frac{\sqrt{B_{IF}/2B_N}}{K_R[T_a + (F-1)T_0]}\right)}\underset{\text{(processing)}}{\left(\frac{1}{\sqrt{S/N}}\right)}\right]^{1/2}. \tag{16}$$

The key parameters are identified by function as

Antenna
- η_A Aperture efficiency
- D antenna diameter
- λ operating wavelength
- η_R radiation efficiency

Target
- A_T projected physical area
- ΔT_T target radiometric contrast

Radiometer
- B_{IF} IF amplifier bandwidth
- B_N scan rate amplifier bandwidth
- T_a antenna temperature of background
- F radiometer noise figure (DSB)
- T_0 standard temperature (290°K)
- K_R radiometer constant

Processing
- S/N power signal-to-noise ratio in the low-frequency bandwidth.

A simpler form of the range equation is frequently used:

$$R = (\eta_R A_T\ \Delta T_T / \Omega_A\ \Delta T_{min} \sqrt{S/N})^{1/2}. \tag{17}$$

Several observations can be made relative to the range equation:

(1) Range increases directly with the antenna diameter and operating frequency (neglecting atmospheric attenuation).
(2) IF amplifier performance influences the range by the $\frac{1}{4}$ power of its bandwidth.
(3) Range is inversely proportional to the square root of the receiver noise figure.
(4) Range is relatively insensitive to the S/N in the output bandwidth.

The range equation in either of the two forms previously derived can be used to predict the performance of any type of passive radiometer that uses a single pencil-beam antenna. Each different type of radiometer has a unique radiometer constant K_R that represents the particular type of processing that is used in measuring target temperature. The constant for a conical scan tracking radio meter is not readily determined because it will vary depending on relative target size and antenna beam shape. An estimate of the conical scan radiometer constant can be made by noting that, at maximum range when a target is being tracked, power is being received from the target at all time during 360° of the scan. However, since the beam squint angle is typically $\frac{1}{2}$ the 3-dB beamwidth, the power received is 3 dB down or $\frac{1}{2}$ of the power that could be received from the main beam if no quint were employed. This amount of power is equivalent to that available in a Dicke square-wave switched radiometer. Therefore, under-tracking conditions,

$$K_R = 2\sqrt{2} \quad \text{(conical scan tracking radiometer when tracking target).} \tag{18}$$

When the target is being acquired, the value of K_R is larger. If the target is displaced from the conical scan axis such that the nose of the beam traverses the target, the beam is only on the target for about $\frac{1}{3}$ of the total scan period. Neglecting the increase in power during this time, a reasonable value for the conical scan radiometer during target acquisition is

$$K_R = 3\sqrt{2} \quad \text{(conical scan tracking radiometer when acquiring target).} \tag{19}$$

This value of K_R holds for targets at angles slightly less than $\frac{1}{2}$ beamwidth away from the conical scan axis. Therefore, for both conditions, the conical scan tracking radiometer is only slightly less sensitive than a Dicke-stabilized design.

B. Conical Scan Active

The active radiometer shown in the block diagram of Fig. 15 differs from the passive radiometer since the active system radiates a short pulse, noiselike signal with a large bandwidth product as typically derived from a chirp pulse IMPATT diode in a cavity. The wide bandwidth is utilized for ground clutter fluctuation reduction. The radarlike properties of the signal are only used to gate out the sidelobe clutter and close-in rain backscatter effects. In principle, the active radiometer is identical to the passive radiometer where a measurement is made of the average value of the signal plus noise and of the average noise alone. The signal is then obtained by forming the difference of the two measurements. The measurements result from a conical scan motion of the antenna.

Similar to the previous derivation for a passive radiometer, the S/N in the low pass filter bandwidth is obtained by forming the ratio of the dc power component of the signal to the total fluctuation noise power present in the final narrow band amplifier. The total fluctuation noise power is the sum of the system thermal noise power, the fluctuating power returned from the terrain, and other sources of fluctuation noise due to scanning and variations in the background terrain effective radar cross section. The S/N in the video filter bandwidth of the active radiometer is

$$S/N = P_T/(P_{SN} + P_G + P_F), \tag{20}$$

where P_T is the target power (dc component), P_{SN} the system thermal noise power, P_G the terrain return fluctuation power, and P_F the fluctua-

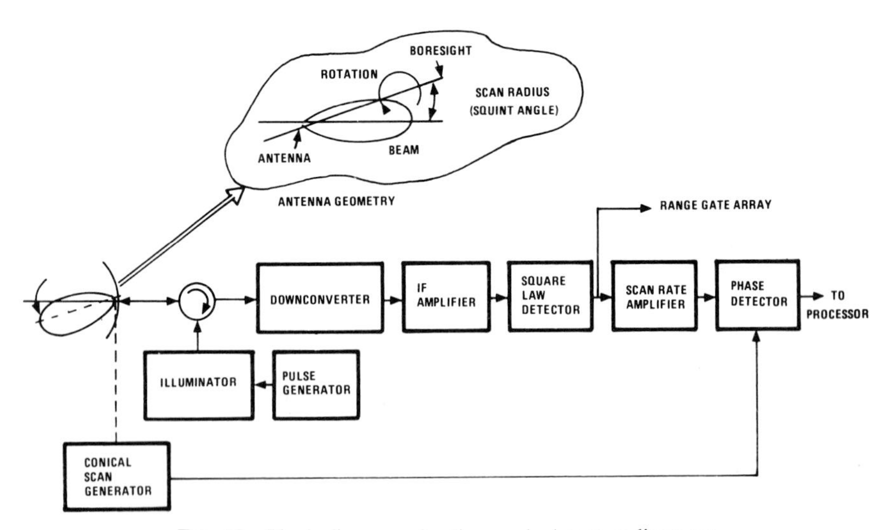

Fig. 15 Block diagram of active conical scan radiometer.

tion power due to other sources. Initially, it will be assumed that P_F caused by scanning and background terrain radar cross-section variations is zero. Also, since the signal-cross-noise and noise-cross-noise components of the video detector are omitted from the above expression, the equation applies when the IF S/N is less than unity and the IF bandwidth is large compared to the video bandwidth. These two conditions are generally satisfied in a radiometer receiver at the low signal levels occurring during target detection and lock on.

To calculate the S/N by introducing the system parameters, the power density of the signal and noise components in the IF amplifier will first be determined and then converted into noise and signal powers at the detector output. Letting P_{AT} be the average target return power at the antenna terminals, the IF amplifier output signal power density will be $GP_{AT}/2B_{IF}$, where G is the overall system rf and IF gain. Similar to the passive radiometer case, the dc signal power at the square-law detector output is

$$S = 4a^2(GP_{AT}/K_{RA}2B_{IF})^2 B_{IF}^2, \tag{21}$$

where the received target power has been reduced by the active radiometer constant K_{RA}. In a similar fashion, the thermal noise power in the low frequency filter B_v at the video detector output is

$$P_{SN} = 4a^2(\tfrac{1}{2}gkGT_{op})^2 B_{IF}2B_v, \tag{22}$$

where g is a gating factor that is the ratio of the time dwell of the main beam return to the interpulse period. In the IF amplifier, the random terrain return has a statistical distribution similar to that of Gaussian noise. Because of this, the terrain return produces a fluctuation component approximately equal to its dc output power component. However, because of the wideband characteristic of the transmitter signal, the fluctuating part of the terrain return will be reduced by $(B_{IF}/B_C)^{1/2}$, where B_C is the correlation bandwidth of the terrain that is approximately equal to $1/\tau_p$, where τ_p is the two-way propagation across the terrain patch. Therefore, the terrain return fluctuation power in the low frequency filter is given by

$$P_G = 4a^2(GP_{AG}/2B_{IF})^2 B_{IF}2B_v/\sqrt{B_{IF}\tau_p}, \tag{23}$$

where P_{AG} is the average power of the terrain return, referenced to the antenna terminals. This derivation has assumed a constant terrain return power as the antenna scans. Using these equations, an expression for the S/N at the low frequency filter is

$$S/N = P_{AT}^2/\{K_{RA}^2[(gk2[T_a + (F-1)T_0])^2 B_{IF}2B_v + P_{AG}^2(2B_v/B_{IF}\sqrt{B_{IF}\tau_p})]\}. \tag{24}$$

From the basic radar equation, the average power received from a point target of radar cross section σ_T at a range R is

$$P_{AT} = (\eta_A^2 \pi D^4/4^3\lambda^2)(\sigma_T/R^4)P_A, \tag{25}$$

where η_A is the antenna aperture efficiency, D the antenna diameter, P_A the average power (peak power $\times$ duty cycle), and λ the operating wavelength. The average power returned from the area on the ground illuminated by the main beam is

$$P_{AG} = (\eta_A^2 \pi D^4/4^3\lambda^2)(\sigma_G/R^4)P_A. \tag{26}$$

This equation is valid since, considering average power, the actual distributed pulse return over the area illuminated by the main beam is equivalent to a single point target of radar cross section σ_G.

Referring to the beam-terrain geometry of Fig. 16, σ_G can be calculated. The illuminated area on the ground is elliptical in shape and has an area

$$A_G = \tfrac{1}{2}\pi(R\theta_b/\sin\delta)^{1/2}R\theta_b = \tfrac{1}{4}\pi(R^2\theta_b^2/\sin\delta). \tag{27}$$

The radar cross section is given by

$$\sigma_G = A_G\sigma_0 = \tfrac{1}{4}\pi R^2\theta_b^2(\sigma_0/\sin\delta) = \tfrac{1}{4}\pi R^2\theta_b^2\gamma_0. \tag{28}$$

An equivalent expression in terms of solid angle Ω_A of the main beam is

$$\sigma_G = R^2\Omega_A\gamma_0, \tag{29}$$

where R is the average range to the terrain patch illuminated by the antenna beam. Substituting σ_G in Eq. (26) for the average power returned from the terrain results in

$$P_{AG} = (\eta_A^2 \pi D^4/4^3\lambda^2)(\Omega_A\gamma_0/R^2)P_A. \tag{30}$$

Introducing the expression for Ω_A gives

$$P_{AG} = (\eta_A\eta_R D^2/16)(\gamma_0/R^2)P_A. \tag{31}$$

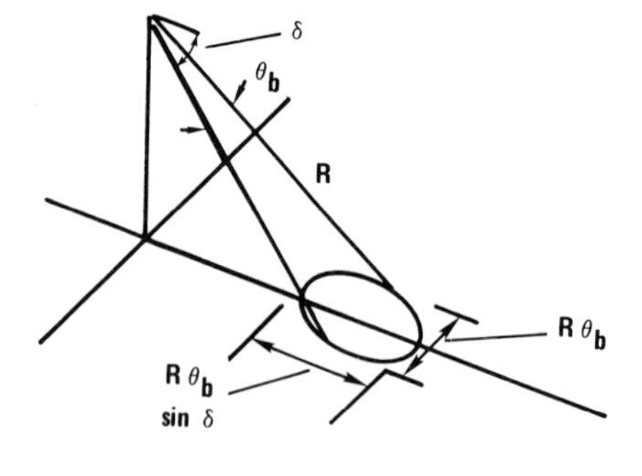

FIG. 16 Main beam geometry for illuminating terrain area.

It is noted that the terrain fluctuation power decreases as the inverse square of the range, while the target power decreases as the inverse fourth power of the range. In the expression for the S/N in the output low frequency filter, the denominator contains the sum of thermal noise power and terrain return fluctuation power. Since, as indicated above, the terrain return power has an inverse square relationship with range and the thermal noise power is constant, at long ranges thermal noise will be dominant, while at short ranges the terrain clutter noise will dominate.

In the region where the receiver thermal noise power is larger, the S/N is

$$\mathrm{S/N} = \frac{1}{K_{\mathrm{RA}}^2}\left[\frac{([\eta_{\mathrm{A}}^2 \pi D^4/4^3\lambda^2](\sigma T/R^4)P_{\mathrm{A}})^2}{(gk[T_{\mathrm{a}} + (F - 1)T_0](^2B_{\mathrm{IF}}2B_{\mathrm{v}}}\right]. \tag{32}$$

Case 1. *Long Range.* Thermal noise power larger than terrain return clutter fluctuation power:

$$R = \left[\underset{\text{(antenna)}}{\left(\frac{\eta_{\mathrm{A}}2\pi D^4}{4^3\lambda^2}\right)} \underset{\text{(target)}}{(\sigma_{\mathrm{T}})} \underset{\text{(active radiometer)}}{\left(\frac{\mathrm{P_A}\sqrt{\mathrm{B_{IF}}}/2\mathrm{B_v}}{gK_{\mathrm{RA}}2K[T_{\mathrm{a}} + (F - 1)T_0]B_{\mathrm{IF}}}\right)} \underset{\text{(processing)}}{\left(\frac{1}{\sqrt{\mathrm{S\ N}}}\right)}\right]^{1/2}. \tag{33}$$

Case 2. *Short Range.* Terrain return clutter fluctuation power larger than thermal noise power:

$$R = \left[\underset{\text{(antenna)}}{\left(\frac{\eta_{\mathrm{A}}\pi D^2}{4\lambda^2}\right)} \underset{\text{(target)}}{(\sigma_{\mathrm{T}})} \underset{\text{(terrain)}}{\left(\frac{1}{\gamma_0}\right)} \underset{\text{(active radiometer)}}{\left(\frac{[\sqrt{B_{\mathrm{IF}}\tau_{\mathrm{p}}}]^{1/2}\sqrt{B_{\mathrm{IF}}}/2B_{\mathrm{v}}}{K_{\mathrm{RA}}}\right)} \underset{\text{(processing)}}{\left(\frac{1}{\sqrt{\mathrm{S\ N}}}\right)}\right]^{1/2}. \tag{34}$$

The key parameters are identified by function as

Antenna	η_{A}	aperture efficiency
	D	antenna diameter
	λ	operating wavelength
Target	σ_{T}	target radar cross section
Terrain	γ_0	normalized terrain return
	τ_{p}	two-way propagation delay across terrain patch

Active radiometer

- P_A average transmitted power
- g gating factor
- K_{RA} active radiometer constant
- B_{IF} IF amplifier bandwidth
- B_V video amplifier bandwidth
- k Boltzmann's constant
- T_A antenna temperature
- F system noise figure, double sideband (DSB)
- T_0 standard temperature (290°K)

Processing

- S/N power signal-to-noise ratio in the low frequency bandwidth

The critical range R_c where the thermal noise power equals the terrain clutter fluctuation power can be derived as

$$R_c = \left[\left(\frac{\eta_A \eta_R D^2}{16}\right)\left(\frac{\gamma_0 P_A}{gk2FT_0B_{IF}}\right)\left(\frac{1}{(B_{IF}\tau_p)^{\frac{1}{4}}}\right)\right]^{1/2}. \tag{35}$$

At this range the active radiometer range performance changes from $\frac{1}{4}$ to a $\frac{1}{2}$ power-law dependence.

The atmospheric attenuation α, one-way attenuation in decibel per unit length, can be included in the active radiometer range equation as follows:

$$R10^{\alpha R/20} = \left[\left(\frac{\eta_A^2 \pi D^4}{4^3\lambda^2}\right)(\sigma_T)\left(\frac{P_A}{gK_{RA}2k[t_a + (F-1)T_0]\sqrt{2B_{IF}B_v}}\right)\left(\frac{1}{\sqrt{S/N}}\right)\right]^{1/4}. \tag{36}$$

IV. Millimeter-Wave Seeker Design

Missile guidance employing millimeter-wave seekers provides a unique solution for a variety of air-to-ground tactical missions. Specific advantages of operation at millimeter-wave frequencies include the following:

(a) adverse weather capability effective against a variety of priority military targets;

(b) a cooperative target not required for detection and tracking by the seeker

(c) seeker hardware amenable to low cost and small size due to a combination of millimeter-wave-integrated circuit (MMIC) components, LSIC processing hardware, and unique gimbal designs;

(d) low CEP due to the true-centroid tracking capability achieved in terminal guidance with a passive terminal tracking mode; and

(e) reduction of target glint, clutter, and multipath effects as a result of utilizing a broadband illumination waveform.

Significant seeker design parameters describe the target, background, propagation, rf seeker head hardware, search geometry, tracking accuracy, and processing effectivness. For many tactical targets of interest, antenna depression angles of 25°–45° from the horizontal provide the largest target RCS. Appropriate rf design tradeoffs include operating frequency, transmitter waveform, tracker configuration, and search rate–volume requirements.

A. Missile Guidance Options

There are basically two modes of operation for a millimeter-wave seeker, active (radar) and passive (radiometric). Current design concepts involve seeker operation as a radar for search, detection, and lock on at an appropriate standoff range. As the missile nears the target, problems of glint, multipath, and other wavelength-dependent phenomena make it prudent to switch to a passive mode in which the target signal is derived from reflected sky radiation. The transmitter or illuminator uses an IMPATT diode for a pulse waveform or a Gunn diode for a cw waveform that can be readily frequency-modulated. The waveform is designed to produce a wideband noiselike energy resembling random noise. The purpose of this is to enhance the radiometric signal in order to achieve longer range and still maintain the advantages of the passive mode with respect to glint, multipath, and clutter decorrelation.

A total energy seeker system such as a millimeter-wave wideband incoherent radar must determine target presence and location from the integrated signal return from a 360° scan. This is unlike a conventional radar that provides a discrete designation as the radar scans past the target. Specifically, within a single scan of the total energy system, spectral background such as streams or smooth-surfaced roads may return as much, or more, energy than a metal target of concern. The signal processing problem is one of devising techniques of enhancing the capability for identifying a metallic target within a diverse background and to discriminate the target signature from other competitive energy returns when radar is operating in the noise illumination mode. The basis of target detection becomes the same as conventional monochromatic radar, i.e., metal targets have greater backscattering (cross-section) properties than the background that results in an increase in target-to-background temperature contrast. Target acquisition will occur when the designated target has been determined and can be maintained within the seeker field-of-view to control the attitude of the missile. The missile control system operates by maximizing the designated target's returning energy as measured by the radar receiver, thereby causing the antenna to point directly at the target.

Emphasis in the design detail that follows is on an autonomous seeker capable of carrying out all of the functions identified in Fig. 17. Key stages

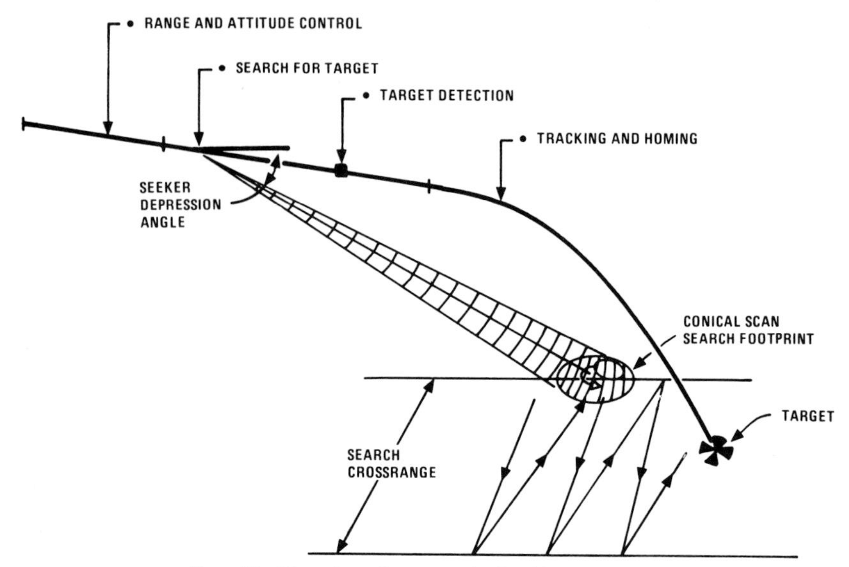

FIG. 17 Functions for a terminal guidance seeker.

include (a) range and attitude control, (b) target search, (c) target detection, and (d) target tracking and homing. In a lock-on-before-launch (LOBL) operation, only the seeker tracking capability would be used. A conical scan antenna, or its equivalent, is required to derive the tracker error signals. Lock-on-after-launch (LOAL) operations require, in addition to a tracking capability, the search, detection, and feature discrimination capabilities identified above. The LOAL seeker concept may also use the radar's ranging capability to provide signals for missile attitude and altitude stabilization.

Two other millimeter-wave guidance modes are currently being evaluated under concept demonstration programs, namely, the semiactive and beamrider systems. The operational concept of the millimeter-wave semiactive concept is that the illuminator will be carried in a mini-RPV, a man on the ground, a penetrating manned aircraft, or a standoff aircraft. The seeker, on the other hand, will be capable of guiding a projectile, missile, or other weapon to the designated target area. System concept studies to date have led to the following conclusions:

(a) A millimeter-wave semiactive guidance system is a viable concept for use against targets in a clutter background.

(b) Semiactive guidance provides a significant increase in standoff range performance over active or passive autonomous mode guidance. The problem of target acquisition for an autonomous guidance system is eliminated since target acquisition and illumination is performed by a separate remote system.

(c) Semiactive guidance must be supplemented by autonomous terminal homing in order to achieve a high probability of kill against an armored target. Direct tracking handover to autonomous guidance can be accomplished without reacquisition.

(d) A 35-GHz semiactive system with a solid-state illuminator will achieve the desired maximum illuminator (2 km) and seeker (4 km) standoff ranges in 4 mm/hr rain with a small illuminator package.

(e) A 94-GHz semiactive system with a solid-state illuminator has a shorter range capability; however, this may be adequate for 2 mm/hr rain. Predicted standoff ranges would be illuminator (2 km) and seeker (2.1 km). A tube illuminator would be required to meet the performance goals under 4 mm/hr rain.

(f) The 6-in. diameter seeker and compatible tracker–illuminator designs have been established for all-aspect angle weapon delivery. The seeker identifies and rejects the direct signal from the illuminator with clever signal processing design and seeks out the reflected energy from the target area. The seeker is configured to operate in a totally autonomous mode if the illumination beam is absent.

In the basic millimeter-wave beamrider concept, the target is acquired by a broad-beam millimeter radar and then tracked by a narrow-beam millimeter radar such that the center of the beam axis forms a line along which it is possible to direct a missile. The missile is launched toward the target along the radar beam, which is deliberately coded to provide information on the position of the missile relative to the center of the beam. A receiver located in the rear of the missile detects the beam signal, which is then decoded to determine the missile position relative to the beam and to generate error signals for missile guidance to beam center. Basic functions of the beamrider system thus include target acquisition, target track, missile capture, and missile guidance. The basic system configuration considered to be desirable for a ground-to-ground antiarmor beamrider requires that the entire system be self-contained, crew-served, and vehicular-mounted package with an antenna aperture diameter less than one meter. Nominal system parameters relating to the requirements are the following:

Operating range: maximum, 2 km; minimum, 0.5 km

Weather and environment: rain, 4 mm/hr; fog, 100-m visibility; temperature, 90° − 100°F; relative humidity, 80% − 100%; smoke; battlefield generated dust and aerosols

Acquisition: search section, 5° × 25°; movable scan sector; search time less than 10 sec

Missile data rate: 50 Hz

Antenna beamwidths: capture, 6°; tracking, 0.5°

Potential operating frequencies: acquisition, 94 GHZ; target tracking and missile guidance 140 or 220 GHz

B. Seeker Search and Track

The typical application for a millimeter-wave seeker is in a cruise or glide trajectory missile delivered to search out targets along a specific ground path. For this application, the seeker employs a linear search pattern generated by a combination of missile forward motion and seeker antenna azimuth scan. This is shown in Fig. 18, where a seeker head look-down angle of 20°–45° is utilized to produce an elliptical projected illumination footprint on the ground. The antenna provides a conical scan, or its equivalent, to enhance target detection and aid in target feature evaluation. The resulting antenna footprint is typically two antenna beamwidths in width and length. Alternate search strategies may delete the conical scan or replace it with an appropriate azimuth dither.

Gimbal design is very important in that a constant-rate search is usually required. An azimuth outer and elevation inner gimbal arrangement will provide the desired search range. A reversal of this arrangement causes the range-to-ground to oscillate as the azimuth search is carried out. This degrades the signal in each range gate and makes range tracking more difficult. A constant sweep rate $\dot{\psi}$ is frequently used to hold the time-on-target (TOT) constant. For point targets, the TOT is just the width of the projected footprint. At the footprint center, assuming a con-

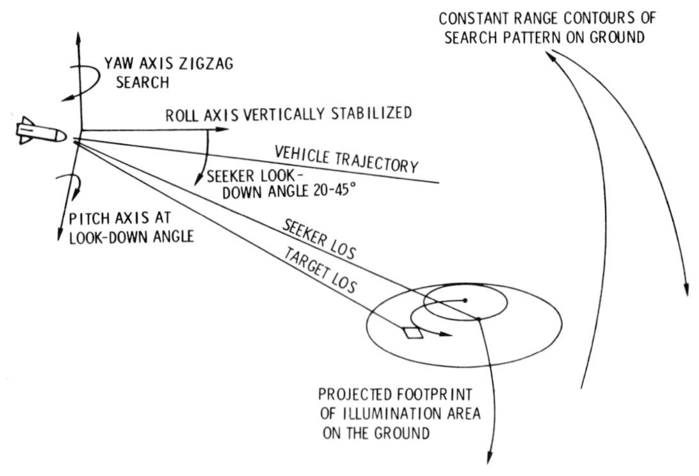

Fig. 18 Seeker search mode operation.

ical scan, TOT is given by

$$\mathrm{TOT} = 2\beta/(\dot{\psi}\cos\theta), \tag{37}$$

where β is the antenna 3-dB beamwidth, θ the depression angle from horizontal, ψ the $\frac{1}{2}$ total search angle, and $\dot{\psi}$ the sweep rate. The optimum search pattern generally has the following characteristics; (a) no coverage gaps, (b) match to the missile maneuver capability, and (c) a constant azimuth scan rate.

The active mode linear search tradeoff equation between the minimum scan rate and the ratio of vehicle velocity (V) to slant range (R) is

$$\frac{\dot{\psi}}{\psi} = \frac{2.5[1 + (\beta/\psi)]\sin(\theta - \gamma)(V/R)}{\beta - 2.5(\psi/\dot{\psi})(V/R)\sin(\theta - \gamma)}. \tag{38}$$

This is based on moving one-half the conical scan footprint length as the antenna beam moves from the center line to one edge and back. Typical angular acceleration are $\ddot{\psi} = 1200°/\mathrm{sec}^2$, and peak angular velocities are $\dot{\psi} = 80°/\mathrm{sec}$. An approximate form of Eq. (38) is plotted in Fig. 19 where the terms β/ψ and $\dot{\psi}/\ddot{\psi}$ are neglected in order to illustrate the frequency

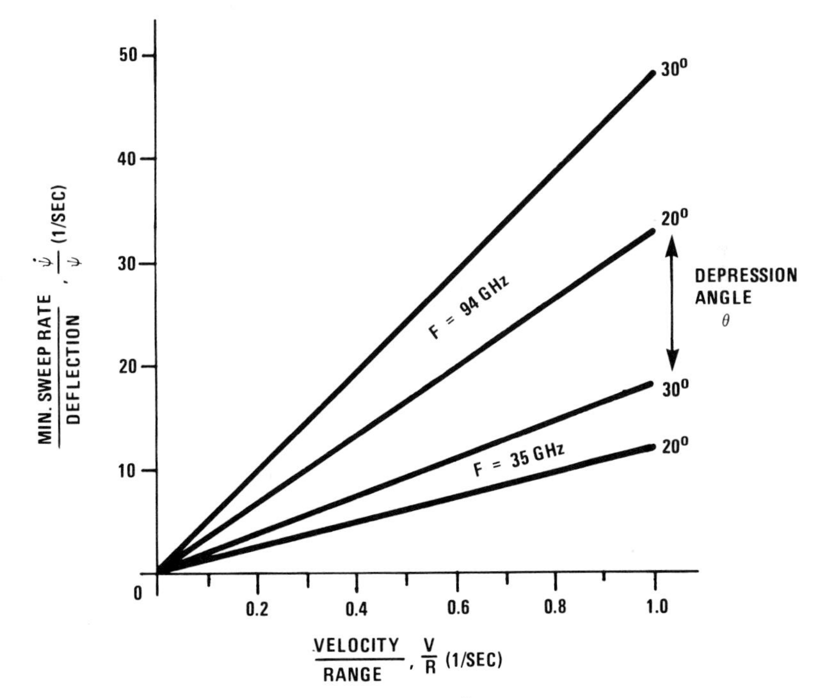

FIG. 19 Search coverage for missile cruise, $\dot{\psi}/\psi = (2.5\sin\theta/\beta)(V/R)$; $D = 150$ mm; $\beta = 4°$ at 35 GHz and 1.5° at 94 GHz.

tradeoff. The coverage cross range (W) is found from

$$W = 2R \cos \theta \sin \psi. \tag{39}$$

An upper bound on V/R ratio is found to be

$$V/R \leq 0.5 \text{ @ 35 GHz; } 0.23 \text{ @ 94 GHz.}$$

Missile maneuver cross-range bounds are approximated by

$$Y = k_1 R^2, \tag{40}$$

where a constant acceleration $\ddot{Y}$ and flight time linear with slant range R are assumed. The search pattern to match this maneuver bound is obtained by letting ψ decrease linearly with R in Eq. (38). The cross-range width has an R^2 shape also since

$$W = 2R \cos \theta \sin(k_2 R),$$

or

$$W \cong 2K_2 R^2 \cos \theta. \tag{41}$$

This $\psi = K_2 R$ concept will provide maximum range and total coverage as the range decreases.

It is easier to hold ψ constant in most applications. In this case $\dot{\psi}$ must be higher than the optimum value resulting in coverage overlap and maximum range and coverage gaps at the minimum ranges. An equation for

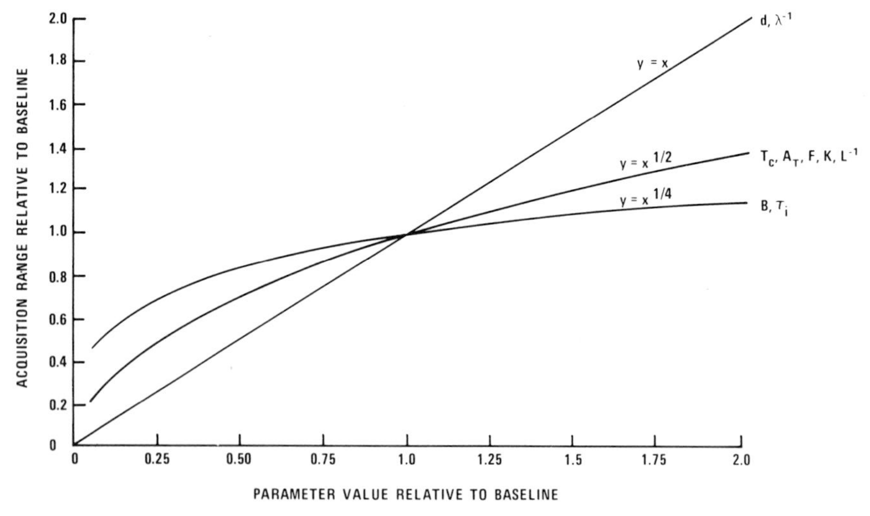

FIG. 20 Passive mode acquisition range. $R^*_{acq} = (T_c \pi d^2 A_T \eta B / 4\lambda^2 k F T_0 L^{1/2}_{SNR})^{1/2} \ (B\tau_i)^{1/4}$. *Receiver noise limited.

the gap length (L_{gap}) is

$$L_{gap} = \left[\underset{\left(\substack{\text{coverage}\\ \text{time}}\right)}{\frac{4(\psi + \beta)}{\dot{\psi}}} + \underset{\left(\substack{\text{turn}\\ \text{around}\\ \text{time}}\right)}{\frac{2\dot{\psi}}{\ddot{\psi}}}\right] \underset{\left(\substack{\text{footprint}\\ \text{velocity}}\right)}{V\frac{\sin(\theta - \gamma)}{\sin\theta}} - \underset{\left(\substack{\text{useful}\\ \text{footprint}\\ \text{length}}\right)}{\frac{1.68^{\beta}\text{R}}{\sin\theta}}. \tag{42}$$

Figures 20 and 21 provide convenient parametric variations for seeker acquisition range relative to any baseline design for passive and active operation, respectively. From these sensitivity curves, it is found that antenna diameter is the most important parameter since range varies linearly with it. The least sensitive range parameter is integration time; however, it still is a very useful parameter for matching S/N and range requirements. The acquisition range is formulated based on a receiver noise-limited condition. Parameter definitions are as follows:

- d antenna daimeter
- λ wavelength
- P_{xmt} transit peak power
- σ_t radar cross section
- F noise figure (DSB)
- τ_i integration time
- τ_p transmit pulse width
- B receiver IF bandwidth
- PRF pulse repetition frequency
- L, L_p RF and processing losses
- η antenna efficiency
- S/N detection signal-to-noise ratio

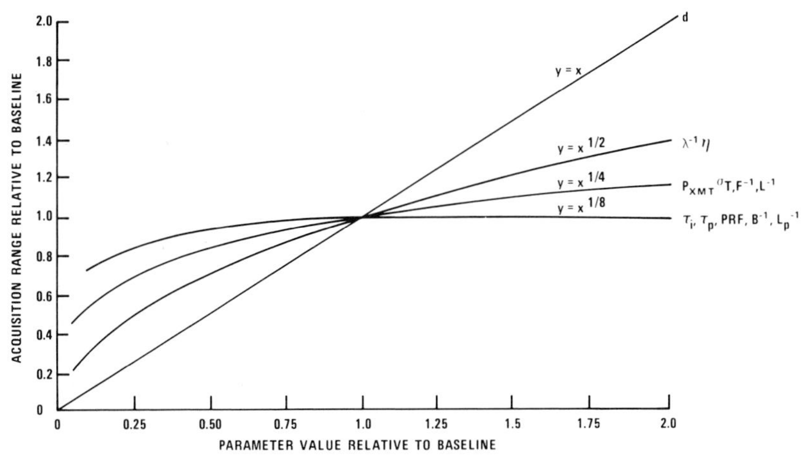

FIG. 21 Active mode acquisition range. $R^*_{acq} = (P_{XMT}\pi d^4\eta^2\sigma_T/64F^2L^2kT_0BF^{1/2}_{SNR})^{1/4} \times (B\tau_p\tau_i PRFL_p^{-1})^{1/8}$. * Receiver nose limited.

The seeker active mode slant range as a function of missile forward velocity is shown in Fig. 22. It is noted that the maximum range decreases as the missile velocity increases due to processing integration time effects. For example, selecting a velocity of 200 m/sec, a 35-GHz seeker can operate at any range between 800 m and 2.5 km. A 6-in. diameter antenna, a 30° antenna depression angle, and a 15° azimuth scan angle are assumed. The conical scan beamwidth is plotted versus antenna diameter in Fig. 23. For example, a 6.5-in. diameter antenna provides a conical scan beamwidth of 7.4° at 35 GHz and 2.18° at 94 GHz assuming a 3-dB scan crossover point.

Automatic transfer into the tracking mode is accomplished after target search and detection. Upon detecting and discriminating a target, azimuth scanning stops and the signal processor initiates a transition sequence. The existing azimuth gimbal scan rate command is reversed to drive the antenna axis to the target line of sight. This command is maintained for a period of time equal to the time on target prior to detection. Use of a gimbal closed-loop, bang-bang control system results in a minimum time convergence of the antenna boresight to the target line of sight, therefore assuring a valid lock on. When convergence is completed, the gimbal control is transferred to azimuth and elevation tracking electronics. Any residual errors are corrected by the closed-loop gimbal system performing as a precision tracker. Upon completion of the settling transient, the tracker pitch and azimuth errors are used as the proportional guidance command.

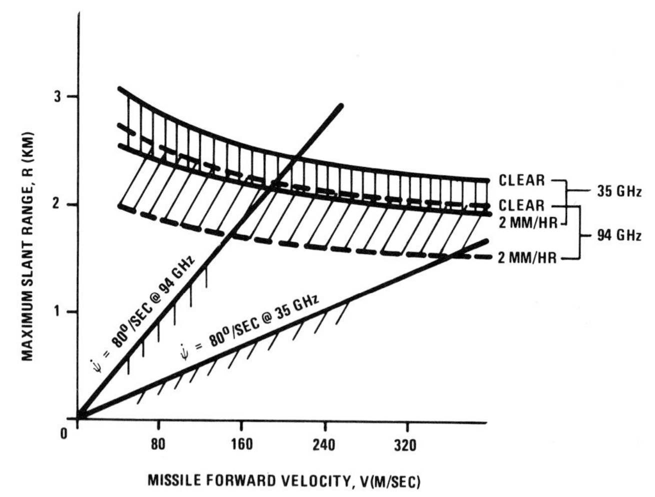

FIG. 22 Active mode range for missile cruise. $\psi V R^7 = kf(20/\mathrm{S/N})(D/150\ \mathrm{mm})^6 \times [0.321/\cos\theta\ \sin(\theta-\gamma)]$. At clear weather, $k_{35} = 27{,}000$, $k_{94} = 12360$, $\theta = 20°$, $\psi = 15°$. $R = [5\ \psi V \sin(\theta - \gamma)/\beta\psi]$.

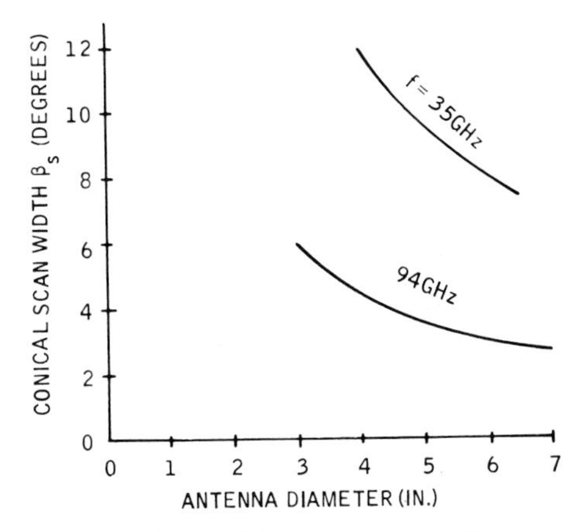

FIG. 23 Conical scan beamwidth versus antenna diameter. $\theta_s^\circ = 142\lambda/d$.

Tracking error signals are generated from amplitude modulation of the main lobe return as it is conically scanned. Because ground return from the entire scan field-of-view must be available for classical conical scan processing, range gating is not employed to the extent used in the target detection mode. Receiver noise is suppressed by gating the receiver on only when a return from the ground is anticipated. Error signals derived from targets relative to antenna boresight are coupled through the platform torquers to move the antenna. Rate gyros mounted on the platform are used to decouple missile body motions from the target tracking function. Position signals give antenna pointing angles relative to the missile body axis. These signals are used to develop the homing guidance.

At long range, tracking performance is limited by receiver noise. Receiver noise is generally suppressed by gating the video on only when the target return is present. Target range information, which is continuously available from a range tracker, is used to center the track gate about the target. Typically, the track gate is twice the pulse width. Receiver noise power is thus reduced by the ratio of the track-gate width to the pulse-to-pulse period. At shorter ranges, where the target begins to fill the antenna beam, glint noise appears. In narrow band radars, glint noise is caused by amplitude and phase interference from two or more target bright spots. It causes the radar aimpoint to wander off the target about 15% of the time. This glint noise is significantly reduced with a wideband, noiselike illumination waveform. Phase variation effects are reduced by averaging over many independent frequencies. The tracking aimpoint will always remain on the target, at the multiple bright spot centroid, although it still wanders as the aspect angle change causes the bright spot amplitude to change.

The variance of glint noise is obtained by assuming bright spot at each end of the target. Let L be the spacing and R the seeker range, then the 3σ value is assumed to be 0.5 L/R. *The* 1σ glint is given by

$$\sigma_g \cong 0.45\ L/R. \tag{43}$$

A more accurate tracking signal can be developed by utilizing the passive radiometric mode. Since the wideband and distributed sky radiation provide the effective target illumination, multipath effects are eliminated and the strongest signal originates from the area near the physical centroid of the target as discussed earlier in this chapter. Tracking accuracy for a conical scan radar can be expressed as

$$\sigma = \beta/\sqrt{\mathrm{S/N}}, \tag{44}$$

where σ is the rms tracking accuracy in radians, β the antenna 3-dB beamwidth, 1.2 λ/D rad, and S/N the signal-to-noise ratio in tracking channel.

Figures 24 and 25 present tracking accuracy as a function of range in fair weather and in 4 mm/hr rain, respectively. Figure 24 shows in clear weather that 94 GHz tracking accuracy is significantly better than that of 35 GHz. However, the tracking accuracy at 94 GHz degrades significantly in moderate rain as shown in Fig. 25.

C. Waveforms and Processing

Figure 26 presents a block diagram illustrating a typical dual-mode ra-

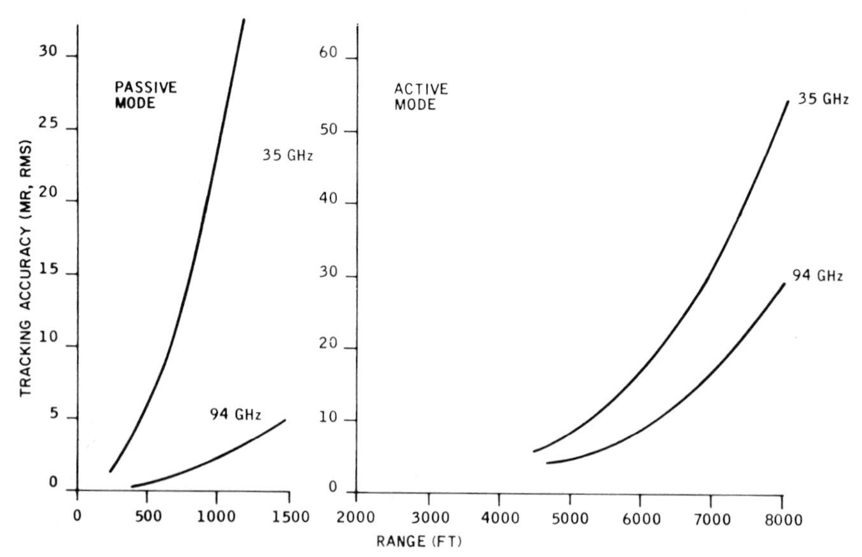

Fig. 24 Tracking accuracy versus target range (clear weather).

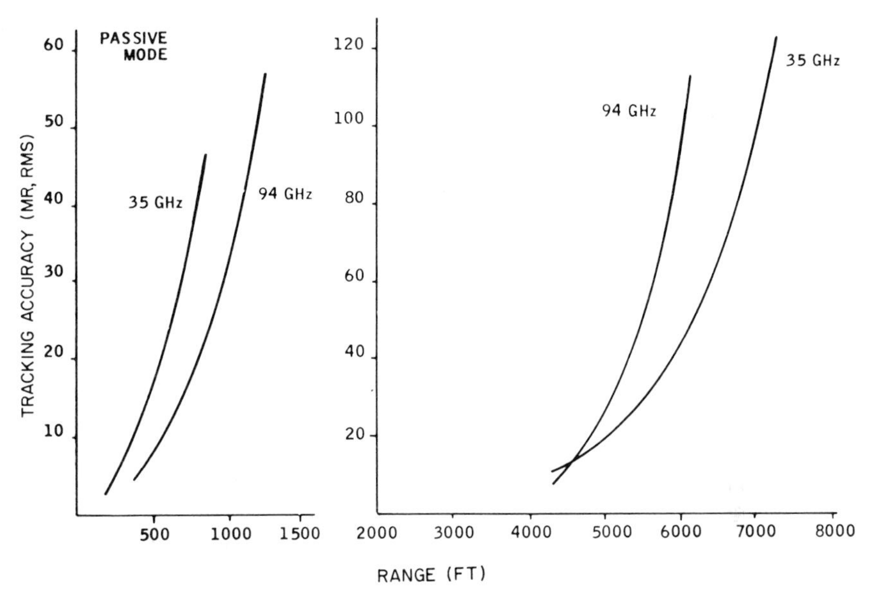

FIG. 25 Tracking accuracy versus target range (4 mm/hr rain).

diometric rf sensor head based on a FMCW illuminating waveform. Sensor parameters are given in Table XI. The receiver front end up through the IF amplifier is common to both radar and radiometric modes of operation. Figure 27 shows the illumination waveform that is triangular. A sample of the transmitted signal is used as the local oscillator, thus the instantaneous IF signal frequency is the difference between the received and transmitted signal frequencies. The IF is a function of time and round-trip delay τ_d, as shown in Fig. 28, where

$$\tau_d = R/c, \tag{45}$$

where R is the range to target, and c the speed of light. The maximum IF signal frequency at a range R is given by

$$f_{max}(R) = \frac{(f_{max} - f_{min})}{T_p} \frac{(2R)}{c}, \tag{46}$$

where T_p is the period of transmitted signal.

The maximum IF signal frequency, which the receiver must process, is determined by the slope of the illumination waveform and the maximum operating range. Assuming an operating range of 10,000 ft and a waveform slope of 10 MHz/μsec ($2B_{XMT}/T_p$), the maximum active mode IF signal frequency is 200 MHz. Since a 500-MHz bandwidth is employed to optimize passive operation, the IF filter is selected to reduce the band-

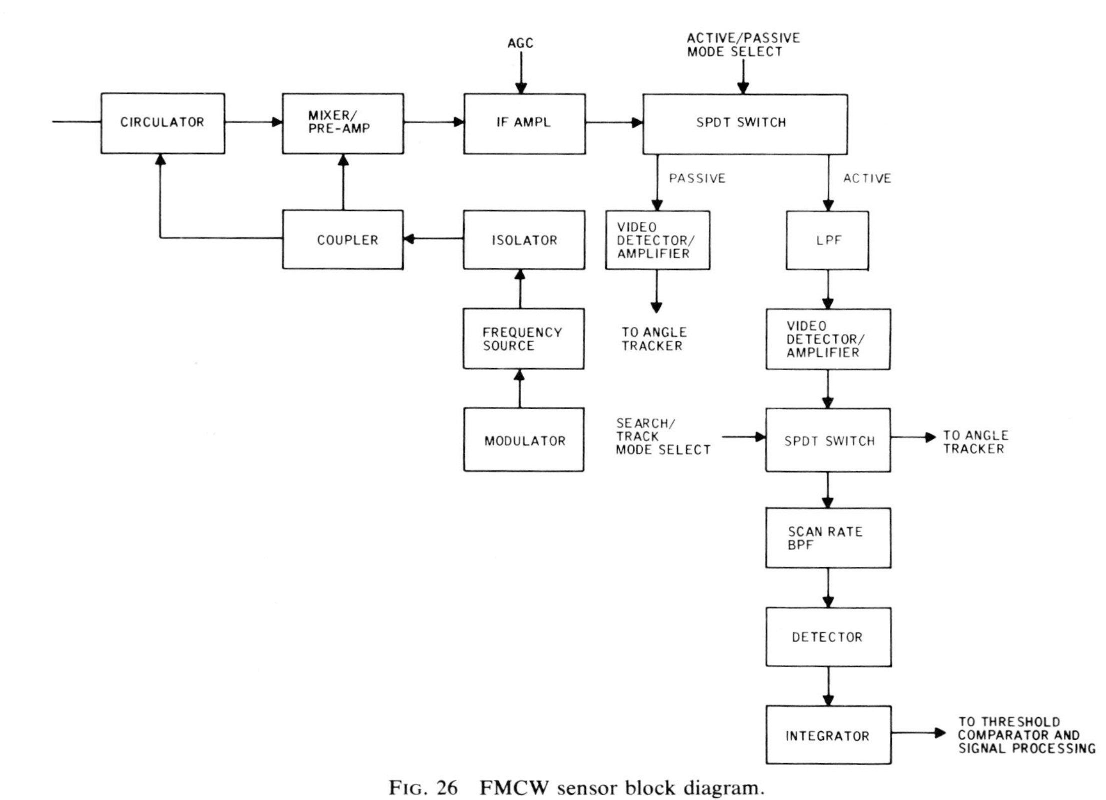

FIG. 26 FMCW sensor block diagram.

TABLE XI

PARAMETERS FOR FMCW SENSOR

Parameter	Value
Average transmitted power	100 mW (35 GHz) 50 mW (94 GHz)
Transmitted bandwidth	500 MHz
Receiver IF bandwidth	200 MHz (active) 500 MHz (passive)
Period of transmitted signal	100 μsec
Antenna diameter	162.5 mm
Antenna beamwidth	3.6° (35 GHz) 1.3° (94 GHz)
Antenna efficiency	0.60
Target RCS	50 m^2
Scanning loss	0.10
Wavelength	8.6 mm (35 GHz) 3.2 mm (94 GHz)
Reference temperature	290°K
Noise figure (SSB)	9 dB (35 GHz) 12 dB (94 GHz)
Integration time	500 msec (35 GHz) 70 msec (94 GHz)
Background terrain reflectivity	−14 dB m^2/m^2

width to 200 MHz for active or radar operation. The power spectrum of the transmitted FMCW waveform is approximately flat, band limited to B_{XMT}, and centered at f_0^2. Thus the target and background terrain are illuminated by a wideband noiselike signal that reduces target scintillation, target glint, and also decorrelates the clutter return similar to a frequency agile radar. Receiver noise and clutter are averaged in a post detection integrator to reduce their respective variances and enhance the sensor target discrimination capability. The active mode S/N for the FMCW de-

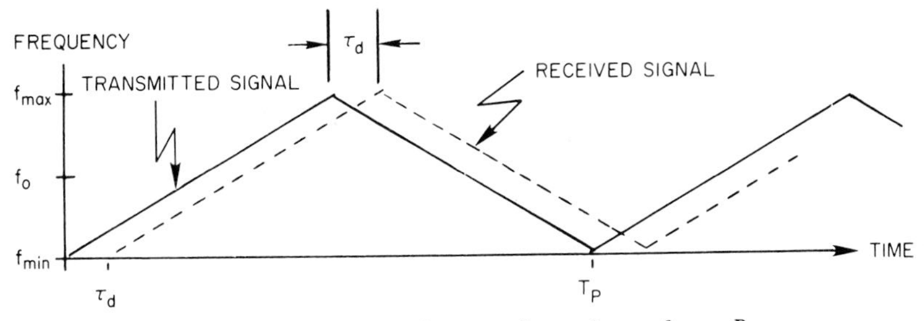

FIG. 27 FMCW illumination waveform; $f_{max} - f_{min} = B_{XMT}$.

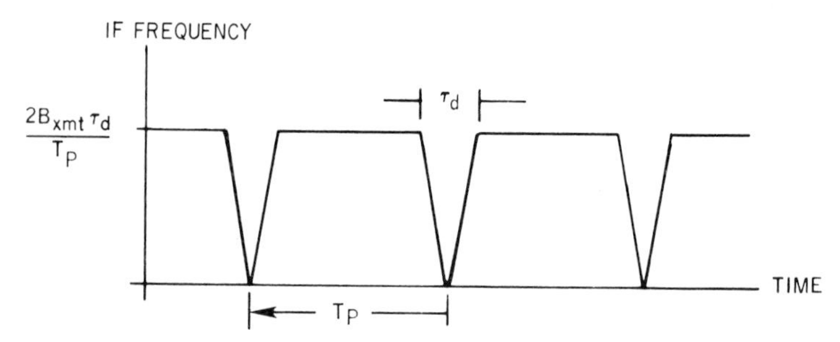

FIG. 28 IF frequency of received signal.

sign is expressed as

$$S/N_{FMCW} = \left(\frac{P_{avg}d^2\pi\eta^2\sigma_t L_t}{64\lambda^2 R^4 K T_0 F B_{LPF}}\right) B_{LPF}\tau, \tag{47}$$

where P_{avg} is the average transmitted power, d the antenna diameter, η the antenna efficiency, σ_t the target RCS, L_T the scanning loss, λ the wavelength, R the range, k Boltzmann's constant, T_0 the reference, F the noise figure, B_{LPF} the active mode IF bandwidth, and τ the integration time.

Figure 29 presents a block diagram illustrating a typical radiometric sensor head based on a pulse illuminating waveform. Sensor parameters are given in Table XII. Figure 30 shows the pulsed illumination waveform. The transmitting source is an IMPATT diode in a cavity that produces a chirp characteristic when modulated with a high current drive pulse. A separate cw source is used as the receiver local oscillator. A SPST switch is used to prevent the receiver main IF amplifier from saturating in response to transmitter pulse leakage through the circulator. The IF preamplifier is allowed to saturate since its gain value is relatively low. Range gating is employed to enhance the signal-to-noise-plus-clutter ratio and also to provide narrow range resolution cells for improved target discrimination. A gate width equal to the transmitted pulse width is selected to maximize the noise and clutter suppression. The active mode S/N at the output of the video integrator can be written as

$$S/N_{pulse} = (P_{avg}d^4\pi\eta^2\sigma_t L_t/64\lambda^2 k T_0 F B_{XMT}) B_{XMT}^{1/4}/d_t, \tag{48}$$

where B_{XMT} is the transmitted signal bandwidth and d_t the duty factor.

A comparison of the S/N achieved for a FMCW and a pulse sensor is shown in Fig. 31, assuming 35-GHz operation and the other parameters of Tables XI and XII. Considering a homogeneous background terrain, the signal-to-receiver noise ratio is essentially equivalent for the two cases. The assumption of homogeneous background terrain permits modeling

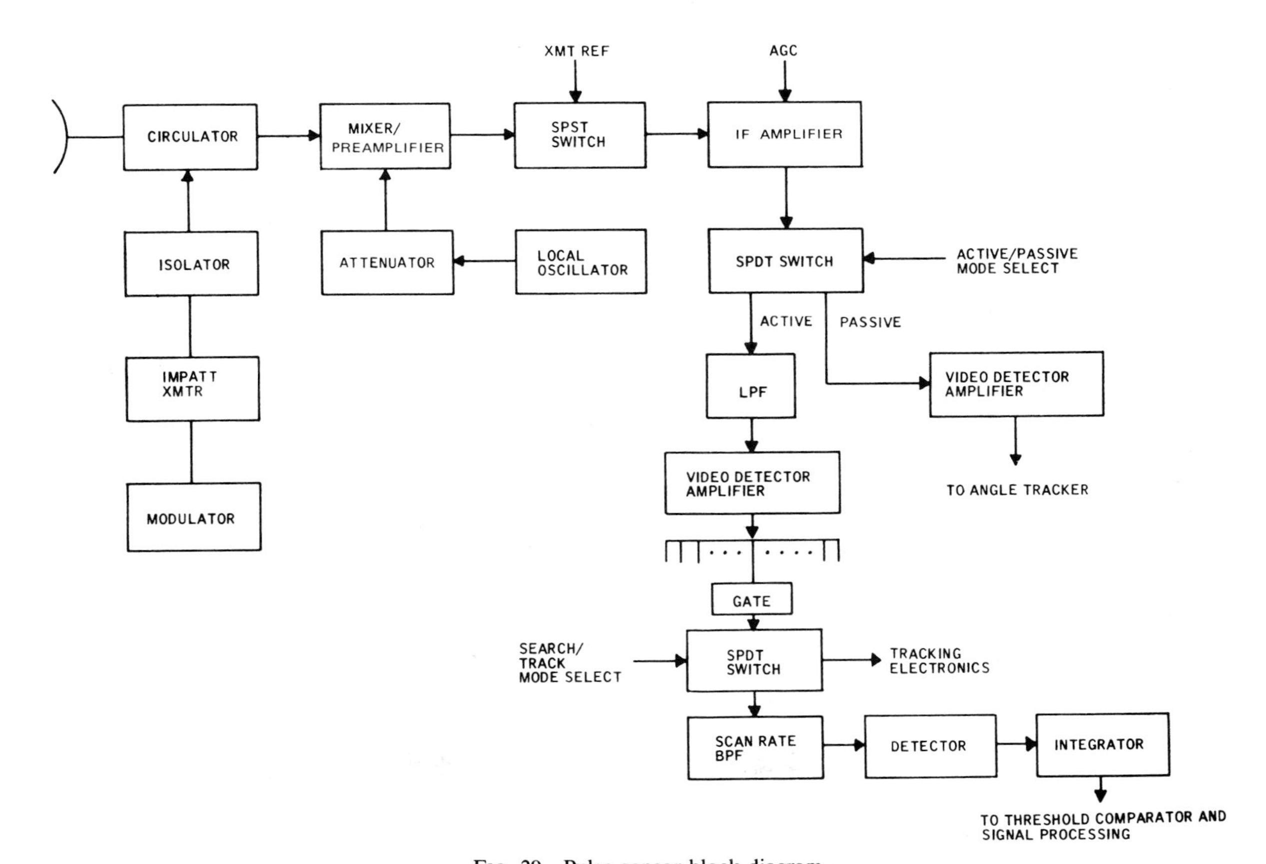

FIG. 29 Pulse sensor block diagram.

TABLE XII

PARAMETERS FOR PULSE SENSOR

Parameter	Value
Peak transmitted power	2 W (35 GHz)
	1 W (94 GHz)
Average transmitted power	10 mW (35 GHz)
	5 mW (94 GHz)
Transmitted bandwidth	300 MHz
Receiver IF bandwidth	300 MHz (active)
	500 MHz (passive)
Transmitted pulse width	0.10 μsec
Pulse repetition frequency	50 KHz
Duty factor	0.005
Antenna diameter	162.5 mm
Antenna efficiency	0.6
Target RCS	50 m²
Antenna beamwidth	3.6° (35 GHz)
	1.3° (94 GHz)
Scanning loss	0.10
Noise figure (SSB)	9 dB (35 GHz)
	12 dB (94 GHz)
Integration time	500 msec (35 GHz)
	70 msec (94 GHz)

the clutter return for a FMCW sensor as a band-limited white-noise process with total received power given by

$$P_c = (P_{avg} d^4 \pi \eta^2 \sigma_c / 64 \lambda^2 R^4), \tag{49}$$

where σ_c is the RCS of the clutter return. The clutter RCS can be expressed as

$$\sigma_c = \tfrac{1}{4}\pi(R^2\theta^2)\gamma_b, \tag{50}$$

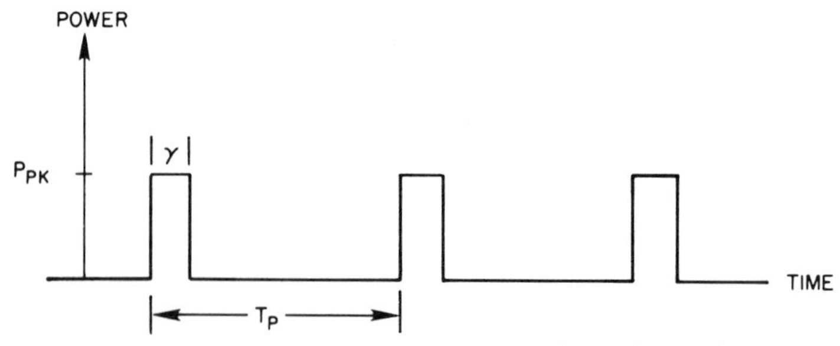

FIG. 30 Pulse illumination waveform; γ/τ_p = duty cycle.

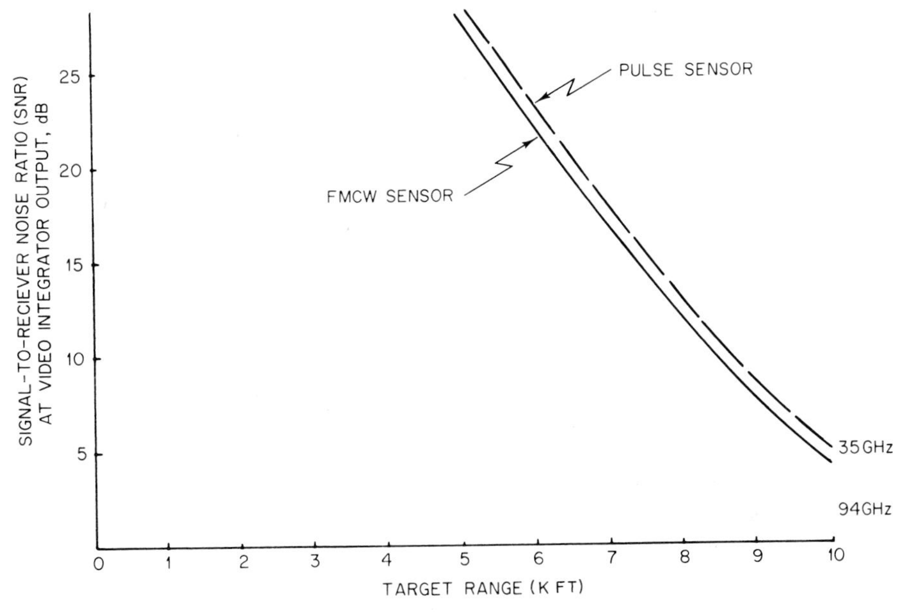

FIG. 31 S/N for FMCW and pulse sensors.

where R is the range to ground, θ the antenna 3-dB beamwidth, and γ_b the background reflectivity.

Clutter RCS is presented in Fig. 32 as a function of range at 35 GHz. A terrain reflectivity of -14 dB m^2/m^2 is assumed. At a range of 5000 ft, the received clutter power is approximately 10^{-12} W, which is less than the

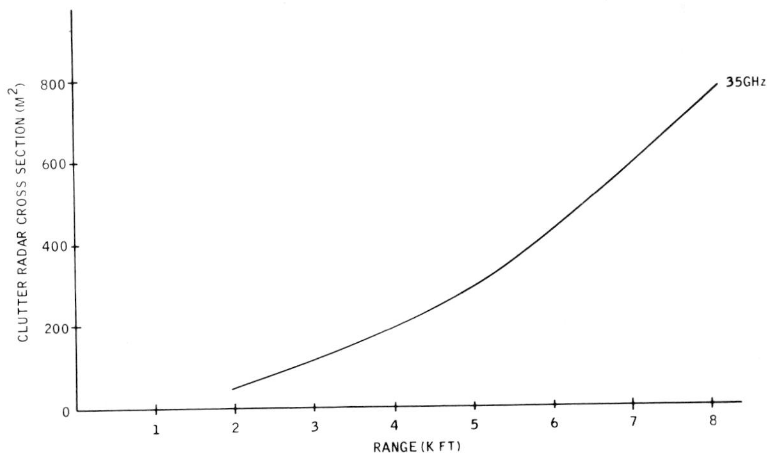

FIG. 32 Clutter RCS versus range for FMCW sensor.

receiver noise power referred to the antenna output of 6.4×10^{-12} W. Therefore, at ranges greater than 5000 ft, the signal-to-noise-plus-clutter ratio is essentially equivalent to the signal-to-receiver-noise ratio.

The effects of clutter on a pulse sensor are similar to the FMCW case. However, the range gating capability inherent to a pulse waveform permits a significant performance advantage in a clutter-limited environment. This advantage results from a smaller clutter RCS relative to the target RCS and from the improved resolution in the downrange direction. The clutter RCS for a pulse sensor with clutter cell limited by pulse length is defined as

$$\sigma_c = \gamma_b \sin(\psi) 2R(C\gamma/2) \tan(\theta/2) \sec \psi, \tag{51}$$

where γ_b is the background, $c\gamma/2$ the gate width (equals transmitted pulsewidth), R the slant range-to-range cell, θ the antenna 3-dB beamwidth, and ψ the antenna depression angle.

The clutter RCS as a function of range and transmitted pulsewidth is presented in Fig. 33. At long range, this clutter RCS is considerably smaller than that encountered with a FMCW sensor. However, as the range is reduced, the pulse length approaches the actual length of the main lobe footprint, thus reducing the clutter rejection advantage.

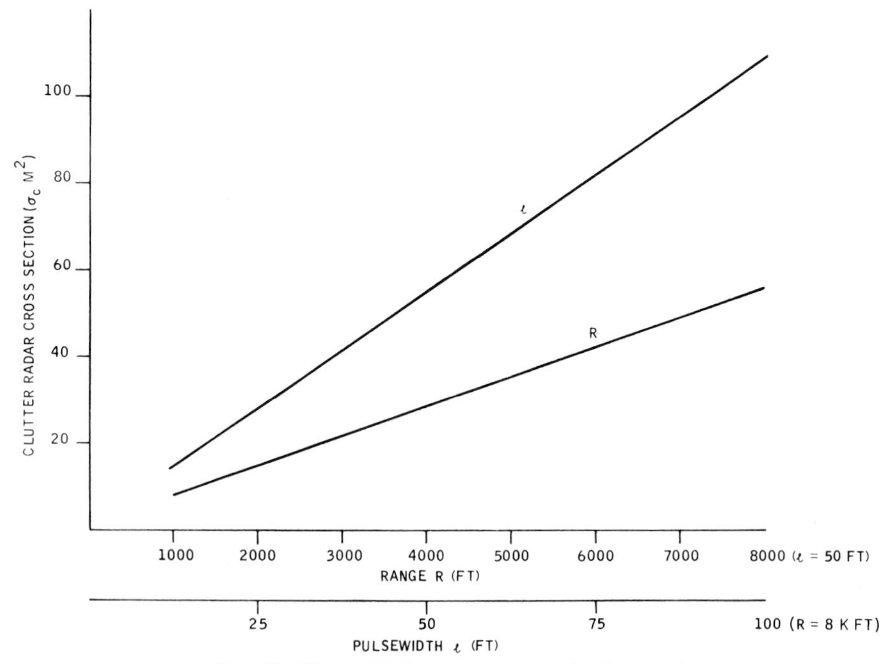

FIG. 33 Clutter RCS versus range and pulse width.

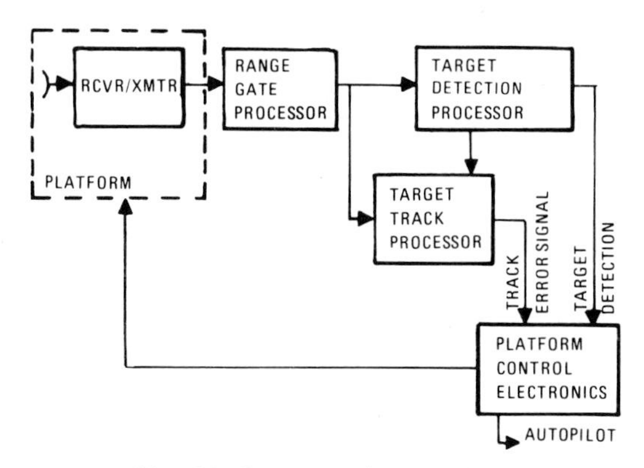

FIG. 34 Processors in the seeker.

Figure 34 shows the three seeker processors used to furnish target detection and tracking information. They all operate on the data available in the receiver video bandwidth and interface directly with the platform control electronics. Figure 35 illustrates the antenna beam motions for both pencil beam and conical scan cases. Amplitude and phase properties are shown for the typical conical scan footprint. There is a double amplitude peak for targets passing near the footprint center and a single peak for those passing through either end of the footprint. The phase signature is

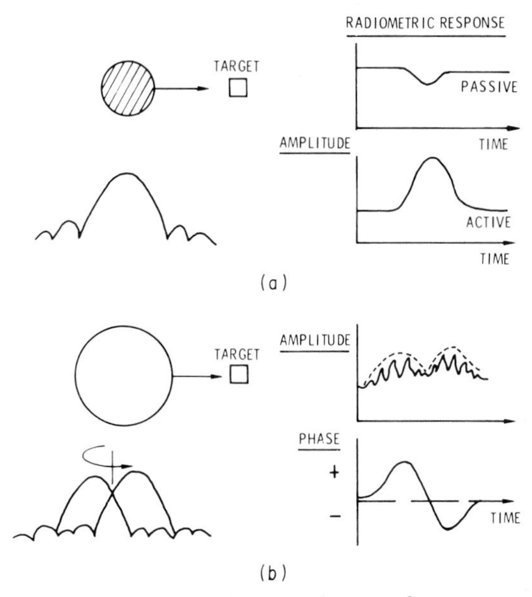

FIG. 35 Antenna beam motion waveforms.

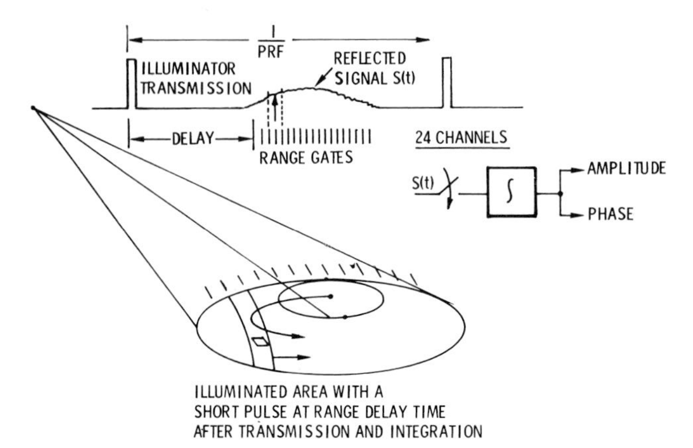

FIG. 36 Short pulse ranged-gated signal processing.

generated by sweeping the conical scan footprint over the target and using a phase demodulator in the receiver video circuitry. The phase signature can provide a measure of target width, using either half of the signature. This is based on the zero crossing of the signature center and the fact that uniform clutter responds 90° out of phase with the reference.

Figure 36 illustrates the short-pulse, range-gated signal processing for a pulse illuminator waveform and its associated return signal. The contiguous range gates ride through the ground return and achieve a resolution proportioned to the gate width. Amplitude and phase response data

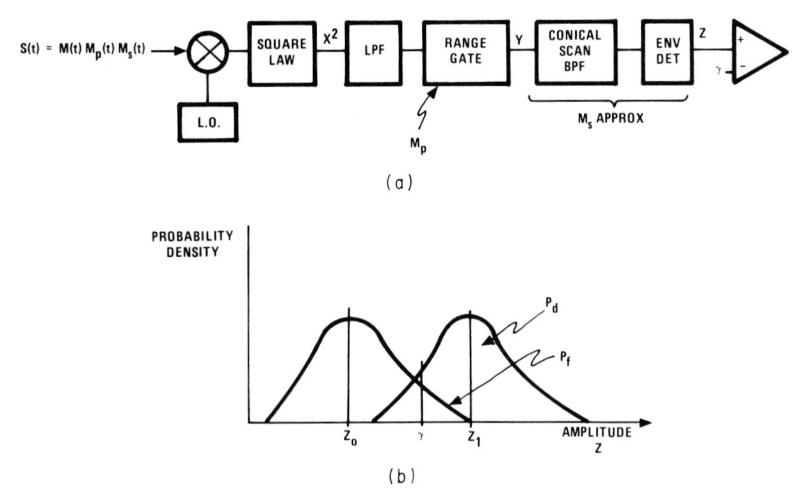

FIG. 37 Target detection processor model for decision threshold. ψ_0 = mean value of noise/clutter (Gaussian distribution); ψ_1 = mean value of target signal (modulated noise); γ = decision threshold.

are tested at the output of each range-gated channel. The seeker performance can be predicted using statistical decision theory techniques. The processor model for deriving an optimum decision threshold is shown in Fig. 37. The Neyman–Pearson criterion is used to derive the optimum processor; this criterion maximizes the probability of detection for a fixed false alarm value. Results of this are plotted in Fig. 38 as a function of the video S/N. The single-channel probability of false alarm P_F is used in Fig. 38 because the false alarm rate (FAR) value depends on other design parameters. The desired probability of detection is usually at the 99% level. A typical probability of false alarm is between 10^{-3} and 10^{-5}. Consequently, the S/N should be 13–15 dB, just ahead of the amplitude comparator threshold. The typical processor is an N-channel design where the FAR is related to the single channel P_F as

$$FAR = P_F/A_T \text{ false alarms/ft}^2, \tag{52}$$

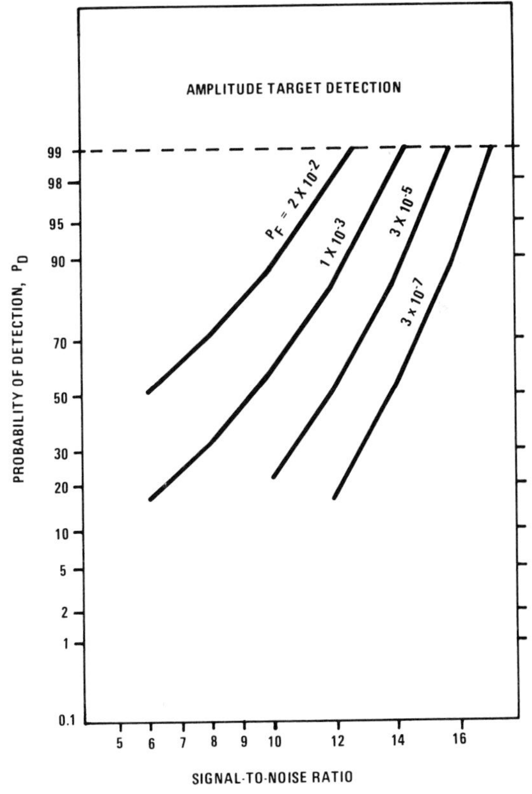

FIG. 38 Detection performance for modulated noise signal in Gaussian noise.

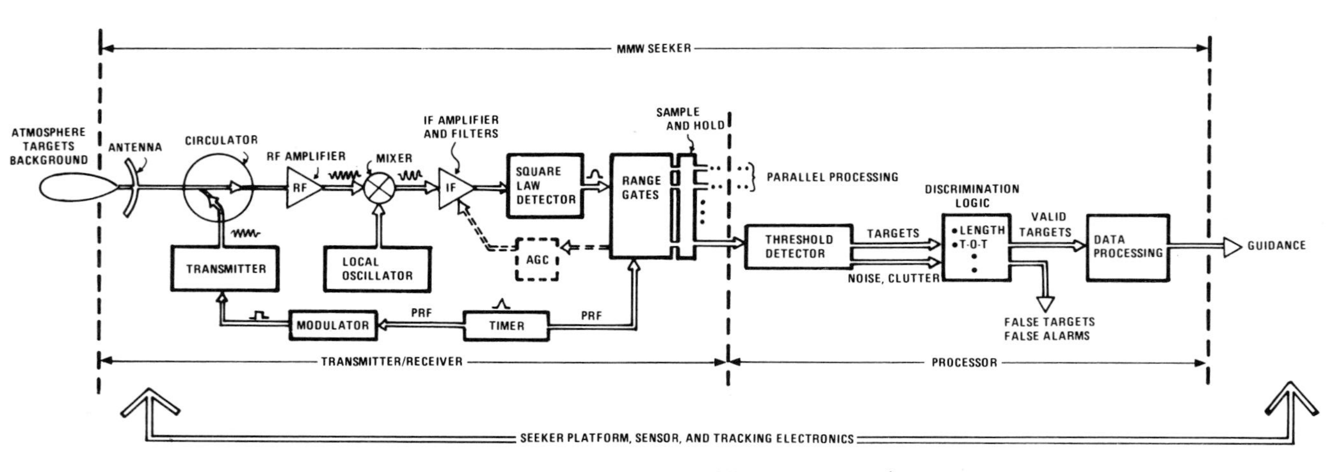

FIG. 39 Block diagram of pulse millimeter wave seeker.

where A_τ is the area covered by one range bin during τ seconds.

$$A_\tau = \tau W_B V, \tag{53}$$

where τ is the integration time, W_B the range bind width on ground, and V the cross range ground velocity.

Figure 38 is based on an amplitude processing decision threshold. It may be used either by starting with a given S/N, resulting in (P_D, P_F) performance values, or by starting with (P_D, P_F) and obtaining S/N values that are then related to other parameters. The signal processing in one range bin utilizes the phase or time characteristic to improve false target rejection and reduce P_F beyond that achieved with amplitude alone.

D. Seeker Hardware

A block diagram of a millimeter-pulse seeker is shown in Fig. 39. The AGC in IF and video is derived from the range-gated signal. An rf amplifier is illustrated between the circulator and mixer input. This is meaningful today only up to about 35 GHz with a TWT or parametric amplifier since no fully solid-state amplifier is available. The basic seeker design utilizes a superhet downconverter as its principal detection noise threshold mechanism; hence considerable emphasis is placed on mixer diodes with very low conversion loss at the incoming rf signal frequency. The seeker gimbal control electronics are shown in Fig. 40. The broad band-

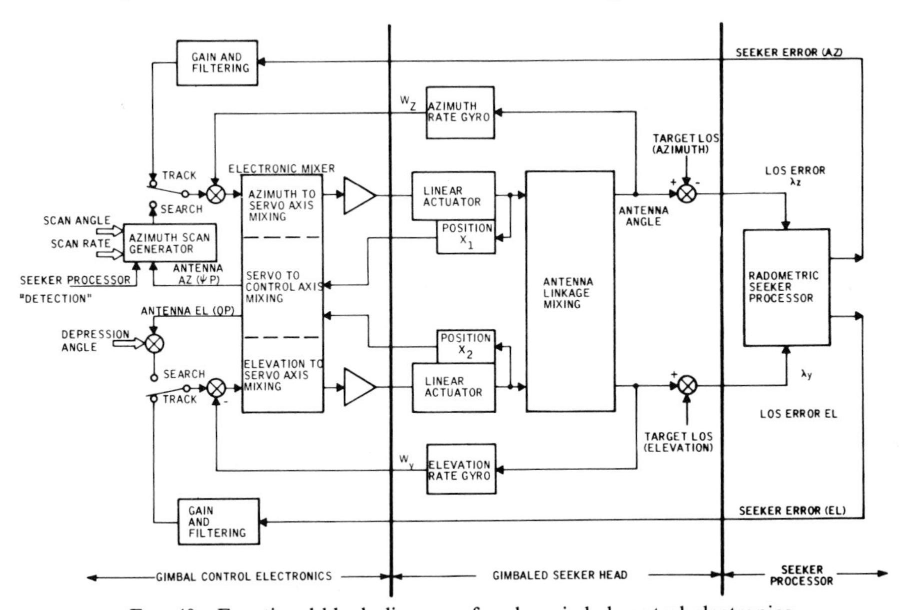

FIG. 40 Functional block diagram of seeker gimbal control electronics.

width of the gyro rate loop swamps out any unwanted friction and spring force effects. This provides the necessary rate damping for the tracker loop and measures line-of-sight rate for the proportional guidance command. The line-of-sight guidance command shown in Fig. 41 is developed from the antenna gimbal angle. Filtering of the tracker signal as a function of range is used to minimize seeker noise particularly at long range.

The typical small-diameter millimeter wave seeker is shown in the line drawing of Fig. 42. Clearly, there is considerable emphasis on hardware miniaturization both in the gimbaled seeker head and the signal processing electronics. A radome with a low transmission loss at the signal frequency is used to protect the seeker rf head and serve as an ogive for aerodynamic purposes. Millimeter-wave-integrated circuits (MMIC) are rapidly maturing at frequencies up to 60 GHz and can be effectively used in the seeker head. A MMIC superheterodyne receiver is shown in Fig. 43 operating at 34–36 GHz. It is complete from rf input to video output including the fundamental local oscillator and AGC, all in a package 2 × 2 × 0.5 in. The electrical performance is as follows:

Operating frequency	34–36 GHz
IF bandwidth	500 MHz
Noise figure	5.2 dB DSB
FR/IF gain	67 dB max
IF gain control	10 db
Video detector sensitivity	1000 mV/mW

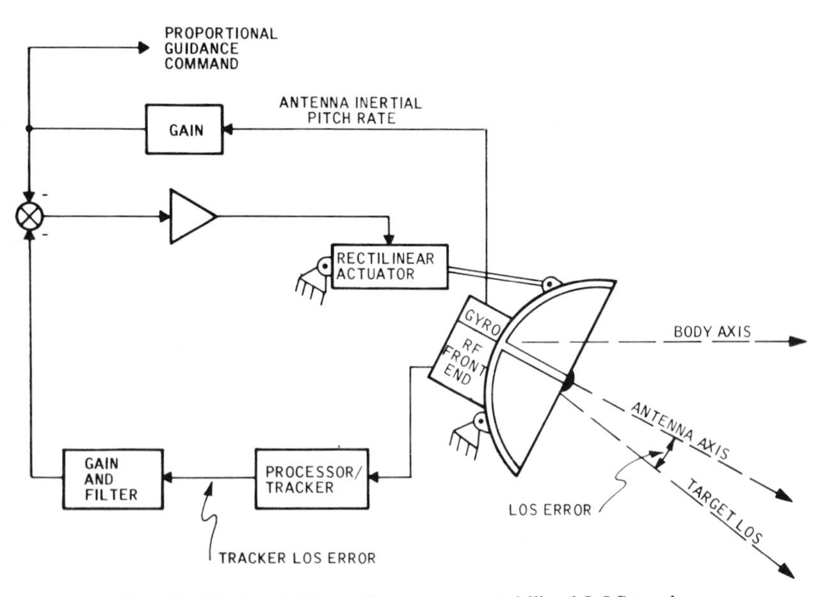

FIG. 41 Pitch axis block diagram, rate stabilized LOS tracker.

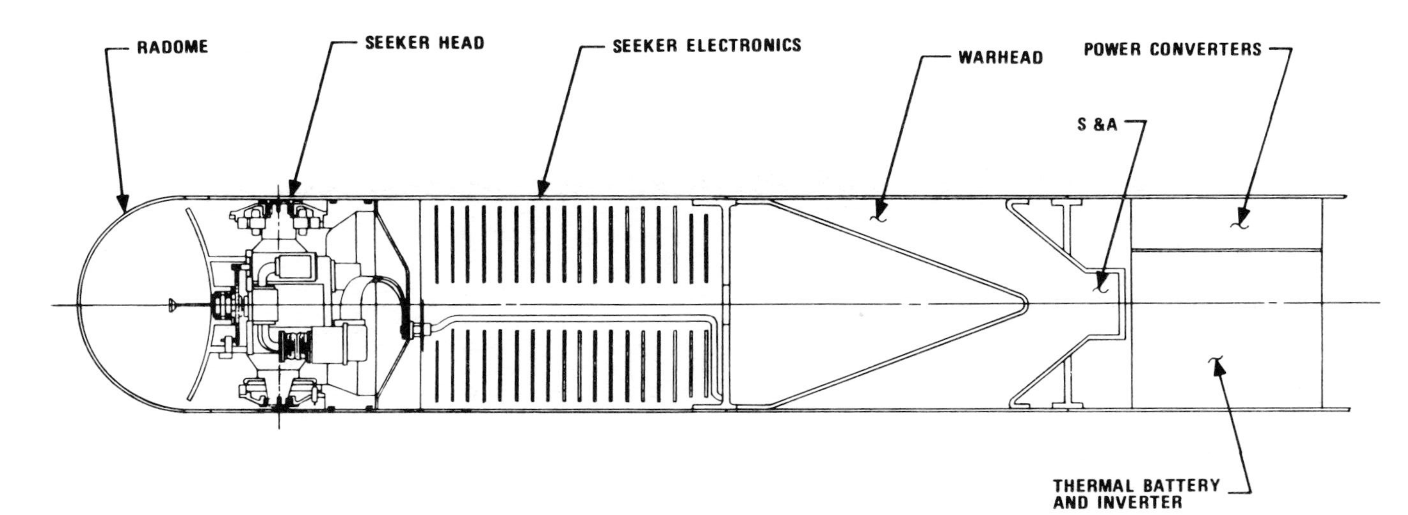

Fig. 42 Line drawing of typical small-dimaeter seeker.

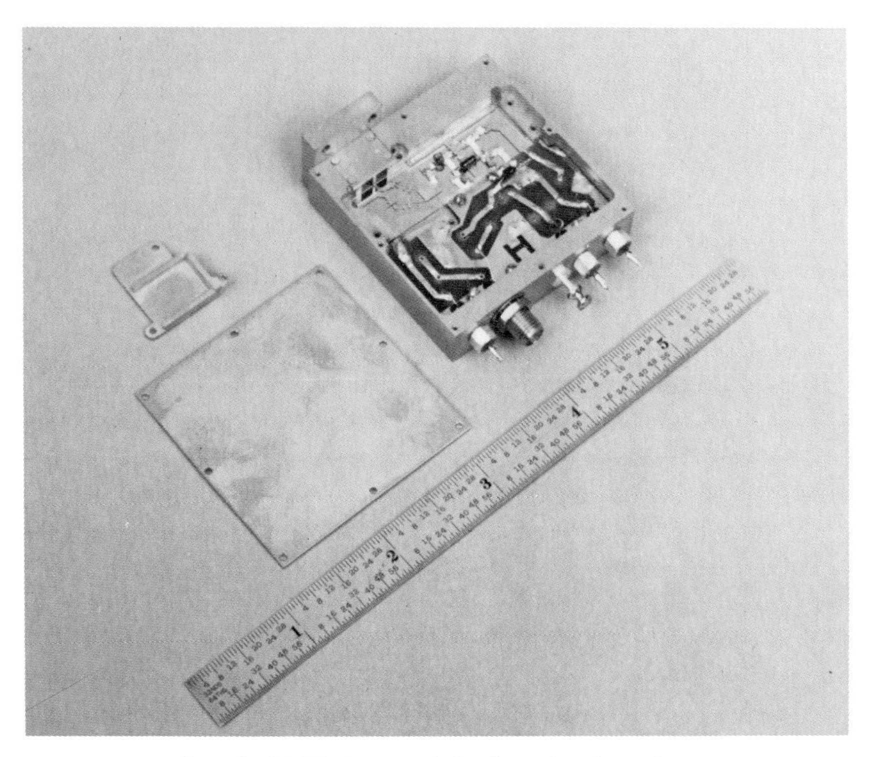

FIG. 43 35 GHz integrated circuit receiver for seeker.

Similar advances in integrated circuit technology are being demonstrated in the transmitter area and in microstrip, flat-profile, multiple-beam antennas.

V. Countermeasures

The subject of countermeasures at millimeter-wave frequencies is becoming of keen interest to the systems designer with the increasing capabilities for generating significant pulse and continuous wave (cw) power at frequencies up to 220 GHz. Techniques of millimeter-wave countermeasures are tending to be extensions of those that have been thoroughly evaluated at microwave frequencies. The reader is referred to the work of Boyd *et al.* (1978), Eustace and Schoniger (1975), and Skolnik (1962) for considerable detail on ECM and ECCM technology.

Table XIII characterizes the inherently useful covert characteristics of millimeter-wave seeker operation as well as indicates the somewhat

TABLE XIII

SEEKER COUNTERMEASURES CONSIDERATIONS

Millimeter-wave covert characteristics	Active–passive countermeasure
Narrow beamwidth	Camouflage
Low side lobes	Absorbers
Low transmitter average power	Coatings
Noncoherent transmitter waveform	Chaff
Passive terminal phase	Decoys (reflectors)
High operating frequency	Jammers
	Sophisticated emitters

standard active–passive countermeasure (CM) techniques. There seems to be general agreement that millimeter wave operation of radar and radiometric sensors offers improved CM immunity when compared to similar operation at microwave frequencies. The very narrow antenna beamwidths, reduced antenna sidelobes, and adequate system performance with low-level and noncoherent transmitter power make the millimeter radar sensor indeed very quiet.

The potential countermeasures for a millimeter-wave seeker fall into three general categories:

(a) target signature reduction;
(b) target signature duplication; and
(c) seeker rf interference generation.

More specific examples of potential countermeasures are as follows:

(a) Target signature reduction.

(i) Absorption blankets: target coverings, possibly using graphite fibers to absorb incident energy for reradiation at longer wavelengths.

(ii) Scattering resonators: use of milled slots in a metallic covering for the target to eliminate specular reflection characteristics.

(iii) Target surface camouflage: use of variable emissivity coatings placed on the target to achieve the millimeter-wave equivalent of visual obscuration. This approach blends the target into the background.

(b) Seeker signature duplication.

(i) Corner reflectors: disposable aluminized plastic structures with millimeter wave RCS approximating those of real targets.

(ii) Reflective sheets: disposable aluminized mylar panels approximately the physical dimensions of the targets.

(iii) Selective use of combat debris, including damaged, destroyed, or abandoned targets, and other metallic structures.

(c) Seeker rf interference generation.

(i) Power jamming: use of high-power transmitters to saturate seeker receivers.

(ii) Function jamming: use of ECM equipment to receive, analyze, and duplicate the seeker signals in any available manner that disrupts the search and track operations.

(iii) Noise jamming: use of multiple low-power transmitters to provide a background in which the real targets cannot be detected from false targets or background.

Acknowledgments

Much of the work described in this chapter was supported by Honeywell IR and D funding over the past six years. I would like to acknowledge the contributions of a number of dedicated engineers from the Millimeter-Wave Guidance Group, including B. Kearns, E. Thiede, A. Hastings, S. Palmquist, M. Razor, and J. Miley.

References

Altshuler, E., Falcone, V., and Wulfsberg, K. (1968). *IEEE Spectrum* **5,** 83–90.

Beebe, M. and Salzman, J. (1976). Final Rep. MSG 65075, pp. 3/10–3/31. Missile Syst. Group, Hughes Aircraft Co. Torrance, California.

Boyd, J., Harris, D., King, D., and Welch, H. (1978). "Electronic Countermeasures," pp. 14/1–14/65. Penninsula Publ., Los Altos, California.

Briscoe, G., Prestwood, F., and Brown, C. (1972). AFATL-TR-71-123. (Confidential.) MICOM, Huntsville, Alabama.

Chen, C. (1975). Rep. R-1964-PR, pp. 1–26. Rand Corp. Santa Monica, California.

Constantine, J., and Deitz, P. (1979). *NAECON 79* **1,** 371–381. Boston, Massachusetts.

Currie, N., Dyer, F., and Hayes R. (1975). Tech. Rep. No. 3. Georgia Inst. of Technol., Atlanta, Georgia.

Eustace, H., and Schoniger, K. (1975). "The International Countermeasures Handbook." EW Communications, Palo Alto, California.

Kulpa, K., and Brown, E. (1979). HDL-SR-79-8, Chap. VI. Harry Diamond Laboratories, DARPA, Arlington, Virginia.

Nelson, R. (1976). Tech. Rep. RE-76-23. MICOM, Huntsville, Alabama. (Confidential.)

Porter, P., and Parker, M. (1960). Tech. Rep. RAD-TR-9-60-20. AVCO Corp. Wilmington, Massachusetts.

Richard, V. (1976). Rep. BRU-MR 2631. U.S. Army Ballist. Res. Lab., Aberdeen, Maryland.

Richer, K., Baurle, R., and McGee, R. (1975). Rep. BRL-MR 2411. U.S. Army Ballist. Res. Lab., Aberdeen, Maryland.

Skolnik, M. (1962). "Introduction to Radar Systems," Chap. 12. McGraw-Hill, New York.

Skolnik, M. (1970). "Radar Handbook," Chapt. 39. McGraw-Hill, New York.

CHAPTER 4

Sources of Millimeter-Wave Radiation: Traveling-Wave Tube and Solid-State Sources

N. Bruce Kramer

Electron Dynamics Division
Hughes Aircraft Company
Torrance, California

I. INTRODUCTION 151
II. TRAVELING-WAVE TUBES 152
A. *Coupled-Cavity Circuits* 154
B. *General Design Considerations* 157
C. *Electron Beams* 160
D. *Electron Guns* 161
E. *Design Formulation* 163
F. *Practical Considerations and Performance* 165
III. SOLID-STATE SOURCES 168
A. *IMPATT Operation* 168
B. *IMPATT Design Considerations* 173
C. *Diode Packages* 180
D. *Diode Performance* 183
E. *Power Amplifiers/Injection-Locked Oscillators* 191
F. *IMPATT Diode Reliability* 194
IV. SUMMARY AND CONCLUSIONS 196
REFERENCES 197

I. Introduction

Over the past ten years both vacuum-tube and solid-state sources for millimeter wavelengths have been intensively developed. A number of traveling-wave tubes for cw and pulsed applications in the 30–100 GHz band have been designed and produced in small quantities. Silicon IMPATT diodes from 30 to 250 GHz have shown good cw and pulsed power, efficiency, and reliability characteristics. Gallium arsenide Gunn diodes have only modest power capability in the 30–100 GHz band but are useful for applications such as local oscillators where low noise is necessary. In this discourse traveling-wave tubes and IMPATT sources will be emphasized because of their expected predominance in power applications. Gunn sources will be discussed, but in less detail.

ISBN 0-12-147704-5

II. Traveling-Wave Tubes

Today, a premier source of high-power millimeter-wave radiation is the vacuum-tube, traveling-wave amplifier. The versatile performance of the traveling-wave tube (TWT) is unmatched, offering high power with good efficiency and high gain over wide bandwidths. One particular type of TWT, the coupled-cavity tube, has emerged to the forefront and is the only design currently being manufactured for wavelengths of 5 mm and less.

Helix tubes at lower (microwave) frequencies are superior to coupled cavity tubes in bandwidth and associated characteristics (gain ripple, group delay, etc.). Techniques are being developed to improve dramatically heat removal from the helix structure using high thermal conductivity diamond supports. Higher power helix tubes can then be made in the 30–50 GHz band that will be competitive with coupled-cavity tubes.

To understand the inherent advantages of a TWT, it is instructive to examine a generalized schematic (see Fig. 1). All TWTs employ an electron gun, which accelerates and focuses electrons from a thermionic cathode to form a small-diameter, high-density electron beam that travels at a high velocity. The small-diameter beam is confined against space-charge repulsion by the radially inward directed $v \times B$ force of an externally generated dc magnetic field to cause it to pass through the rf interaction structure with little interception. In an efficient tube design, much of the energy contained in the high-voltage electron beam has been converted into rf energy and is extracted via a waveguide at the end of the interaction structure. In all cases, however, a significant amount of energy remains in the beam that can be partially recovered by using an electron-collecting structure that decelerates the electrons before they collide with the metal walls of the collector and their energy is converted into heat.

It is apparent from the schematic of Fig. 1 that a basic advantage of the TWT for generating high power is the spatial separation of the functions of generating, extracting rf energy from, and dissipating the waste energy

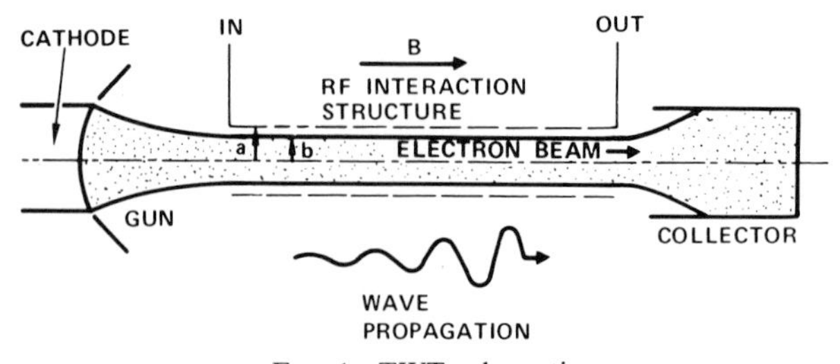

FIG. 1 TWT schematic.

of the electron beam. As pointed out by Forster in an earlier description of practical high-power millimeter-wave generators, this physical separation of functions is a key element in the success of the TWT (Forster, 1968).

Generalized TWT theory (Pierce, 1950) shows that if a transmission line structure can be devised that both propagates an electromagnetic wave at a velocity slightly slower than the velocity of a nearby dense electron beam and also provides strong longitudinal rf electric fields in the region occupied by the beam, strong amplification of the wave propagating on the transmission line will occur. This imposes two basic requirements: The wave must travel at a fraction of the velocity of light since reasonable electron accelerating voltages must be used for reliable power supply design, and the beam must flow close to the transmission line circuit since, in general, propagating electromagnetic fields decay exponentially away from the metal wave supporting structure. A helically wound wire (helix) in the form of a common tensioning spring was the first practical structure that was employed for microwave TWTs. Subsequent analysis (Pierce, 1950) showed that the helix was a very fine slow-wave structure in regard to the above two criteria and possessed enormous bandwidth as well since the phase (and therefore group) velocity of the wave remained nearly constant as a function of frequency.

In order that the longitudinal electric field strength remain high, the helix must be wound tighter and tighter as frequency increases and consequently the wire diameter must become smaller and smaller. The helix must be supported by materials that are electrical insulators to confine the propagating wave. Normally the helix is supported inside a stainless steel vacuum shell by dielectric rods. The support problem becomes difficult at wavelengths shorter than 1 cm, because of the necessity to remove heat from the fine helix wire. Heating is caused by electron beam interception and rf losses.

Techniques for bonding or brazing helices to metallized type IIa diamond supports are being developed and, if successful, may result in formidable helix–tube performance at 1 cm to 5 mm wavelengths. Type IIa (high-purity) diamond is a good electrical insulator and has thermal conductivity two times higher than copper at typical tube operating temperatures (250°C). It is currently being used very effectively to improve IMPATT diode heat sinking.

The investigation of other types of slow-wave structures in the late 1940s and early 1950s for handling more power than the helix resulted in the coupled-cavity structure being developed. The coupled-cavity slow-wave structure is comprised of a stack of reentrant cylindrical cavities with coupling holes to adjacent cavities being arranged 180° apart in opposing cavity walls as shown in Fig. 2. Its principal advantage over the helix is that a direct, all-metal path is provided for removal of heat through the webs of the cavities to an external heat sink.

A. Coupled-Cavity Circuits

The coupled-cavity interaction structure was the "chosen" design to be taken up in frequency in the first millimeter-wave TWT amplifier to be developed and demonstrated in the early 1960s (Forster, 1968). This basic design philosophy has proven out in practice; all successful high-power TWTs developed and manufactured for operation at wavelengths of 5 mm and less have been variants of the first millimeter-wave coupled-cavity design.

A beam passing through a hole in the center of a stack of iris-coupled cylindrical cavities is depicted in Fig. 2. Note that the structure resembles a folded waveguide that has obvious slow-wave propagation relative to a normal waveguide, i.e., the developed waveguide is much longer than the total electron beam length. Thus the slow-wave property of the coupled-cavity distributed interaction circuit is easily visualized. Returning for a moment to the coupled-cavity visualization, it might be expected that two types of modes could propagate if the coupling slots are relatively large (which they always are). There are modes associated with the cylindrical cavity resonances and modes associated with the coupling slot resonances (slot length equals an integral number of half-wavelengths). Millimeter-wave TWT amplifiers operate in the lowest order cavity mode. The bandpass behavior (bandwidth and phase-

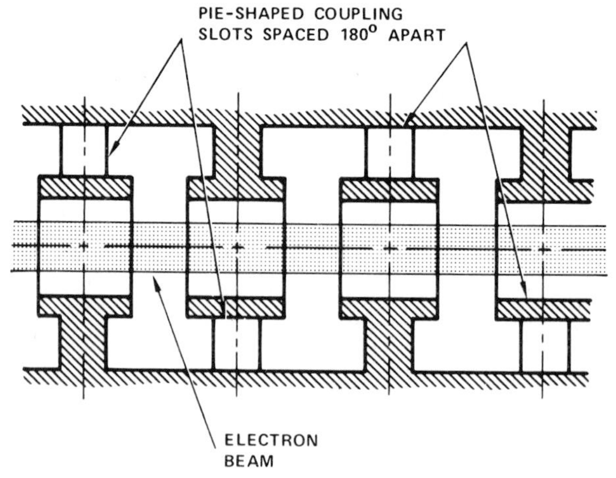

Fig. 2 Coupled-cavity interaction structure.

frequency or ω–β characteristics), however, depends to first order only on the coupling slot dimensions.

Now let us consider how the rf fields vary with z in a lossless structure of the type shown in Fig. 2. After moving one spatial period L down the axis, everything is the same looking forward and backward in an infinite structure except for the 180° rotation of the coupling slots. Thus the propagating fields near the axis should be the same except for a phase shift represented by a factor $\exp(-j\beta L)$. Each time a movement in z of amount L takes place, the fields are multiplied by this factor. A function that has this property is obviously $\exp(-j\beta z)$. The most general function that has this property is the product of $\exp(-j\beta z)$ and an arbitrary periodic function of z of period L, which does not change as z is increased by L, $2L$, $3L$, etc. This periodic function can be represented by a Fourier series with terms $E_n \exp(-j\beta_n z)$, where $\beta_n = \beta + (2n\pi/L)$. Now we make a simplifying assumption that the amplitude of the longitudinal electric field E_{gap} in the gap of length g between ferrules is constant (which is accurate for close spacing of ferrules); then

$$\begin{aligned} E(r, z) &= \sum_n A_n \exp(-j\beta_n z) I_0(\gamma_n r), \\ \gamma_n^2 &= \beta_n^2 - \beta_0^2, \\ A_n &= E_{\text{gap}}[\sin(\beta_n g/2)/(\beta_n g/2)][1/I_0(\gamma_n a)]. \end{aligned} \tag{1}$$

In these equations $I_0(\gamma_n r)$ is a modified Bessel function and γ_n is chosen to make Eq.(1) satisfy Maxwell's equations with $\beta_0 = 2\pi/\lambda_0$, the free-space propagation constant. From the relation $\beta_n = \omega/v_p + (2n\pi/L)$, where v_p is the phase velocity of the fundamental ($n = 0$) wave component, it can be seen that the group velocity $v_g = (\partial\beta_n/\partial\omega)^{-1}$ is the same for all n, i.e., for all space harmonics. The exact dependence of β on ω is determined primarily by the coupling slot, but in general will have the shape shown in Fig. 3. Note that the extended ω–β plot shows the propagation constants β_{-1}, β_0, β_1, etc. As β_n becomes large, γ_n does also [Eq. (1)]; the function $[I_0(0)/I_0(\gamma_n a)]$ becomes very small as γ_n gets large, thus the longitudinal electric field on the axis becomes very small making high-order space-harmonic operation impractical.

A beam velocity line that typifies millimeter-wave operation is drawn in Fig. 3. The first space harmonic is used at the $\beta L = 3\pi/2$ point. In this way a match of phase velocity ($v_p = \omega/\beta$) to beam velocity can be obtained with reasonable accelerating voltages.

The dispersion characteristic of Fig. 3 can be deduced in a physical way by examining the folded-waveguide model of the coupled-cavity structure. Consider a rectangular waveguide periodically loaded with reactive irises as shown in Fig. 4a. By changing the proportions of this drawing we

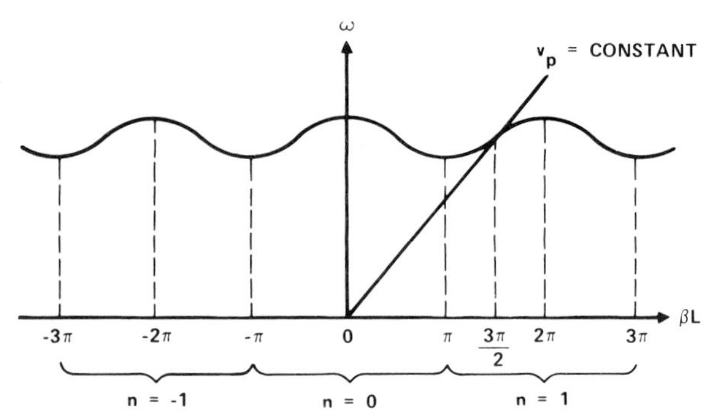

FIG. 3 Phase-frequency characteristic of coupled-cavity structure.

get a folded waveguide structure (see Fig. 4b). Note the difference in the spatial period in the z direction. As is well known, the dispersion characteristics of the periodically loaded rectangular waveguide can be obtained as a perturbation of the unloaded waveguide dispersion curve. Due to the reflections from the discontinuities spaced L apart, there is a stop band that occurs when $\lambda_g/2 = L$. So the $\omega-\beta$ diagram for the lowest order mode of the periodically loaded guide appears as shown in Fig. 5.

Now, for the periodically loaded guide, consider the transverse electric field variation with z at two points: $\beta L = 0$ (lower cutoff frequency) and

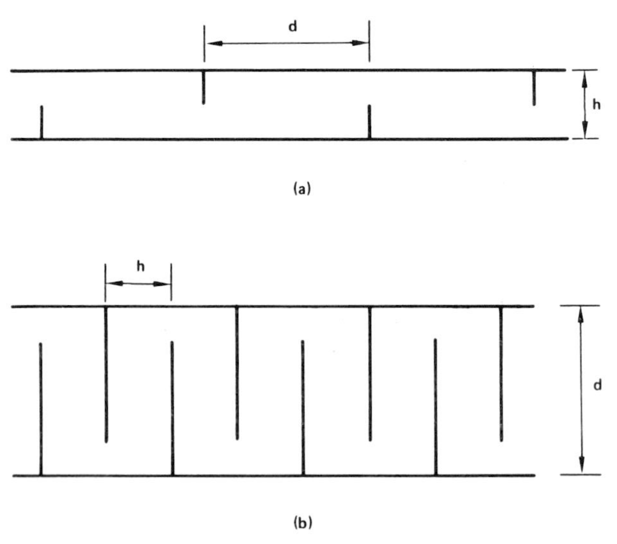

FIG. 4 Reactively loaded waveguide (a) becomes folded waveguide (b) with interchange of dimensions.

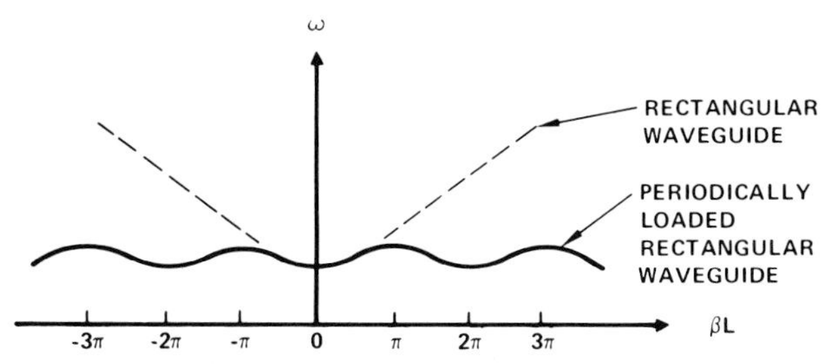

FIG. 5 Dispersion curve for lowest-order mode in a periodically loaded rectangular waveguide.

$\beta L = \pi$ (upper cutoff frequency). At $\beta L = 0$, $\lambda_g \to \infty$ and at $\beta L = \pi$, $(\lambda_g/2) = L$, thus the field patterns are as shown in Fig. 6a and b. If we look at the folded waveguide in the same way, we see that the respective field patterns for $\lambda_g \to \infty$ and $(\lambda_g/2) = L$ appear as shown in Fig. 6c and d. In terms of the period $L = h$ of Fig. 4b, however, we have $\beta h = \pi$ at the lower cutoff frequency and $\beta h = 0$ at the upper cutoff frequency. This leads to the inverted (or backward wave) ω–β characteristic for the coupled-cavity structure shown in Fig. 3.

B. General Design Considerations

The effectiveness of the interaction of the rf fields propagating down a slow-wave structure (circuit) with the electron beam can be measured by the strength of the peak longitudinal rf electric field on the axis of the beam relative to the rf power flow P down the circuit. This can be cast in the form of an impedance K, using E/β as the rf voltage:

$$K = E^2/2\beta^2 P. \tag{2}$$

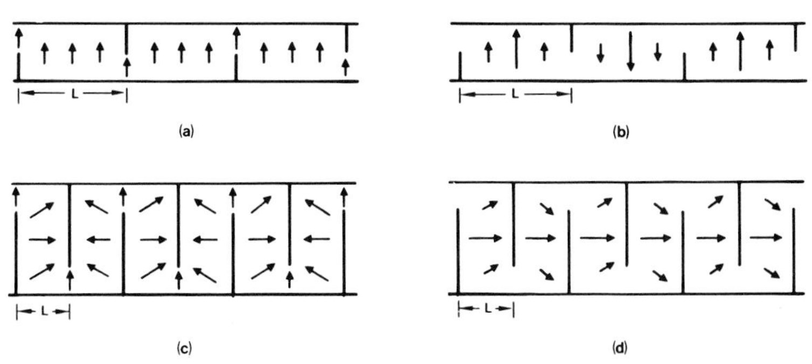

FIG. 6 Electric field patterns in (a)–(b) periodically loaded and (c)–(d) folded waveguides. (a) lower cut off, $\beta L = 0$; (b) upper cut off, $\beta L = \pi$; (c) lower cut off, equivalent to (a), $\beta L = \pi$; (d) equivalent to (b), upper cut off, $\beta L = 0$.

K is known as the Pierce interaction impedance and is a good figure-of-merit for slow-wave circuit capability. What about the electron beam? A measure of the beam capability is the peak ac current i_{peak} relative to the beam power I_0V_0. After an electron beam has drifted through the interaction structure, if things have gone well, it is tightly bunched into concentrated charge packets. In the limit, a narrow spike of charge passes the output cavity once per rf period. In this case the peak value of the fundamental frequency component of current is twice the dc current $i_{\text{peak}} = 2I_0$. Therefore, as a measure of the beam capability, analogous to K, we have

$$i_{\text{peak}}^2/2I_0V_0 = 2I_0/V_0, \tag{3}$$

which has the form of an admittance. An overall dimensionless figure-of-merit that makes sense is $2I_0/V_0$ folded against K, noting that both factors must be nonzero for rf power to be generated.

Theoretically and experimentally (Cutler, 1956), it has been shown that a good approximation for the dc to rf conversion efficiency η of a TWT is given by the cube root of this figure-of-merit:

$$\eta \approx [2(I_0/V_0)K]^{1/3}. \tag{4}$$

Usually this is expressed in terms of the Pierce coupling parameter C, $\eta \approx 2C = 2(KI_0/4V_0)^{1/3}$. C is also a measure of gain per wavelength of the tube.

In order to understand how K varies with various tube design parameters, a simple model can be analyzed. First we can write $P = Wv_g$, where W is the energy stored per unit length of the circuit. To evaluate W, consider the cavity geometry shown in Fig. 7. The cavity is represented by two opposing capacitor plates containing beam holes and coupling slots spaced by the circuit period L. The energy stored in the cavity capacitance is approximated simply as $WL = (\varepsilon A/2L)(E^2/\beta^2)$. For reentrant cavities $D \approx \lambda_0/2$ so that $A \approx \pi\lambda_0^2/16$. From Fig. 3 the operating point is $\beta L = 3\pi/2$, which means that $L = 3\pi v_p/2\omega$. Thus we can express K ex-

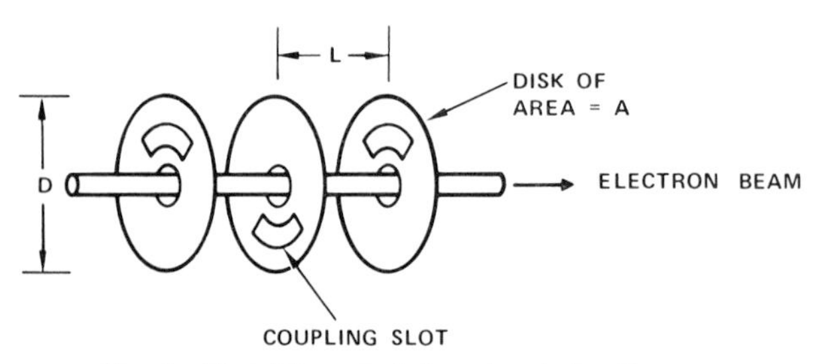

FIG. 7 Model for calculating stored circuit energy.

plicitly as

$$K = E^2/2\beta^2 W v_g = (9/\pi \varepsilon c^2)(v_p/v_g)v_p, \tag{5}$$

where c is the velocity of light. Now as stated before, the $\omega-\beta$ diagram (cold pass band) and, therefore, v_p/v_g do not depend on circuit period (only on the coupling slot); we see that K varies as v_p, which means that at a given operating frequency K varies as the circuit period L.

Using a simple straight-line model for a millimeter-wave $\omega-\beta$ diagram shown in Fig. 8, the value of v_p/v_g can be roughly determined in terms of the cold pass band $\Delta\omega/\omega_0$.

$$v_p/v_g \approx \tfrac{2}{3}(\omega_0/\Delta\omega). \tag{6}$$

As hinted by Eq. (6), the interaction impedance can be increased by increasing v_p/v_g only at the expense of hot bandwidth since large v_p/v_g implies rapid departure from beam-wave synchronism as frequency is changed.

Characteristic of coupled-cavity structures are the "ferrules" shown in Fig. 2. The ferrules result in relatively narrow gaps in which the circuit electric field is concentrated, separated by longer drift tubes in which the beam is "hidden" from the circuit wave. This is desirable for space harmonic operation because the "gap" factor $\sin(\beta_1 g/2)/(\beta_1 g/2)$ in Eq. (1) deteriorates as g becomes large. Also note that the reentrant resonator

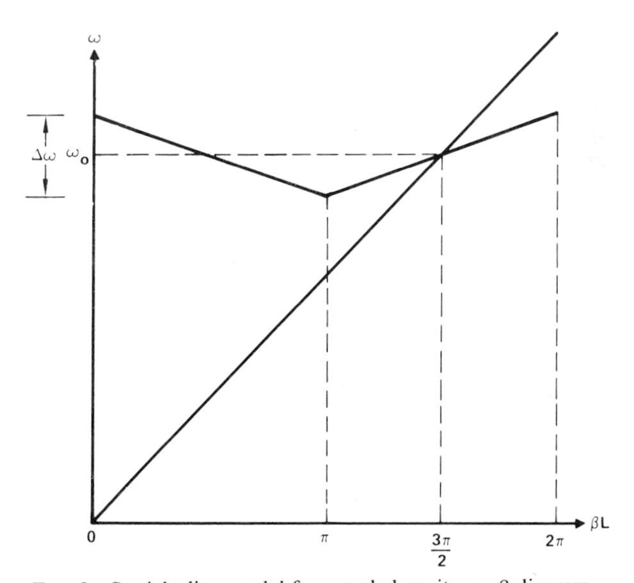

FIG. 8 Straight-line model for coupled-cavity $\omega-\beta$ diagram.

has a lower cavity capacitance so that W is reduced for the same value of E^2.

C. Electron Beams

The beam figure of merit $2I_0/V_0$ indicates that high beam currents at low accelerating voltages are desirable. There are constraints placed on I_0 and V_0 by the strength of the magnetic field that is available for confining or focusing the beam. To examine the beam limitations, consider Brillouin flow (Brillouin, 1945), which is the most efficient means (requires least magnetic field) of producing a smooth beam. The cathode from which electrons are emitted is shielded from the magnetic field. Upon entering the magnetic field, the electrons are caused to rotate about the beam axis by the radial magnetic field component at the entry point, and a stable flow condition can be reached in which the centripetal, space-charge repulsion and $v \times B$ forces balance. The magnetic field B required to establish Brillouin flow for a beam of radius b is

$$B = \left(\frac{\sqrt{2}}{\pi\varepsilon(\mathrm{e/m})^{3/2}}\right)^{1/2}\left(\frac{I_0}{V_0^{1/2}}\right)^{1/2}\frac{1}{b}. \tag{7}$$

The key question in evaluating relation Eq. (7) is how small must the radius of a millimeter-wave beam be made? First of all, b must be smaller than the beam hole a; at millimeter wavelengths with tight tolerances and difficult alignment requirements (to minimize interception), it is prudent to assume that $b \approx a/2$. Returning to Eq. (1), the electric field strength on the beam axis for the first space harmonic is proportional to $I_0(0)/I_0(2\gamma_1 b)$. $I_0(0) = 1$ and $\mathrm{I}_0(x) \approx \exp x/(2\pi x)^{1/2}$ for $x > 1$; thus it is desirable to have $2\gamma_1 b < 1.5$. For slow waves $\gamma_1 \approx \beta_1$, so that $a < (\lambda_0/4)(v_p/c)$ and $b < (\lambda_0/8)(v_p/c)$ will give good interaction, with low beam interception.

The condition on b can be written $b < v_p/8f$, where f is the operating frequency. For beam-wave synchronism we have $v_p = (2(\mathrm{e/m})\mathrm{V}_0)^{1/2}$, so that the required magnetic field from Eq. (7) becomes

$$B = 8[\sqrt{2}\pi\varepsilon(e/m)^{5/2}]^{-1/2}(I_0/V_0^{3/2})^{1/2}f, \tag{8}$$

so that B is given by the product of the square root of the electron beam perveance $I_0/V_0^{3/2}$, the operating frequency, and a collection of constant factors. The magnetic field can be produced by a large permanent magnet or a solenoid. Periodically reversed magnetic fields created by small permanent magnets arrayed along the outside of the tube will also focus the beam (Mendel *et al.*, 1954). The magnetic field created by the periodic permanent magnet (ppm) structure will be sinusoidal in nature, and the beam will expand and contract radially with a period equal to one-half the magnet period. With ppm focusing, Eq. (8) for the required magnetic field applies, with the rms value of the sinusoidal magnetic field replacing the steady field. There are obvious limits on the magnetic field strengths that are practical: Solenoids are heavy and consume dc power (usually more than the dissipated dc beam power); ppm focusing is lightweight and requires no power, but the pole pieces cannot be made an integral part of the coupled cavities at millimeter frequencies as they can at lower frequencies because iron is a poor heat conductor. So that even with samarium cobalt magnets the ppm-focused beam-power levels are an order of magnitude lower than those achievable with solenoid focusing.

Given a limitation on the magnetic field strength, it is seen from Eq. (8) that the beam perveance $I_0/V_0^{3/2}$ must decrease as $1/f^2$, which is not favorable for the ratio I_0/V_0. Thus, it is to be expected that the Pierce C parameter and, therefore, gain and efficiency will suffer badly at high frequencies. Gain, however, can be recovered by lengthening the tube, and efficiency can be recovered by using single or multistage depressed collectors to recover the unused beam energy (if the electronic efficiency is low, then the rf modulation on the beam is small and the collector can be depressed more without running the risk of returning slow electrons to the rf interaction structure).

D. Electron Guns

The most commonly used device to form a cylindrical electron beam for a millimeter-wave TWT is a spherically convergent Pierce-type gun. Such a gun is depicted in Fig. 9. Note that the gun can be crudely modeled as a portion of two concentric spheres, one of radius r_c the cathode, the other of radius r_a the anode, with a hole in the anode sphere for the beam to exit. Space-charge limited flow between spheres (Langmuir and Blodgett, 1924) is given by

$$I_0 = F(\log r_a/r_c) \sin^2 \tfrac{1}{2}\theta V_0^{3/2}, \tag{9}$$

where $F(\log r_a/r_c)$ is a tabulated function (see, e.g., Brewer, 1967) and θ is the half-angle of convergence. Obviously $I_0/V_0^{3/2}$ is invariant with geometrical scaling and is called the perveance of the gun. Pierce showed that a straight focus electrode at an angle of 67.5° with the beam edge can be used to bend the equipotential lines in the region outside of the beam so as to achieve a potential distribution at the beam edge that simulates the missing space charge of the rest of the diode (Pierce, 1940). The hole in the anode is a lens that bends the beam and also reduces the field near the center of the cathode causing nonuniform emission. The effects of the hole can be compensated for by a modified focus electrode that deviates from the Pierce design with an extension or inward bent tip near the anode (Brewer, 1957) (see Fig. 9).

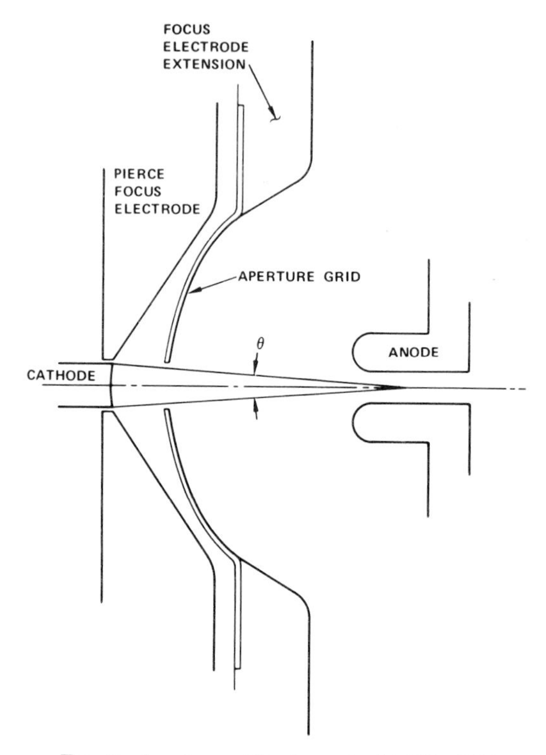

FIG. 9 Aperture-gridded gun for 94-GHz TWT.

In the previous section it was pointed out that magnetic field limitations require low values of perveance at high frequencies. This means that θ can be made small and consequently the hole in the anode can be made small, which minimizes the lens effect and gives uniform cathode emission. This results in ordered electron flow, which is necessary for the stringent magnetic focusing required to transmit an electron beam through a long interaction structure with a tiny hole. θ cannot be reduced indefinitely because cathode-emission current densities should be kept under 6 A/cm^2 for reliable operation. The cathode–anode spacing $(r_c - r_a)$ must be sufficient to prevent voltage breakdown between the anode and focus electrode. This results in high area compression [(the ratio of cathode area to focused-beam cross-section area $\approx (\theta r_c/2b)^2$)]. The transverse electron thermal velocities increase as the area compression does (the equivalent electron temperature in the beam is the cathode temperature times the area compression factor), and the magnetic field must be increased to confine these radially energetic electrons. Even so, a dispersion of the beam edge takes place, which results in a beam that has approximately a Gaussian current density distribution in the interaction region.

E. Design Formulation

The previous sections have discussed features and limitations of the various functional elements of a millimeter-wave TWT. It is useful to step through a design exercise to see how these elements interrelate. Frequency and rf power output are two basic specifications. Normally maximum electronic efficiency will be a goal because it is desirable to minimize the complexity of (or do without) depressed collectors, which add to power supply requirements and introduce possible voltage-breakdown problems between collector electrodes and the rf interaction structure. Gain will also be specified. Maximizing gain per unit length is desirable. The tube can then be made short, which makes beam focusing easier (with less dc power for solenoid focusing) and reduces interception. Also a shorter tube means fewer cavities, which reduces the number of parts and hence weight and fabrication costs. Thus, on many counts, it is desirable to maximize the Pierce coupling parameter C.

Returning to the basic specification of rf output power, a first-order estimate of electronic efficiency can be made and from this the required beam power I_0V_0 can be estimated. Power supply complexity goes upward and reliability downward as the beam voltage is increased. Therefore, an upper limit on V_0 is usually set, which gives a value for $I_0 = I_0V_0/V_0$. The magnetic focusing field B is then estimated from Eq. (8). If B is low enough, ppm focusing can be used, if not, a solenoid must be used at a severe penalty in size, weight, and dc power consumption (which may or may not be a problem). If the magnetic field required is not available, then the question of increasing the operating voltage must be reopened. When values of I_0 and V_0 have been determined that meet the beam power and focusing field requirements, the efficiency can be estimated from Eq. (4) once a value for K has been gauged.

At millimeter frequencies absolute bandwidths are large so v_p/v_g may be made large to improve K [see Eq. (5)]. Also the ferrule (gap) design may be adjusted to improve K. The ferrule design must take into account heat removal since the tip of ferrule on the side nearest the coupling slot will normally be the hotest point in the rf structure. Finally, the remaining factor in K is $v_p = [2(e/m)V_0]^{1/2}$. Based on the optimization of K and the previously determined values of I_0 and V_0, Eq. (4) tells us if the original electronic efficiency estimate was reasonable. If not, an iteration of the process is easily done to arrive at values for I_0, V_0, and B compatible with required tube performance.

Because $I_0/V_0^{3/2}$ goes as f^{-2}, the C parameter for a millimeter-wave

TWT will invariably be small compared to lower-frequency amplifiers. This means that 150 cavities will be used in a high-gain tube (~50 dB). Also a multistage depressed collector will be used if efficiency and/or cooling are important. Multistage depressed collectors effectively recover the unused beam energy electrically, before it is dissipated as heat so that the overall beam efficiency is equal to that of a microwave tube (~30%).

Detailed dimensional design of a device as complicated as a millimeter-wave TWT might be considered as a rather hit-or-miss affair. Such is not the case. A schedule showing a systematic design procedure for a high-power TWT is displayed in Fig. 10. Every block on the chart, except one, represents a computerized analysis or output. The one step that must be done experimentally is cold-test measurement of the phase-frequency characteristic ($\omega-\beta$ diagram) and axial electric field

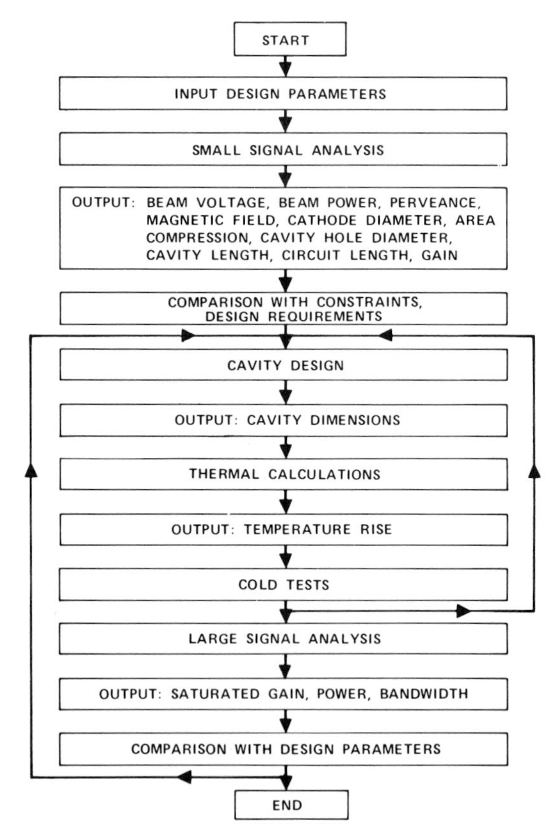

FIG. 10 Design schedule for coupled-cavity TWT. (From Kramer, 1979. Reprinted from *1979 IEEE NTC Proceedings*. © 1979 IEEE.)

strength (interaction impedance) on a cavity stack. Numerical techniques do not as yet give sufficient accuracy to obtain the cavity electrical parameters needed to do a strictly paper design. All other design steps can be done numerically with a high degree of confidence, considerably shortening the design time.

F. Practical Considerations and Performance

Millimeter-wave circuits in general are more lossy than microwave circuits; therefore regenerative effects are reduced. This is beneficial in suppressing spurious oscillations due to reflections from output mismatches. In high-gain tubes, however, a lossy circuit sever (or two) must be used to suppress oscillations. Because of the small cavity dimensions, loss buttons incorporated directly into the cavity are small and difficult to match and may overheat at high-power levels. Thus standard practice for microwave amplifiers must be reexamined for use at millimeter wavelengths. For high-power millimeter-wave tubes it is often best to provide internal terminations in the form of inverted cones that fill an entire cavity before and after a sever.

Previous discussion on operating voltage has concentrated on power supply problems. Interaction circuit design can benefit from increased voltage that in some cases may offset power supply complications and costs. As V_0 is increased, v_p increases so the circuit period can be increased. This allows the cavity walls to be made thicker for better heat removal and reduces tolerances on the ferrules (gaps) because g can be increased while maintaining the gap factor $\sin(\beta_1 g/2)/(\beta_1 g/2)$. This reduces costs since the ferrules are difficult to machine with tolerances being held to 0.0001 in. Finally, the beam hole can be made larger without changing the factor $I_0(0)/I_0(\gamma_1 a) \approx I(0)/I_0(\beta_1 a)$, which means the rf field strength in the beam remains the same while allowing a larger passageway for the hard-to-focus electrons.

Inspection of the coupled-cavity tube cross section in Fig. 2 reveals a rather complicated structure that must be made from solid copper. The elemental disks with coupling holes and ferrules are expensive to machine (electrical discharge machining must be used for some operations) and are a major factor in overall tube fabrication cost. Under development is a simpler coupled cavity (*ferruless circuit*) shown in Fig. 11. It can be made up of elemental "disks" and "washers," which can be fabricated by a coining technique known as *fine-edge blanking*. the interaction of the circuit wave with the beam is weaker for such a circuit, so a larger number of "ferruless" cavities must be employed to achieve the same amount of gain. The lower cost of the cavity parts, however, will more than offset the increased number of parts required. The penalty in

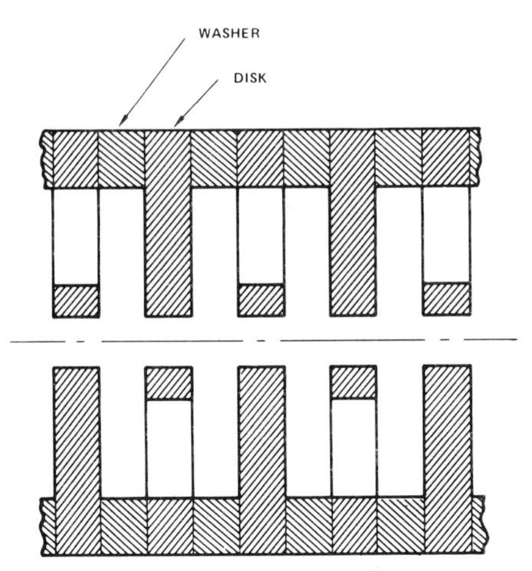

FIG. 11 Ferruless coupled-cavity circuit stack. Compare with Fig. 2. (From Kramer, 1979. Reprinted from *1979 IEEE NTC Proceedings*. © 1979 IEEE.)

electronic efficiency is expected to be about 1 dB, which can be recovered in depressed collector designs.

Regardless of the fabrication techniques used for making the individual cavity parts, because of the tight tolerances and accurate alignment required, standard brazing techniques for assembling the circuit stack are not permissible. Instead, the parts are assembled by diffusion bonding to achieve interfaces with low electrical resistance. The vacuum integrity is then ensured by enclosing the assembled stack in a stainless steel barrel.

Performance of millimeter-wave TWTs is shown in Fig. 12. Estimated upper limits for optimized designs based on present technology are also shown as dashed (ppm) and solid (solenoid) curves (Grant and Chris-

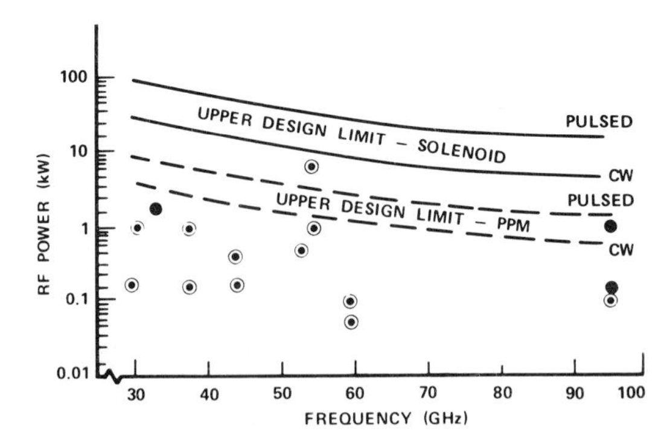

FIG. 12 Demonstrated TWT output power and projected limits. ●, Pulsed; ◉, cw. (From Kramer, 1979. Reprinted from *1979 IEEE NTC Proceedings*. © 1979 IEEE.)

TABLE I

OPERATING CHARACTERISTICS OF 915 H[a]

RF characteristics	
Frequency band	43.0–45.0 GHz
Power output	200 W
Gain	50 dB
Duty factor	cw
Electrical characteristics	
Cathode voltage	−22.0 kV
Cathode current	0.088 A
Collector voltage	−12 kV
Collector current	0.078 A
Body current	0.010 A
Heater voltage	7.0 V
Heater current	1.0 A

[a] From Kramer, 1979. Reprinted from *1979 IEEE NTC Proceedings*. © 1979 IEEE.

tensen, 1979). As specific examples, characteristics of two millimeter-wave TWTs are presented in Tables I and II. Photographs of these tubes are shown in Figs. 13 and 14. The first tube is a ppm-focused 200-W cw tube operating at 44 GHz. The second is a solenoid-focused 1-kW peak 10% duty factor tube operating at 94 GHz.

TABLE II

OPERATING CHARACTERISTICS OF 980 H[a]

Rf characteristics	
Frequency band	93.25–94.75 GHz
Power output	1.0 kW
Gain	50 dB
Duty factor	10%
Electrical characteristics	
Cathode voltage	−39 kV
Cathode current	0.390 A
Collector voltage	−27 kV
Collector current	0.386 A
Body current	0.004 A
Grid voltage	1.7 kV
Grid bias	−1.0 kV
Heater voltage	4.8 V
Heater current	2.0 A
Solenoid voltage	160 V
Solenoid current	23 A

[a] From Kramer, 1979. Reprinted from *1979 IEEE NTC Proceedings*. © 1979 IEEE.

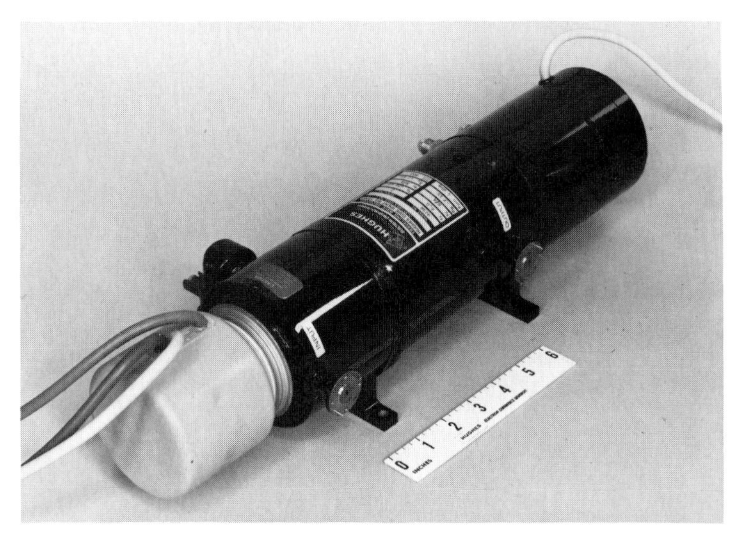

Fig. 13 Photograph of 915H, 200-W cw, 44-GHz TWT. (From Kramer, 1979. Reprinted from *1979 IEEE NTC Proceedings*. © 1979 IEEE.)

III. Solid-State Sources

A. IMPATT Operation

Truly enormous rf power per unit active volume can be produced by millimeter-wave double-drift silicon IMPATT diodes, perhaps unrivaled by any other nonself destructive source of electromagnetic radiation. At 94 GHz, 18 W that correspond to nearly 10^{10} W/cm^3 can be produced in 100-nsec pulses at a 50-kHz repetition rate. This power capability is due to several factors: The device operates at maximum electric field strength

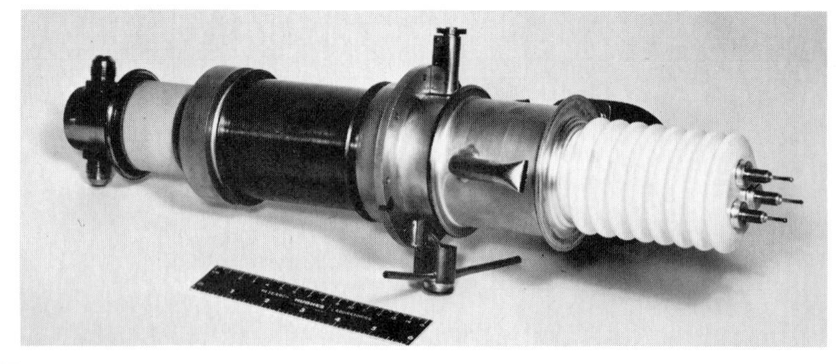

Fig. 14 Photograph of 980H, 1-kW peak power, 10%-duty factor, 94-GHz TWT.

(dielectric breakdown ~500 kV/cm) and enormous current densities (>10^5 A/cm^2), reasonably good electronic efficiencies can be achieved (~10%), and silicon has good thermal conductivity for efficient heat removal. It is interesting to note that a current density of 10^5 A/cm^2 transiting a drift region of 0.3 μm width (typical of 94-GHz diodes) produces a space-charge field approaching the breakdown field.

The electric field profile of a reverse biased $p - n$ junction diode is shown in Fig. 15. As the reverse bias voltage is increased, some electrons and holes in the region of the peak field will gain enough energy, before suffering other types of diverting collisions, to impact and knock electrons off, or ionize, atoms in the crystal lattice. An impact ionization event creates an electron–hole pair. The electron and hole thus created are accelerated by the field and can cause additional impact ionization events resulting in large multiplication of current (avalanche breakdown). In Fig. 15 the multiplied hole current will exit to the left of the avalanche zone and the multiplied electron current will exit to the right. If the doping levels and p and n layer thicknesses are properly selected, the transit time delay associated with the holes and electrons moving across their respective drift regions (p and n layers) will generate rf energy in a chosen frequency band.

Two high electric field effects are important for IMPATT operation: impact ionization and slow transit velocities. If the transit velocities were not low, the semiconductor layers would have to be thick, which increases thermal resistance, and excessive temperatures would cause alloying or other destructive processes to occur. (Practical techniques for heat-sinking diodes from both sides have not yet been developed. One-sided heat sinking results in the metallized back side of the diode where

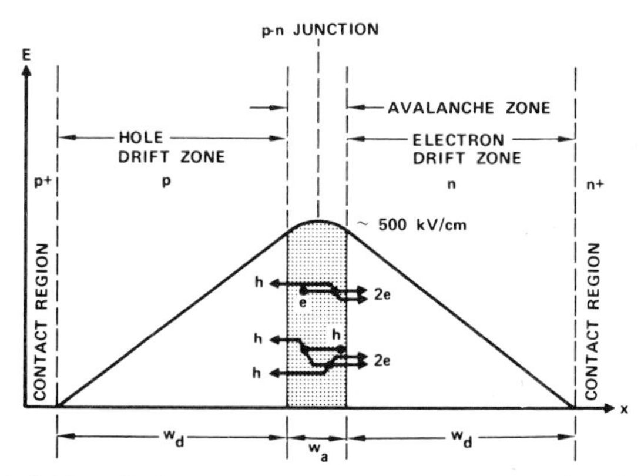

FIG. 15 Electric field profile in an avalanching reverse biased p–n junction diode designed to operate as a double-drift IMPATT.

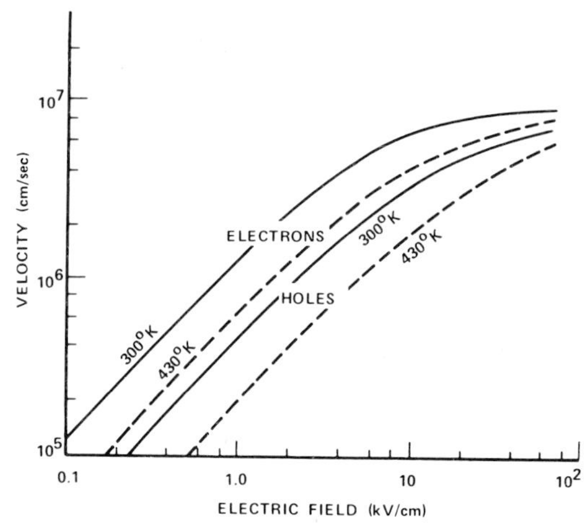

FIG. 16 Drift velocities of electrons and holes in silicon (Canali *et al.*, 1975). IMPATT diodes normally operate at temperatures between 400 and 550°K.

the ribbon connection is made getting nearly as hot as the avalanche zone. Therefore, even with excellent heat sinking on the junction side of the diode, the gold supplied by the ribbon will alloy through from the back side and cause the junction to short if the active region is too thick.

The ionization process can be measured by an ionization rate α, which is the number of ionizing collisions produced per unit length of travel of an accelerated carrier. The theoretical dependence of α on electric field E can be expressed in the form $\alpha = a \exp(-b/E)$, where a and b are constants. This is sort of a "threshold function" because E must reach very large values before α becomes nonnegligible; at low fields electrons and holes will suffer collisions with phonons and ionized impurities before gaining enough energy for an ionizing collision. The electron and hole velocities tend to saturate with electric field for the same reason; as the field is increased, most carriers suffer optical phonon collisions before significant energy (velocity) can be gained. Thus the velocity-field characteristic flattens out at high electric field values, as is shown in Fig. 16 (Canali *et al.*, 1975). From the electric field profile in Fig. 15 it is seen that throughout the avalanche zone and most of the two drift zones the electron and hole velocities will be at the saturated value v_s, somewhat below 10^7 cm/sec. Also note that because of nearly equal drift velocities of electrons and holes the two drift zones are of equal width w_d.

In the avalanche zone (assuming no variation with x) the time dependence of the total particle current I is given by (Read, 1958)

$$\frac{1}{2}\tau\frac{dI}{dt} = I\left(\int_0^{w_a} \alpha\, dx - 1\right) + I_s, \tag{10}$$

where $\tau = w_a/v_s$ is the particle transit time across the avalanche zone of width w_a, and I_s is the current due to thermally generated electron–hole pairs. Note that $\int_0^{w_a} \alpha \, dx$ is the number of ionizing collisions suffered as the particle transits through the avalanche zone. As $\int_0^{w_a} \alpha \, dx \to 1$, the avalanche becomes self-sustaining. From Eq. (10) note that for the steady-state condition the current "multiplication" is $I/I_s = (1 - \int_0^{w_a} \alpha \, dx)^{-1}$.

Ignoring the reverse saturation current I_s, which is normally quite small, the behavior of the double-drift diode of Fig. 15 can be deduced as follows. A positive variation of the electric field will cause an increase in α, and $(\int_0^{w_a} \alpha \, dx - 1)$ will be positive, resulting in an exponential increase in current, whereas a negative variation will cause an exponential decrease. This can be illustrated for an applied ac voltage waveform $V_1 = Ew_a$, as shown in Fig. 17. The generated particle current increases while the voltage is positive, peaking at the voltage waveform zero-crossing and then decaying. The particle current is $I = I_e + I_h$, so at the right side of the avalanche zone the multiplied electron current exits and $I \approx I_e$ since at this location the hole current $I_h < I_s$, and at the left side the multiplied hole current exits and $I \approx I_h$ since here the electron current $I_e < I_s$. The electron current becomes a packet of negative charge drifting across the n-doped region, and the hole current becomes a packet of positive charge drifting across the p-doped region. The induced current in the external circuit due to these injected charge packets is shown in Fig. 17 for a transit angle $(\omega w_d/v_s) = \pi$. Note that the induced current due to the drifting electrons and holes is nearly a square wave that is 180° out of phase with the applied ac voltage, which amounts to an ac negative resistance. Some of the important details of IMPATT operation, however, are not revealed

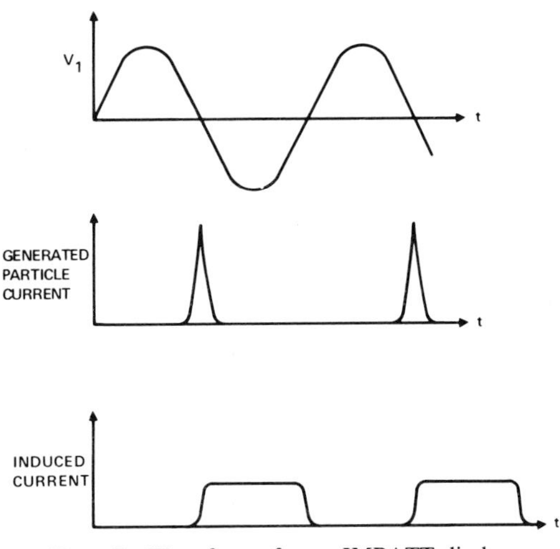

FIG. 17 Waveforms for an IMPATT diode.

by the simple waveforms of Fig. 17. Displacement current in both the avalanche and drift zones must be taken into account to complete the physical picture.

Equation (10) and Fig. 17 show the inductive behavior of the avalanche. Let L_a represent the "inductance" of the avalanche, then for small amplitude signals the generated ac particle current i_p will lag the applied ac voltage by 90°, $i_p = V_1/(j\omega L_a)$. The avalanche zone also has a displacement current $i_d = j\omega(\varepsilon A/w_a)V_1$. Everywhere in the diode at a given instant, the total ac current i_t is the same and is the sum of the particle and displacement currents $i_t = i_d + i_p$. Thus the avalanche zone represents a parallel LC circuit with an "avalanche" resonance at $\omega_a = (L_a \varepsilon A/w_a)^{-1/2}$. As would be expected from Eq. (10), the inductance depends on the dc current level and on the functional dependence of α on electric field. Small-signal analysis gives an expression for the avalanche resonance frequency of a thin avalanche zone (Gilden and Hines, 1966):

$$\omega_a^2 = 2\alpha' v_s J_0/\varepsilon, \tag{11}$$

where $\alpha' = d\alpha/dE$ and J_0 is the dc current density. Note that $i_p/i_d = 1/(1 - \omega^2/\omega_a^2)$, so that below avalanche resonance i_p, i_d, and i_t are all in phase, but above resonance i_p is 180° out of phase with i_d and i_t. The terminal (external) current I_{ex} of a drift zone is obtained by integrating the sum of the particle and displacement currents over the zone.

$$I_{ex} = \frac{1}{w_d}\int_0^{w_d}\left(I_p + \varepsilon A\frac{\partial E}{\partial t}\right)dx. \tag{12}$$

Here the drift zone particle current, for saturated drift velocities, can be expressed in terms of the avalanche zone particle current simply as $I_p(x, t) = I(t - x/v_s)$. The ac version of Eq. (12) for an injected particle current $i_p \exp(j\omega t)$ is then

$$i_{ex} = \frac{1}{w_d}\int_0^{w_d} i_p \exp\left(-\frac{j\omega x}{v_s}\right)dx + \frac{j\omega\varepsilon A V_d}{w_d}. \tag{13}$$

For $(\omega w_d/v_s) = \pi$, integration of Eq. (13) gives $i_{ex} = -j(2/\pi)i_p + j\omega C_d V_d$, where C_d is the capacitance of the drift zone. Figure 18 contains a phasor diagram for a double-drift IMPATT diode operating just above avalanche resonance at a drift transit angle $\omega\tau_d = \pi$. This diagram illustrates in vector form the ac particle current waveforms, plus it shows the displacement currents in the two drift zones with the avalanche and drift zone voltages summed to give the total voltage. The diode impedance Z is simply $Z = (|V_t|/|i_t|)\exp(-j\phi)$, where ϕ is the phase angle between V_t and i_t. It is seen that $\pi/2 < \phi < \pi$. Thus the diode impedance has a negative resistance component and a capacitive reactance component.

The diode will not operate below avalanche resonance because i_t and i_p

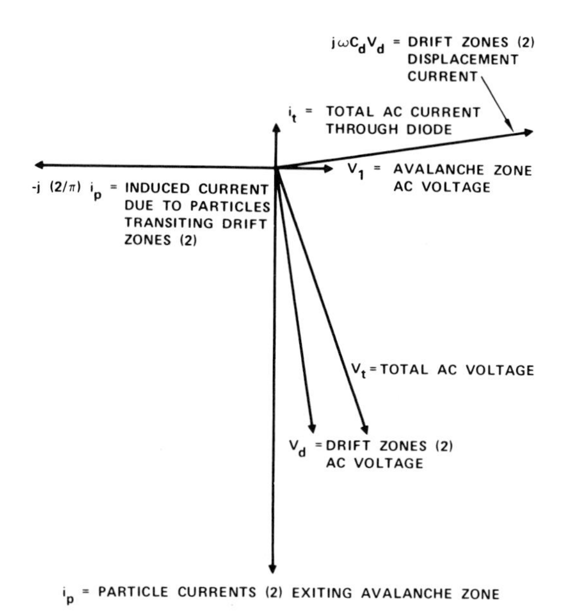

FIG. 18 Phasor diagram for a double-drift IMPATT diode operating just above avalanche resonance at a transit angle $\omega\tau_d = \pi$.

are in phase and no negative resistance results. This can be seen by constructing a diagram similar to Fig. 18. Also note that as ω increases above ω_a, i_p/i_d steadily decreases causing ϕ to approach $\pi/2$ and consequently the negative resistance to decrease. A simple conclusion is that the device should be operated in a narrow frequency band just above the avalanche resonance frequency.

For the simple device model discussed above, at $\omega = \omega_a$ there is a singularity and the negative resistance becomes very large. There are negative and positive damping effects in the avalanche zone due to space charge, unequal electron and hole ionization rates, and finite I_s, which remove the singularity in more detailed device models (Gummel and Scharfetter, 1966; Rucker *et al.*, 1978). For these more accurate device models a useful figure-of-merit is the small-signal negative Q, i.e., $\omega/2Q = -(\omega/2) \cot \phi$ is the rate of growth of oscillation amplitude for the diode mounted in a lossless cavity. Numerical analyses of IMPATT diode structures have shown that the maximum value of $-1/Q$ is obtained for $\omega\tau_d \simeq \pi$ and for $\omega \simeq 1.2\, \omega_a$ (Scharfetter and Gummel, 1969).

B. IMPATT DESIGN CONSIDERATIONS

The double-drift IMPATT (Scharfetter *et al.*, 1970), fabricated from silicon, with uniformly doped p and n regions (see Fig. 15) is a superior device for millimeter-wave operation. The active region of a double-drift diode is approximately twice as thick as a single-drift diode, which provides twice the impedance per unit area. Circuit losses, which will be discussed later, are very important at millimeter wavelengths and increase rapidly as device impedance is decreased. It has been shown (Seidel *et al.*, 1974) that double-drift diodes of up to four times the area of single-drift devices can be used before measured (circuit) efficiency begins to fall off. The basic electronic efficiency of double-drift diodes is also higher than single-drift devices. To show this, large-signal operation must be discussed.

The model examined in Fig. 18 shows that when ω decreases toward ω_a, the phase angle between i_t and V_t approaches 180°; the generated rf power will be $-i_t V_t/2$. If we assume full current modulation as shown in the square-wave of Fig. 17, then $i_{t_{max}} = (2/\pi)2I_0$ and the maximum efficiency is $(2/\pi)V_t/V_0$, where I_0 and V_0 are the dc current and voltage, respectively. Note that the charge packet of total charge Q transiting the drift zone will depress the field in the drift zone by an amount ΔE just behind the packet, e.g., $\Delta E = Q/\varepsilon A = J_0/\varepsilon f$, in the case of full current modulation. The depression in electric field will turn the avalanche off, prematurely degrading the phase relationship between voltage and current in the avalanche zone. The minimum in the ac voltage occurs when the packet is half-way across the drift region; thus ΔE must be smaller than the minimum ac field by about 100 kV/cm or the electrons will begin to drop out of velocity saturation (see Fig. 16) and the phase relationship in the drift zone will also be degraded. Note that for high power it is desirable that J_0 and, therefore, ΔE be as large as possible.

Study of various silicon models shows that Read's original suggestion (Read, 1958) that the total ac field modulation not exceed one-half the average drift zone field E_d is appropriate for high power and efficiency. Thus $E_{ac} = E_d/2$ and $V_t = (E_d/2)(w_a + 2w_d)$. Then we have for the maximum efficiency: $(2/\pi)V_d/(V_a + 2V_d)$, if w_a is small compared to w_d. Here V_a and V_d are the dc voltage drops across the avalanche and the individual drift zones, respectively. Note that in Fig. 15, for example, w_a can be small with $V_a = E_a w_a$ quite large. Even though w_a may be small, it cannot be zero because of the avalanche condition $\int_0^{w_a} \alpha\, dx \approx \alpha w_a = 1$. In fact, when the electron and hole ionization rates are not equal ($\alpha_e > \alpha_h$), then $\alpha_e w_a > 1$; this is because the holes, with a lower ionization coefficient, are likely to leave the avalanche zone without suffering an ionizing collision, which means the electrons must travel further and produce more than one ionizing collision to provide the same current multiplication, and this requires that the avalanche zone be wider.

Not surprisingly, large-scale computer simulations (Scharfetter and

Gummel, 1969) have shown that the small-signal growth factor correlates to efficient large-signal operation. The two conditions for a large growth factor can be combined to give $(\pi/\tau_d) = 1.2\omega_a$. If a power-law approximation $\alpha = \alpha_0(E/E_c)^p$ (where $p \approx 6$ for silicon) is used, then $\alpha' = p\alpha/E$. The current limitation, as discussed above, is $J_0 = \varepsilon E_d f/2$; therefore $\omega_a^2 = (3p/2)(\alpha_e w_a/V_a)v_s(E_d f/2)$.† Then since $\pi/\tau_d = \pi v_s/w_d$ and $f = 1/(2\tau_d)$, we have $V_a/V_d = (1.2/2\pi)^2(3p/2)(\alpha_e w_a)$. At millimeter frequencies E_a is large so that, in silicon, α_e and α_h are more nearly equal than at lower frequencies with the result that $\alpha_e w_a \approx 2.0$. Thus $V_a/V_d \approx 0.66$ and the maximum efficiency is $(2/\pi)V_d/(V_a + 2V_d) = 0.24$. Efficiencies exceeding 10% have been measured at frequencies near 100 GHz in pulsed-oscillator operation. Considering that diode parasitic and circuit losses are high near 100 GHz and that phase degradation due to space charge is not accounted for, the predicted maximum electronic efficiency is in reasonable agreement. For the diode of Fig. 15 the fact that the avalanche voltage V_a must be comparable to the drift voltage V_d is a direct consequence of the requirement that $w_a \approx 2/\alpha_e$ be large enough to provide sufficient carrier multiplication to supply the large pulses of injected current required for high-power, efficient operation. The efficiency relationship demonstrates the efficiency enhancement with two drift spaces connected to opposite sides of a single avalanche zone; for a single-drift diode the maximum efficiency is $(1/\pi)V_d/(V_a + V_d)$.

Another IMPATT structure, the Read diode fabricated in n-type GaAs (see Fig. 19), has proven to be a good performer at X-band frequencies with efficiencies routinely exceeding 20% for diodes manufactured in small quantities. GaAs has an electron velocity-field characteristic (Ruch and Kino, 1967) shown in Fig. 20. Because of the Gunn effect, represented by the negative slope of the velocity-field curve between 3 and 10 kV/cm, the field in the drift region can drop to a relatively small value without the electron velocity dropping out of saturation (compare to Fig. 16). This means that nearly 100% field modulation can be achieved, i.e., $J_0 = \varepsilon E_d f$ so that the maximum efficiency with an electron drift region only (single drift) is $(2/\pi)V_d/(V_a + V_d)$; similarly, the voltage relation is modified, $V_a/V_d = (1.2/2\pi)^2(6p)(\alpha w_a)$. In GaAs $p = 5$, and electron and holes have nearly equal ionization rates so that $\alpha w_a = 1$. Thus $V_a/V_d = 1.1$, and the efficiency is predicted to be 30%. Figure 19 shows that in the

† The effective value of α for different electron and hole ionization rates is given by $\bar{\alpha} = (\alpha_e + \alpha_h)/2$ (see, e.g., Blum and Kramer, 1970). Therefore, if $\alpha_e \gg \alpha_h$, as is generally true in silicon, we have $\bar{\alpha}' \approx \alpha_e'/2$. When space charge in the avalanche zone is taken into account, we have $\omega_a^2 = 3\bar{\alpha}' v_s J_0/\varepsilon$ instead of Eq. (11) (Gummel and Blue, 1967). Thus, for silicon an approximate formula for the avalanche frequency is $\omega_a^2 = 3\alpha_e' v_s J_0/2\varepsilon$.

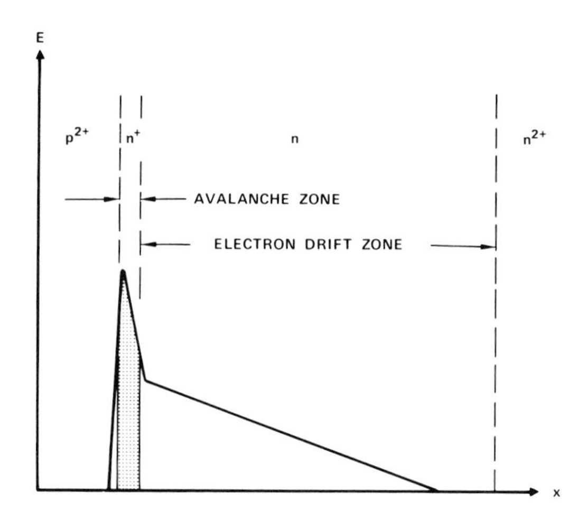

FIG. 19 Electric field profile in a GaAs Read diode.

Read structure w_a can be used as a design parameter to achieve the optimum value of V_a/V_d by adjusting the doping level in the avalanche zone. Note that in Fig. 19 the depletion layer does not extend to the n^{2+} contact. This allows for rf depletion layer width modulation attendant with high-efficiency operation.

It is reasonable to assume that similar GaAs IMPATT performance might be achieved at millimeter wavelengths by design scaling. Such is not the case. There is a time delay associated with the Gunn effect in GaAs; electrons suffer a number of diverting collisions before gaining enough energy to transfer to a subvalley in the conduction band that has a higher electron effective mass. Thus the electrons require a finite amount of time to come into equilibrium with the field, of the order of picoseconds. So as the frequency is increased into the millimeter-wave band, the velocity-field characteristic becomes more and more like that of silicon until above 100 GHz the Gunn effect essentially disappears. This means that the field modulation is reduced to $E_d/2$ as in silicon. An additional advantage that GaAs has over silicon is also removed; at 10 GHz $\alpha_e \gg \alpha_h$ in silicon, requiring that $\alpha_e w_a \approx 3.5$ and consequently that $V_a/V_d > 1.1$, whereas $\alpha_e \approx \alpha_h$ for GaAs. For higher-frequency designs with flat doping profiles, the breakdown fields become larger because w_d, and consequently w_a, must become smaller. Then, since $\alpha_e \rightarrow \alpha_h$ in silicon at higher fields, $\alpha_e w_a \rightarrow 1$ and the V_a/V_d ratio is similar for both materials.

The relative change in the ionization rates is sensitive to the exact functional dependence on electric field, i.e., $\alpha = a \exp(-b/E)$ gives a dif-

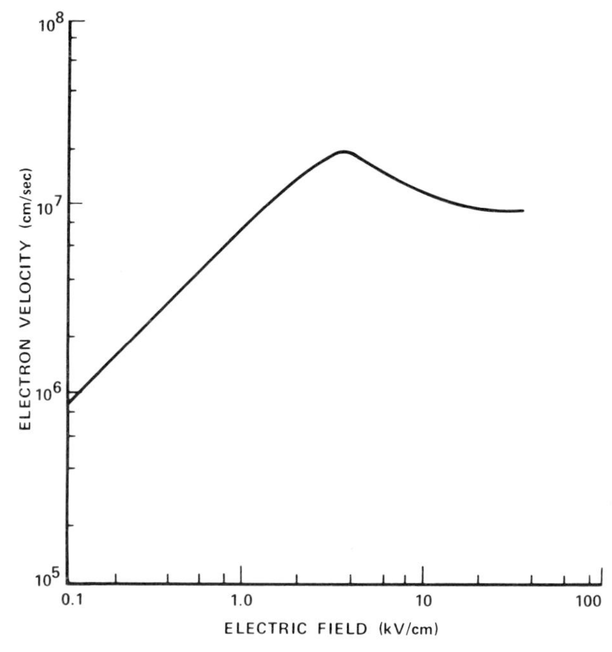

FIG. 20 Electron velocity-field characteristic for n-type GaAs.

ferent result than the approximating formula used by Read (1958): $\alpha = \alpha_0(E/E_c)^p$. For the Read formula a given fractional change in field $\Delta E/E$ produces the same fractional change in the ionization rate $\Delta\alpha/\alpha$ independent of the value of E; for this case the equations describing the diode dynamics can be converted into universal forms that have no dependence on diode width by proper normalization of variables (Misawa, 1972). Misawa showed that the more accurate expression for ionization rates results in a larger (degraded) avalanche response time as the frequency goes higher (avalanche zone becomes narrow), i.e., since $\Delta\alpha/\alpha$ decreases relative to $\Delta E/E$, the generated particle current becomes less spiky and the phase relationship in the drift zone is degraded causing electronic efficiency to suffer. The onset of this effect was calculated by Misawa to occur at approximately 50 GHz for GaAs and 100 GHz for silicon. This is another reason why no efficiency improvement is expected, or has been observed, with GaAs at the higher millimeter-wave frequencies.

Other, practical factors are important. The thermal conductivity of silicon is several times larger than that of GaAs, which is advantageous for high-power/high-reliability operation. Silicon also has the benefit of a superior processing technology based on high-precision, extremely high-rate production for computer and consumer electronics. On the other hand, the electron low-field mobility in GaAs is several times larger than in silicon, which means the rf loss in the parasitic bulk resistance is lower in GaAs.

A deleterious effect, minority charge storage is avoided in double-drift diodes. In single-drift ($p^+ nn^+$ or $n^+ pp^+$) diodes minority carriers generated by the avalanche can back diffuse against the field into the heavily doped, low-field contact region adjacent to the avalanche zone and then rediffuse into the avalanche zone when the field reverses. This effect provides a large starting current for the multiplication [equivalent to a large value of I_s in Eq. (10)]. As a result, the maximum-induced current level $2J_0$ can be reached well before the end of the positive portion of the ac voltage cycle (Misawa, 1970a, b), with space charge quenching the ava-

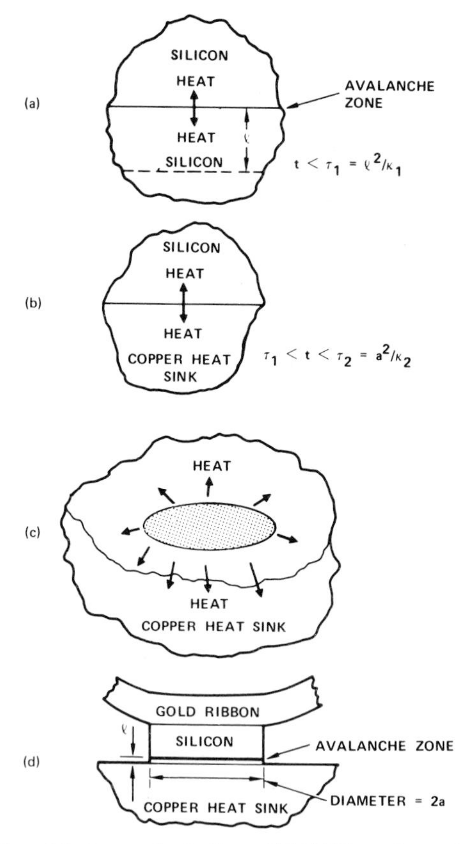

FIG. 21 Thermal models for a pulsed-silicon IMPATT are shown in (a) infinite solid, (b) infinite composite solid, (c) infinite half space with heat supplied over disk of radius = a. A schematic of a diode chip is shown in (d), actual diode.

lanche prematurely. This results in a phase shift of $\Delta\phi$ away from the ideal value of 180° between current and voltage. Efficiency is then proportional to $\cos(\Delta\phi)$. In the double-drift diode the always large field in the avalanche zone and immediately adjacent drift regions inhibits charge storage and therefore, effectively, $I_s = 0$.

Pulsed IMPATTs can produce well over an order of magnitude more power in short pulses (~100 nsec) than cw devices. This is because the peak temperature reached with short pulses at moderate duty factors (<0.5%) is more than an order of magnitude below the temperature reached in steady operation. Figure 21 shows three thermal models for a pulsed IMPATT that can be tied together in time sequence to approximate the temperature behavior at the center of the silicon–metal interface where alloying will initiate and cause device failure. For small values of t, Fig. 21a shows the avalanche zone, where heat is generated, layered in a one-dimensional silicon structure in which heat flows equally in both directions. The distance from the avalanche zone to the heat sink interface is taken to be l. For a diode on a copper heat sink, Fig. 21b shows a one-dimensional model for time greater than the silicon thermal time constant $\tau_1 = l^2/\kappa_1$, where κ_1 is the thermal diffusivity of silicon. Because the avalanche zone is very close to the copper heat sink for millimeter-wave devices relative to other device dimensions (l is small), the power dissipation is modeled as occurring at the heat sink interface for $t > \tau_1$. Finally, the three-dimensional model shown in Fig. 21c is used for time greater than the heat sink thermal time constant $\tau_2 = a^2/\kappa_2$, where a is the diode radius and κ_2 is the thermal diffusivity of copper. As $t \to \infty$, a steady temperature determined by the model of Fig. 21c is reached, given by $I_0V_0/\pi aK_2$, where K_2 is the thermal conductivity of copper.

In the following, the results of Carslaw and Jaeger (1959) are used. For small values of t the temperature rise at the silicon–heat sink interface is described by

$$T(t) = (I_0V_0(\kappa_1 t)^{1/2/}\pi a^2 K_1) \operatorname{ierfc}[l/2(\kappa_1 t)^{1/2}]$$

corresponding to Fig. 21a. Since only the values of the thermal properties of silicon appear in this equation, it is obvious that improved heat sink materials such as diamond will not help reduce the initial temperature rise in pulsed operation. For $\tau_1 < t < \tau_2$ the maximum temperature at the interface is approximated by

$$T(t) = (2I_0V_0/\pi a^2)[(\kappa_1\kappa_2 t)^{1/2}/(K_1\sqrt{\kappa_2} + K_2\sqrt{\kappa_1})](1/\sqrt{\pi})$$

corresponding to Fig. 21b. Then for the model of Fig. 21c the temperature at the center of the disk defining the silicon–copper interface is $(I_0V_0/\pi aK_2)[1 - a/2(\pi\kappa_2 t)^{1/2}]$ for $\tau_2 \ll \tau \to \infty$.

Based on the above equations, the time dependence of temperature at the silicon–metal interface at the diode center is plotted in normalized form in Fig. 22, assuming $a/l = 50$, which is typical for millimeter-wave diodes. Note that the choice $a/l \sim 50$ results in $\tau_2 \sim 1000\tau_1$ for silicon mounted on copper.

Numerically 5 nsec $< \tau_1 <$ 50 nsec, which is typical of tactical radar pulsewidths. From Fig. 22 it is seen that the temperature rise at the end of a 50-nsec pulse is less than 10% of the temperature rise in a cw diode operated at the same input power level, which explains, in part, the extremely high-peak power outputs achievable with pulsed millimeter-wave IMPATTs.

C. Diode Packages

The usefulness of an rf diode depends not only on the fundamental electronic processes that generate millimeter-wave energy, but, to a large degree, on the techniques for packaging and/or circuit mounting of the basic semiconductor chip. Reliability is often determined by packaging and mounting since bond integrity, exclusion of contaminants, and heat removal are known to be among the most important factors in achieving long operating life.

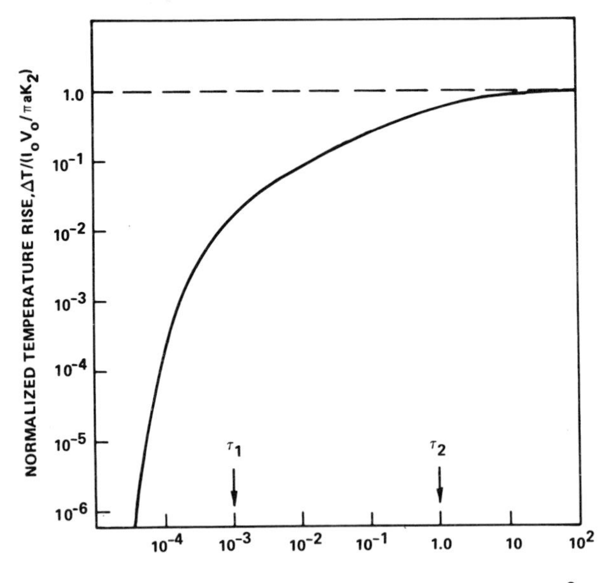

FIG. 22 Temperature rise with time for a pulsed-silicon IMPATT diode on a copper heat sink. The ratio of diode chip radius a to junction depth l is $a/l = 50$. The temperature is at the center of the disk forming the silicon–metal interface.

The importance of mounting, connecting, and protecting the semiconductor ship is often overlooked. At millimeter wavelengths package parasitics usually are a limitation even in narrow-band circuits. It is often useful to be able to tailor the mount/package design to achieve the desired performance. To be able to limit adequately and control the parasitics, however, it is necessary to work within extremely small physical volumes.

It is often stated that an optimum package design has minimum parasitics. This is not altogether true. It is desirable to minimize resistive loss associated with the chip and its mounting connections. The skin effect becomes important in millimeter-wave diode chips and significantly increases parasitic series resistance (De Loach, 1970; Midford and Bernick, 1979). The solution is to thin the diode chip to a total thickness of 10–20 μm to reduce the "cylinder" resistance to a small fraction of an ohm. As a dimensional reference, double-drift diode chips range in diameter as follows (typically): 35 GHz, 100 μm; 60 GHz, 70 μm; 94 GHz, 50 μm; 140 GHz, 30 μm; 240 GHz, 15 μm. In the package equivalent circuit of Fig. 23, it is also desirable to minimize the fringing capacitance C_f that shunts the chip in order to reduce circulating rf currents that are dissipated in the resistive parasitics represented by R_s. However, L_p, the inductance of the connecting lead(s), often has an optimum nonminimum value for use in a particular circuit application at a particular frequency. In other words, in the equivalent circuit of Fig. 23, L_p and C_p form an important part of the network that is used in conjunction with external stepped coax or waveguide transformers to match resistively and reactively the low diode impedance (a few ohms) to a typical circuit impedance of 100 Ω. A package, or mount, should normally be designed to have a total shunt capacitance value that is a small fraction of the operating diode capacitance and to have a means for varying the inductance of the connecting lead(s). Such a package, suitable for operation up to 100 GHz, is shown in Fig. 24.

For diodes in which the heat due to a large amount of dc input power

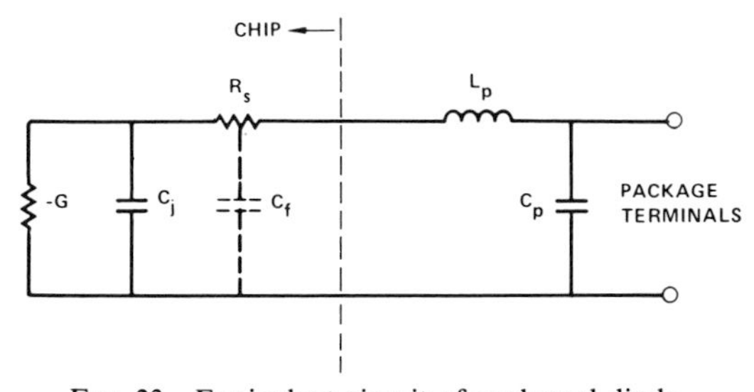

FIG. 23 Equivalent circuit of packaged diode.

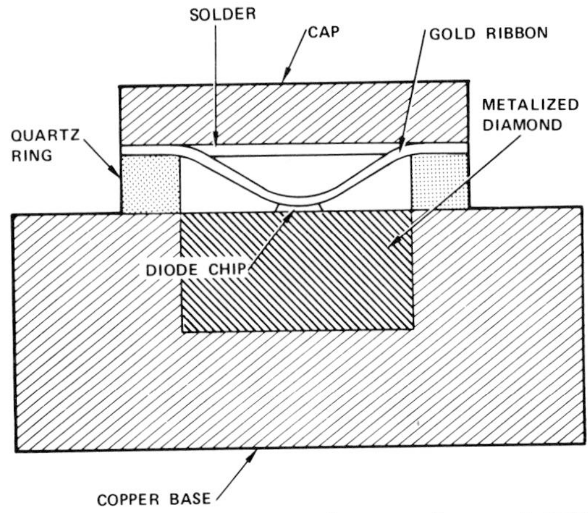

FIG. 24 Millimeter-wave IMPATT package. The cap diameter is 0.75 mm. The drawing is approximately to scale. (From Kramer, 1979. Reprinted from *1979 IEEE NTC Proceedings*. © 1979 IEEE.)

must be dissipated, a mounting scheme is required that provides adequate heat removal and also has the necessary mechanical strength to withstand thermal expansion and contraction. The package of Fig. 24 has these features. The insulating rings are ultrasonically cut from a single-crystal quartz plate. The quartz crystal can be oriented to match closely the thermal expansion characteristics of the copper package base. The low dielectric constant of quartz contributes to the reduced shunt capacitance of this package.

The miniature metal–ceramic package shown in Fig. 25 has been used with good results for Gunn diodes up to 100 GHz. For frequencies above

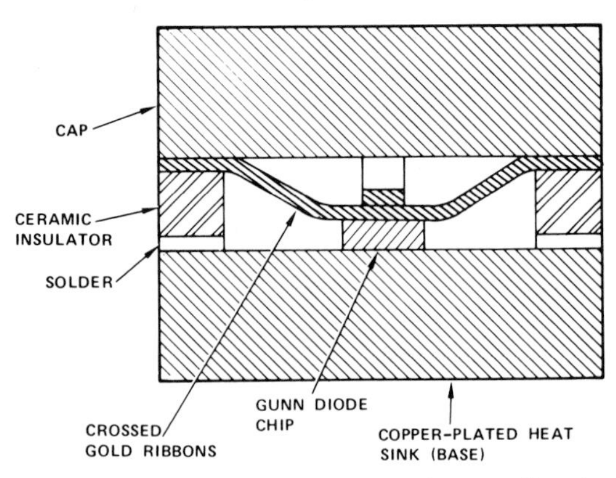

FIG. 25 Millimeter-wave Gunn diode package. The cap diameter is 0.9 mm. The drawing is approximately to scale.

100 GHz, the shunt capacitance of any type of ring insulator becomes intolerable. One approach is to use quartz stand-offs (posts) as shown in Fig. 26. This type of mounting arrangement is used for IMPATT diodes designed to operate from 100 to 250 GHz. At the higher frequencies the position and geometry of the stand-off(s) and the connecting ribbon(s) are critical factors in circuit performance. With a single stand-off mount, the parasitic shunt capacitance can be reduced to approximately 0.01 pF, which is sufficiently low for satisfactory performance at 200 GHz. However, the optimum value for the inductance of the connecting strap may be smaller than can be achieved. Inductance values measured for millimeter-wave packages and stand-off mounts are shown in Fig. 27 for various strapping configurations.

Encapsulated millimeter-wave IMPATT diodes with integral diamond heat sinks, as shown in Fig. 24, are capable of producing roughly twice the cw power and efficiency of similar IMPATT diodes with copper heat sinks when operated at the same junction temperature. High-purity (type IIa) diamond has a thermal conductivity value more than twice that of copper at 200°C, making it useful as a diode heat-sink material. The cost of material may limit the use of diamond heat sinks, but diamond heat sinking is a proven technique with a clear performance advantage for cw operation.

D. Diode Performance

Device area should be as large as possible to minimize heating, thereby ensuring long life operation. This results in a low value of rf impedance since both resistance and capacitive reactance are inversely proportional

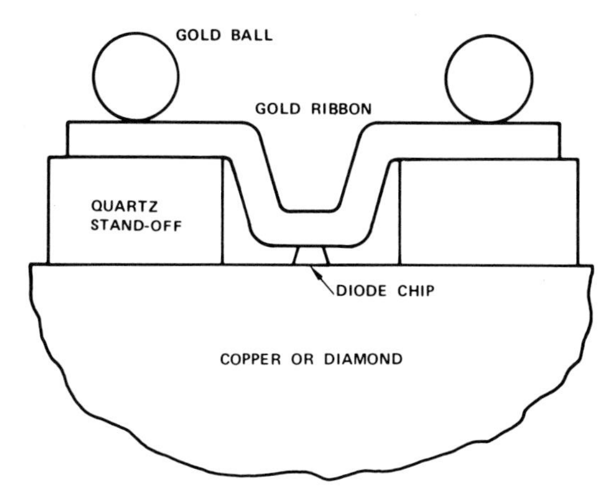

Fig. 26 Stand-off mount for 100–250-GHz IMPATT diodes. The stand-off is 0.003 in. high. The drawing is approximately to scale.

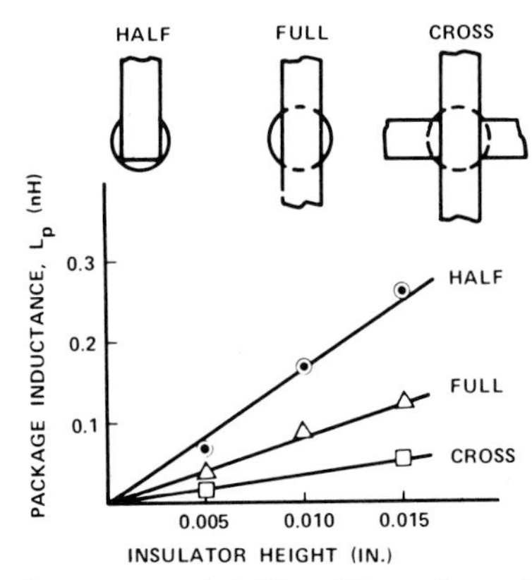

FIG. 27 Inductance values measured at X band for various millimeter-wave diode strapping configurations.

to device area. This means that in general the rf impedance of a diode is much lower than the characteristic impedance of the transmission line or cavity circuit in which it must be used. There are two undesirable byproducts of low device impedance: (1) parasitic impedance of the package elements has a strong influence on overall circuit performance, and (2) circuit impedance must be reduced to match the device impedance by using transformers. Both contribute to making broadband circuit design difficult. Low circuit impedance Z_0 produces large currents flowing in the cavity walls resulting in a disproportionally larger rf power loss since $P_{\text{lost}} = I^2R_{\text{walls}}/2 = P_{\text{RF}}R_{\text{walls}}/2Z_0$. To summarize the situation, low device impedance is a pervasive, deleterious fact-of-life at millimeter-wave frequencies that must be dealt with by compromising performance.

How does impedance level affect the performance of devices used for power generation? At the low end of the millimeter-wave band the power generation capability of Gunn and IMPATT diodes at moderate junction temperatures is comparable even though Gunn efficiency is lower. This can be attributed to the higher "active" impedance (impedance under operating conditions) of the Gunn device. A Gunn diode with several times the area of an IMPATT diode is conveniently used in a simple cavity resonator for operation in the same frequency band. (Above 50 GHz Gunn performance drops off rapidly because of the electron-transfer time constant mentioned earlier.)

In order to understand active impedance quantitatively, equivalent circuits have been devised that are based on simple physical models as well as on detailed large-signal computer simulations. Most of the theoretical studies have been appropriate to the physical parameters and response times of diodes operating in the neighborhood of 10 GHz. For both Gunn and IMPATT diodes, the derived equivalent circuit in the frequency band, where highest power output and efficiency is available, is a negative conductance shunted by a capacitive susceptance that is several times larger. This equivalent circuit is a practical representation for millimeter-wave IMPATT diodes but does not appear to be so for Gunn diodes. Swept-frequency impedance measurements made on Gunn diodes usually correspond to a series RLC circuit characteristic. Since the Gunn diode can be operated near the LC resonance by tailoring the mount inductance, the operating impedance presented to the rf circuit is nearly a pure negative resistance. Thus the net Gunn diode impedance is large compared to that of an IMPATT diode with its large shunt capacitive susceptance.

Higher impedance per-unit-area and series-RLC-circuit behavior of the Gunn diode are beneficial to oscillator circuit design for noise reduction and for such applications as varactor tuning. Amplifier and power combiner design is also benefited by the relatively low Q of Gunn devices.

Temperature rise due to dc power dissipation is a limitation that must be considered in any power device. In a semiconductor device there are chemical-rate processes associated with the semiconductor material and applied metallizations that are accelerated at elevated temperatures and ultimately cause failure. For this reason it is appropriate to compare power generation capability on the basis of a fixed temperature rise.

Figures 28 and 29 contain a cw power and efficiency comparison of Gunn diodes on copper heat sinks and double-drift silicon IMPATTs on diamond heat sinks at a temperature rise not exceeding 225°C. The comparison is based on best data obtained from Hughes's silicon IMPATTs and GaAs Gunn diodes. It should be noted that Gunn diodes normally exhibit a saturation in output power as dc input power is increased. For the size of diodes normally used, the power maximum usually occurs at a temperature rise of less than 225°C. Thus, a power/efficiency comparison of Gunn and IMPATT diodes at equal temperatures would be more favorable for Gunn diodes than is indicated by Figs. 28 and 29. A relatively new material, indium phosphide, shows promise of enabling the construction of Gunn diodes and perhaps IMPATTs that will be more competitive with silicon IMPATTs in power and efficiency at higher frequencies.

The relation between the area of a cw diode and its power and efficiency can be characterized as follows. With higher diode impedance (smaller-diode area), circulating currents in the cavity walls can be decreased resulting in lower rf losses and higher circuit efficiency. A larger

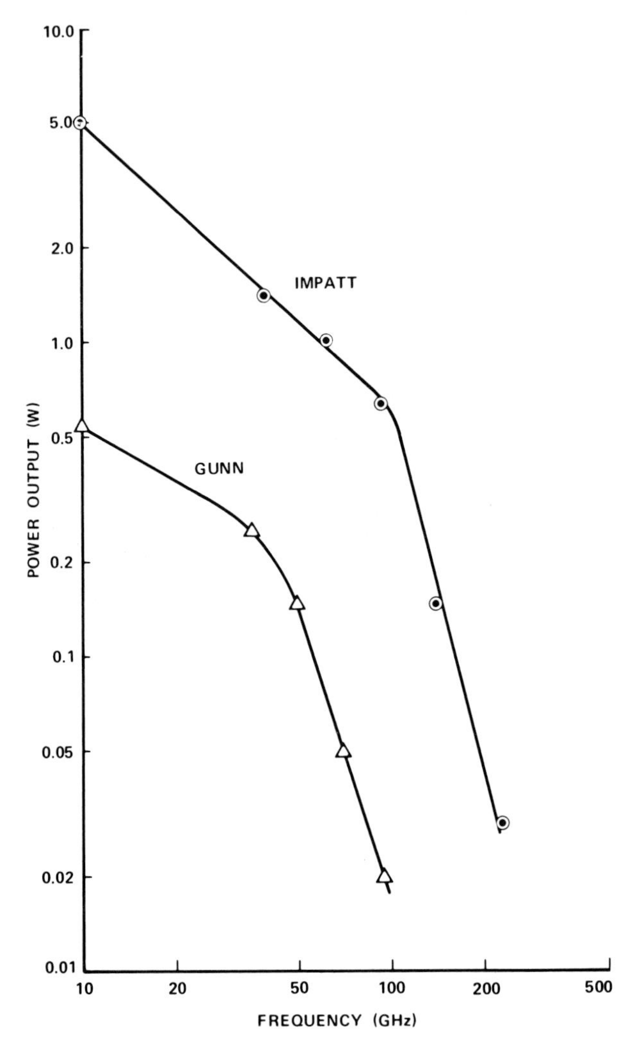

FIG. 28 Cw power output vs frequency for double-drift silicon IMPATT diodes on diamond heat sinks and GaAs Gunn diodes on copper heat sinks at a temperature rise not exceeding 225°C.

diode will produce more power, but not in direct proportion to its area because of the increased thermal impedance per-unit-area and lower circuit efficiency. Electronic efficiency is dependent on the input power density for both Gunn and IMPATT diodes. Since larger diodes present a higher thermal impedance per-unit-area, electronic efficiency tends to decrease with increasing diode area as well.

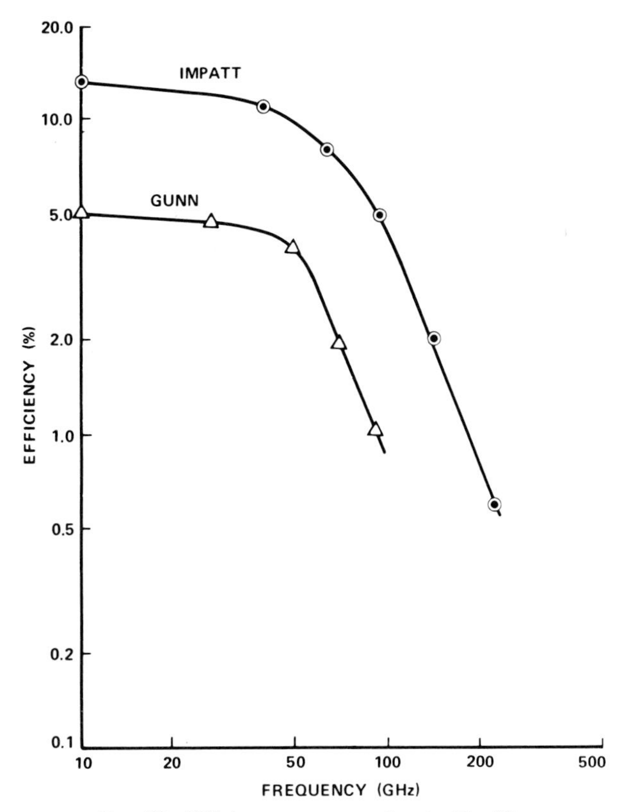

FIG. 29 Efficiency corresponding to Fig. 28.

1. *Pulsed IMPATT Oscillators*

IMPATT diodes can be designed to produce much more peak power in short pulses than is available on a continuous basis. As an example, at 94 GHz about 0.5-W cw is available from double-drift silicon IMPATTs at reasonable junction temperatures (~250°C), whereas 18 W has been obtained in pulsed operation.

Millimeter wavelengths offer a solution to the problem of building small short-range radars that can resolve small targets in a dense clutter environment. For this purpose it is desirable to use a radar pulse of about 100-nsec duration, with a frequency chirp during the pulse of the order of 1 GHz. A pulsed IMPATT oscillator matches these requirements quite well. As shown in Fig. 30, peak power levels an order of magnitude higher than cw levels can be achieved if the pulse width is small (see Fig. 22). The design of a pulsed diode is somewhat different than a cw diode to

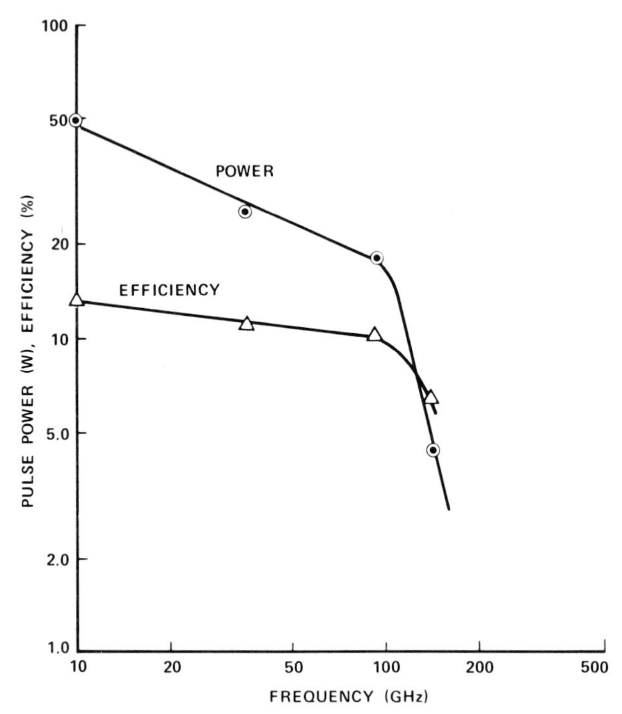

FIG. 30 Pulse-power capability of double-drift silicon IMPATTs. Pulse width is 100 nsec, and PRF is 50 kHz.

accommodate the very high current densities and resulting space-charge region widening.

Downward frequency chirp occurs as a natural consequence of heating during the pulse and can be reduced (or increased) by ramping the top of the bias current pulse. Chirp should be minimized by up-ramping if, for example, injection locking of an IMPATT source to obtain a signal with pulse-to-pulse coherency is desired.

2. *Frequency Tunable Diode Oscillators*

Modern micro/millimeter-wave systems are heavily dependent on electronically tunable solid-state oscillator technology. Tuning is currently accomplished by several methods. The YIG resonator is a useful technique, but is slow and will probably be limited to frequencies no higher than the low end of the millimeter spectrum due to the large magnetic fields required. Gunn and IMPATT Varactor-controlled oscillators (VCOS) can provide high power over narrow tuning ranges and local-oscillator power levels over much wider tuning ranges. In laboratory Gunn VCO units power—bandwidth products as high as 160 mW—GHz have been obtained at 40 GHz and 65 mW—GHz at 54 GHz. It is possible to maintain a relatively constant power output over a wide frequency band with varactor tuning. VCOs are useful in building phase-locked millimeter-wave sources for stabilized or accurately programmable signals.

Monotonic tuning of an IMPATT oscillator can be achieved by sweeping the diode bias current. Today millimeter-wave sweep generators using this principle are used in a wide variety of instrumentation and measurement applications. In general, as bias current is increased, power output as well as frequency increases. This type of behavior is shown in Fig. 31. A 10-GHz sweep bandwidth is available for center frequencies between 45 and 95 GHz. Three IMPATT oscillator cavities can be combined via an electromechanical waveguide switch to make a sweep generator covering a full waveguide band (see Fig. 32).

3. *Diode Noise*

The rf noise power generated in a negative-resistance semiconductor device is dependent on many factors. Detailed noise analyses, including diode doping-profile parameters, bias conditions, rf signal level, and parametric interactions show that significant improvements in noise performance can be expected with alterations in the doping profile and in operating conditions.

The impedance relationship between the oscillator, or amplifier circuit, and the diode also strongly affects noise output. Not a great deal of flexibility for noise optimization exists because of the difficulty in making subtle doping-profile and circuit changes at millimeter-wave frequencies. Therefore, with a given type of diode it is observed that samples from different wafers, when placed in a simple waveguide or coaxial cavity reso-

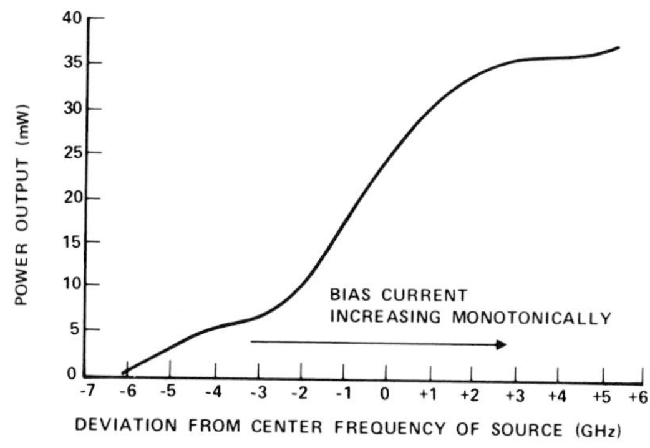

FIG. 31 Characteristic of a millimeter-wave IMPATT sweep generator.

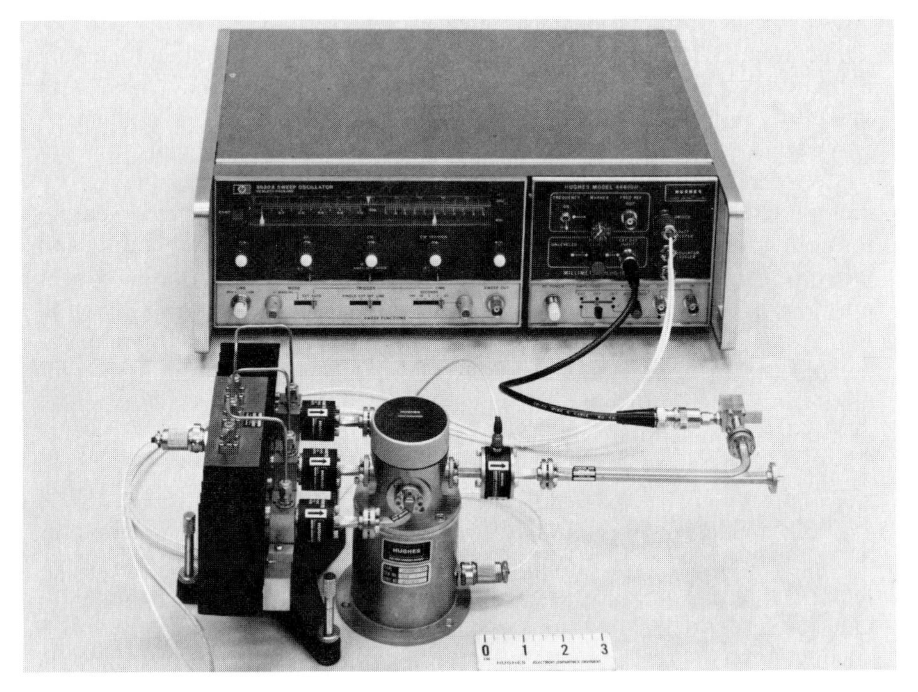

FIG. 32 Full waveguide bandwidth (50 to 75 GHz) sweep generator comprised of three electromechanically switched IMPATT oscillators.

nator, will usually produce about the same noise-to-carrier ratio when tuned for minimum noise at a specified power output level.

For if frequencies less than 1 GHz it is impractical to use IMPATT local oscillators because of high AM noise. IMPATTs are relatively noisy because of the inherent randomness of the avalanche process that results in

TABLE III

94-GHz GUNN OSCILLATOR NOISE

	AM noise	FM noise
Frequency from carrier (kHz)	Double sideband noise-to-signal ratio in a 1-kHz bandwidth (dB)	Δf_{rms} in a 1-Hz bandwidth (Hz)
1	−105	60
10	−116	10
100	−122	10

a variation in the build-up time and launch of the current pulses into the drift zones. If a 1-GHz or higher if frequency is used, an rf filter can be used on an IMPATT LO to reduce sufficiently the AM noise sidebands.

Gunn diodes consistently perform satisfactorily in local oscillators for pumping low-noise mixers if the if frequency is above 100 kHz; rf filters need not be used. Data on a 94-GHz Gunn oscillator is contained in Table III.

E. Power Amplifiers/Injection-Locked Oscillators

Solid-state rf amplifiers at millimeter wavelengths are negative-resistance diode devices that are circulator-coupled and are either operated in a stable mode or as an injection-locked oscillator (ILO). Gunn amplifiers are useful where moderate power levels over wide bandwidths are required. Today, for millimeter-wave frequencies, the IMPATT amplifier is most commonly used. It offers the highest power level available from a negative-resistance diode amplifier or ILO. The power and efficiency capability of Gunn and IMPATT injection-locked oscillators/amplifiers is nearly equal to the free-running oscillator power and efficiency curves (Figs. 28 and 29). Neither the GaAs Gunn nor IMPATT device can be used for low-noise amplification. GaAs Gunn amplifiers exhibit noise figures in the 20–30dB range, whereas IMPATTs exhibit 35–50dB noise figures. Indium-phosphide Gunn devices may provide low-noise millimeter-wave amplification in the future. For now, most power applications will require silicon IMPATTs.

Low-impedance waveguide and resonator circuits are possible to construct but are very lossy at millimeter wavelengths so they are not used. As a result of large IMPATT chip susceptance, the Q of the packaged device is relatively high, in the range of 20 to 30 or more. Impedance transformation is required outside of the package to match the relatively high circuit impedance so the total circuit Q may be even higher. The result is that IMPATT amplifiers and ILOs are relatively narrowband.

Stable amplifiers and ILOs can be crudely compared as follows. Theory predicts that the small-signal voltage gain—half-power bandwidth product of a stable negative-resistance amplifier must be greater than $2/Q_{\mathrm{ss}}$, where Q_{ss} is the small-signal Q. The theoretical voltage gain—locking-bandwidth product of an ILO is given by $2/Q_{\mathrm{LS}}$, where Q_{LS} is the large-signal Q. The large-signal negative conductance of an IMPATT diode tuned for maximum efficiency decreases to a value approaching $\frac{1}{4}$ that of the small-signal negative conductance. Thus $Q_{\mathrm{LS}}/Q_{\mathrm{ss}} = G_{\mathrm{ss}}/G_{\mathrm{LS}} \approx 4$, and the ratio of the gain-bandwidth products is $(2/Q_{\mathrm{ss}})/(2/Q_{\mathrm{LS}}) \approx 4$. Based on this comparison, we would expect stable amplifiers to offer approximately four times the bandwidth of an ILO at the same gain level. For single diode-cavity combinations (see Fig. 33), voltage gain—bandwidth products of 16.6 and 4.4 GHz have been achieved for stable amplifiers and ILOs, respectively, at 60 GHz (Kuno and English, 1973).

An ILO can be truly power optimized by impedance matching, whereas the stability requirement for amplifiers dictates a nonoptimum load impedance so that amplifier power output (and efficiency) is somewhat less. Moreover, in practice, gain per stage of stable amplifiers must be limited to 10 dB or less to avoid spurious instabilities caused by ambient temperature variations. These two factors result in injection-locked oscillator chains producing more power at higher efficiencies than stable amplifier chains.

In order to increase output power levels with solid-stage devices, it is necessary to do power combining. Cylindrical resonant combiners (see Fig. 34) with individual diode modules mounted to couple to the periphery of the cavity (Harp and Stover, 1973) have proven capable of combining up to 18 diodes at *X* band with 90% combining efficiency and 3% injection-locking bandwidth. At millimeter wavelengths, however, the cavity diameter shrinks in proportion to wavelength, so only a few modules can be mounted on the periphery. Thus a more attractive approach is to use a rectangular cavity combiner of the type shown in Fig. 35 (Kurokawa and Magalhaes, 1971). The diode modules are spaced a half-wavelength apart, which makes this configuration feasible at 100 GHz and higher. Heat removal is easily facilitated because of the linear disposal of the modules.

Resonant power combining using cylindrical and rectangular cavities is a proven and practical concept. The diodes tend to lock together due to the intermodule coupling provided by the resonator. Therefore, elaborate diode selection and module tuning procedures are not required. Injection-locking bandwidth, however, becomes a problem. Not only are the individual diodes (and therefore modules) relatively high *Q*, but the dispersal in the individual module's free-running oscillator frequencies due to unavoidable statistical variation in diode and matching circuit

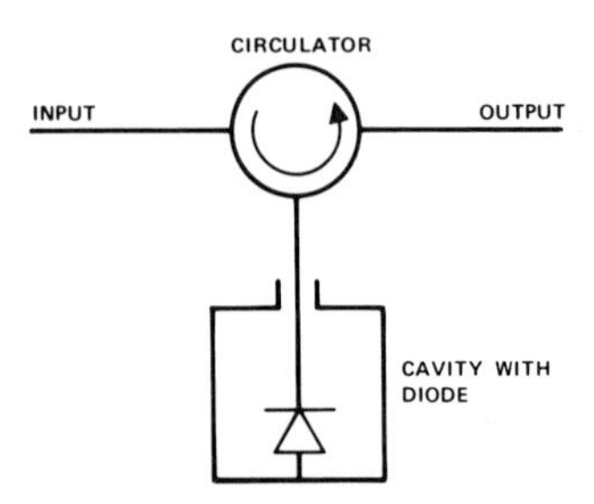

FIG. 33 Single diode reflection amplifier/injection-locked oscillator schematic. (From Kramer, 1979. Reprinted from *1979 IEEE NTC Proceedings*. © 1979 IEEE.)

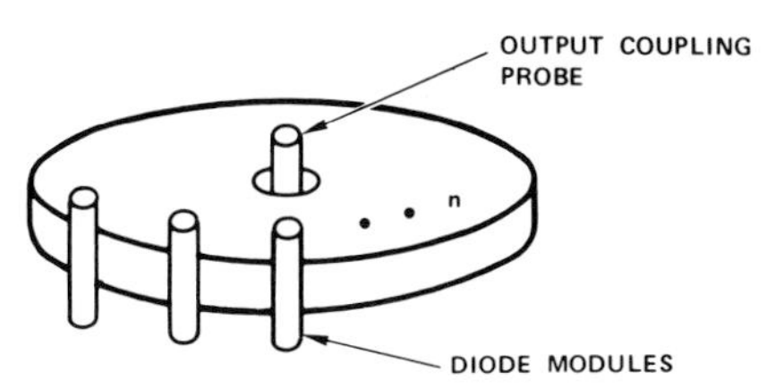

FIG. 34 Power combiner using cylindrical resonator. (From Kramer, 1979. Reprinted from *1979 IEEE NTC Proceedings*. © 1979 IEEE.)

parameters causes the measured external Q of manufactured high-power IMPATT combiners (combiners tuned for maximum power output at a precise operating frequency) to vary approximately as n, the number of modules being combined. This is in spite of the fact that theoretically, and in some carefully controlled laboratory experiments, the external Q of resonant combiners with well-matched diodes (and modules) has been shown to vary as $1/n$ (Aston, 1979; Bayuk and Raue, 1977; Ma and Sun, 1980). So resonant combining is attractive for injection-locked amplification of angle-modulated signals as long as sufficient bandwidth is available. Rectangular combiners at 94 GHz (see Fig. 35) have produced 20-, 40-, and 48-W pulsed power with 2, 4, and 6 modules, respectively (Chang *et al.*, 1979; Chang and Ebert, 1980). The individual diodes were capable of 10–13 W.

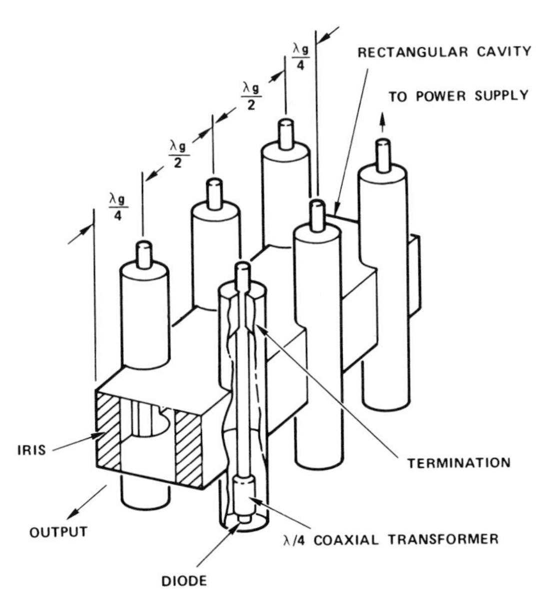

FIG. 35 Power combiner using rectangular resonator. (From Kramer, 1979. Reprinted from *1979 IEEE NTC Proceedings*. © 1979 IEEE.)

Hybrid combining (see Fig. 36) is another approach for increasing output power that has worked. Hybrids provide a high degree of isolation between diode modules so that the diodes may not lock together. For this reason it is better to use hybrid combining for stable amplifiers. The modules are loaded so as to prevent oscillation; however, when a signal is coupled into the input, the individual devices contribute nearly equally to the amplification of the signal and the output powers combine approximately in proportion to n. Because the modules must be stabilized in the quiescent condition, the Q is lower, allowing broader bandwidth operation in a stable-amplifier hybrid combiner. Obviously as the number of diodes to be combined becomes large, the hybrid "tree" required becomes cumbersome and difficult to implement in the wave-guide circuitry that is required for low loss at millimeter frequencies.

To summarize the status of millimeter-wave combiners, n diodes can be combined to get nearly n times the power output shown in Figs. 28 and 30. As n gets large, however, bandwidth and/or circuit complexity may be a limitation.

F. IMPATT Diode Reliability

The preceding discussions have examined the features of millimeter-wave diode construction and operation that are important for practical applications. High-temperature operation of semiconductor materials and small diode geometries with delicate electrical lead attachments are common characteristics that are potential reliability problems. Tests on samples of diodes, devised to focus on the potential problem areas, have been conducted.

Estimates of the operating life of millimeter IMPATT diodes have been obtained from high-temperature (accelerated) life testing. Due to the rise of IMPATT power and efficiency with dc input power, it is desirable, in most cases, to operate IMPATT diodes at the maximum junction temperature consistent with operating life requirements. For this reason accelerated life testing under operating bias conditions has been employed for millimeter-wave IMPATT diodes, allowing a quantitative trade-off

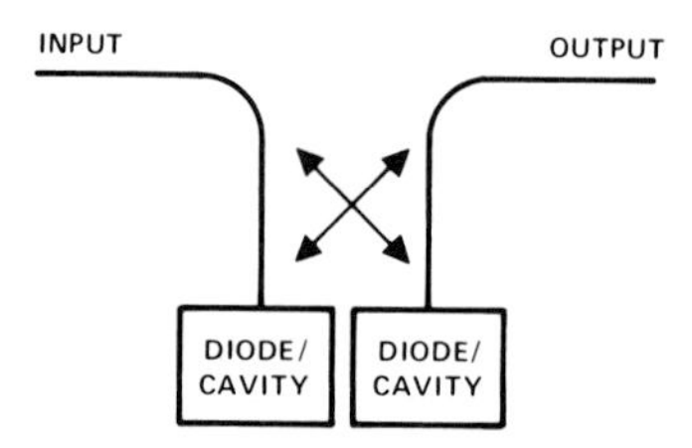

Fig. 36 Power combiner using 3-dB hybrid coupler. (From Kramer, 1979. Reprinted from *1979 IEEE NTC Proceedings*. © 1979 IEEE.)

between power/efficiency and operating life. For a double-drift silicon IMPATT diode mounted in a diamond heat-sink package (see Fig. 24) with an arsenic-doped substrate and chromium–platinum–gold metallization on the junction side of the chip, the median-time-to-failure is given by an Arrhenius relationship:

$$\tau = \tau_0 \exp(E_a/kT). \tag{14}$$

The determination of this expression for diode operating life was obtained by plotting median-time-to-failure data of sample diode lots burned at 325°, 350°, and 375°C, as shown in Fig. 37. The failures occur as a metal alloy spike originating from the heat sink and shorting the junction. The life distribution is log-normal at any given temperature. The line plotted in Fig. 37 is fit to data points on completed tests at 375° and 350°C and a conservative estimate of the data point for the 325°C test that is partially completed. The slope of the line (activation energy E_a) is matched to the activation energy of 1.6 eV obtained for more extensive tests on single-drift silicon IMPATTs of similar construction mounted on copper heat sinks (Kramer, 1976). An operating life of over 10^6 hr at 200°C is predicted. On–off cycling (30 sec on, 30 sec off) of cw silicon IMPATT diodes has resulted in the accumulation of 30.3 million diode-cycles with only six fail-

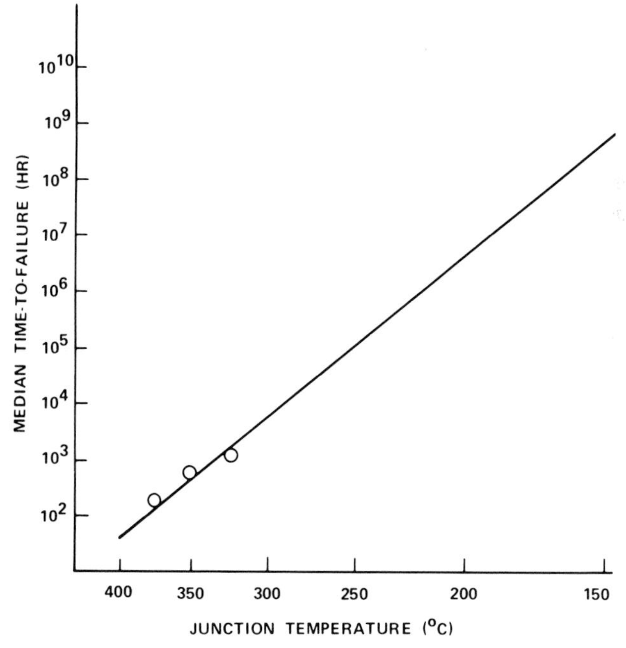

FIG. 37 Accelerated life test results for 40-GHz double-drift silicon IMPATTs on diamond heat sinks.

TABLE IV

SUMMARY OF 50–75-GHz IMPATT RF LIFE TEST

Number of diodes	Running time
2	8 yr 1 month
2	7 yr 2 months
4	5 yr 2 months
1	2 yr 10 months (failed)

ures. A test of ten sample 94-GHz pulsed diodes operating with 3-A pulses, 100-nsec pulsewidth, and a 40-kHz pulse repetition frequency has produced no failures after 24,000 hr.

Nine single-drift IMPATT diodes have been life tested operating as cw free-running oscillators producing nominally 100 mW in the 50–75-GHz band with 2.5% efficiency. The junction temperatures have been maintained at 200°C. Eight of the nine diodes are still operating, and a summary is contained in Table IV.

Five double-drift IMPATT diodes on diamond heat sinks are being life tested as cw free-running oscillators producing nominally 1.3 W at 40 GHz with 11% efficiency. The junction temperatures are being maintained at 250°C. After seven months one diode failed due to a power outage. The failed diode was replaced, and the test has been going on for 17 months with no additional failures.

IV. Summary and Conclusions

Traveling-wave tubes have been advanced in design and in manufacturing techniques so that multikilowatt average power levels with efficiencies exceeding 30% can be obtained over the 30–100-GHz band. No drastic departures from proven lower-frequency (1–10-GHz) tube technology are required so that similar amplifier signal quality and reliability can be expected. Primarily this is due to the excellent suitability of the coupled-cavity slow-wave structure for use at millimeter wavelengths.

IMPATT diodes (with circuit combining) show promise of producing both pulsed and average power levels in the 30–100-GHz band that will make vacuum tube transmitters unnecessary for many applications. IMPATTs are used as laboratory instrumentation sources and in short-range radars in this frequency band today. Gallium arsenide Gunn sources produce modest power levels with low noise over the 30–100-GHz band but will not work at all much above 100 GHz. Thus a critical future need is to develop low-noise solid-state sources capable of tens of milliwatts for operating at 100 GHz and higher frequencies.

REFERENCES

Aston, R. (1979). *IEEE Trans. Microwave Theory Tech.* **27,** 479–482.
Bayuk, F. J., and Raue, J. E. (1977). Final Rep. AFAL-TR-77-16, Contract F33615-74-C-1051. TRW, Redondo Beach, California.
Blum, F. A., and Kramer, N. B. (1970). *IEEE Trans. Electron Devices* **17,** 983–986.
Brewer, G. R. (1957). *IRE Trans. Electron Devices* **4,** 132–140.
Brewer, G. R. (1967). *In* "Focusing of Charged Particles" (A. Septier, ed.), Vol. 2, pp. 23–72. Academic Press, New York.
Brillouin, L. (1945). *Phys. Rev.* **67,** 260–266.
Canali, C., Majni, G., Minder, R., and Ottaviani, G. (1975). *IEEE Trans. Electron Devices* **22,** 1045–1047.
Carslaw, H. S., and Jaeger, J. C. (1959). "Conduction of Heat in Solids." Oxford Univ. Press, London and New York.
Chang, K., and Ebert, R. L. (1980). *IEEE Trans. Microwave Theory Tech.* **28,** 295–305.
Chang, K., Ebert, R. L., and Sun, C. (1979). *Electron. Lett.* **15,** 403–405.
Cutler, C. C. (1956). *Bell Syst. Tech. J.* **35,** 841–876.
De Loach, B. C. (1970). *IEEE Trans. Microwave Theory Tech.* **18,** 72–74.
Forster, D. C. (1968). *Adv. Microwaves* **3,** 301–346.
Gilden, M., and Hines, M. E. (1966). *IEEE Trans. Electron Devices* **13,** 169–175.
Grant, J. E., and Christensen, J. A. (1979). Unpublished data, Hughes Aircraft Co.
Gummel, H. K., and Blue, J. L. (1967). *IEEE Trans. Electron Devices* **14,** 569–580.
Gummel, H. K., and Scharfetter, D. L. (1966). *Bell Syst. Tech. J.* **45,** 1797–1827.
Harp, R. S., and Stover, H. L. (1973). *IEEE Int. Solid-State Circuits Conf. Digest,* 118–119.
Kramer, N. B. (1976). *IEEE Trans. Microwave Theory Tech.* **24,** 685–693.
Kramer, N. B. (1979). *1979 NTC Conference Proceedings,* November 1979, Washington, D.C.
Kuno, H. J., and English, D. L. (1973). *IEEE Trans. Microwave Theory Tech.* **21,** 703–706.
Kurokawa, K., and Magalhaes, F. M. (1971). *Proc. IEEE* **59,** 102–103.
Langmuir, I., and Blodgett, K. (1924). *Phys. Rev.* **24,** 49–59.
Ma, Y., and Sun, C. (1980). *IEEE MTT-S International Microwave Symposium Digest,* 65–66.
Mendel, J. T., Quate, C. F., and Yocum, W. H. (1954). *Proc. IRE* **42,** 1–11.
Midford, T. A., and Bernick, R. L. (1979). *IEEE Trans. Microwave Theory Tech.* **27,** 483–492.
Misawa, T. (1970a). *Solid-State Electron.* **13,** 1363–1368.
Misawa, T. (1970b). *Solid-State Electron.* **13,** 1369–1374.
Misawa, T. (1972). *Solid-State Electron.* **15,** 457–465.
Pierce, J. R. (1940). *J. Appl. Phys.* **11,** 548–554.
Pierce, J. R. (1950). "Traveling-Wave Tubes." Van Nostrand, Princeton, New Jersey.
Read, W. T. (1958). *Bell Syst. Tech. J.* **37,** 401–446.
Ruch, J. G., and Kino, G. S. (1967). *Appl. Phys. Lett.* **10,** 40.
Rucker, C. T., Hill, G. N., Amoss, J. W., Harris, H. M., Cox, N. W., and Covington, D. W. (1978). Final Rep. AFAL-TR-78-63, Contract F33615-75-C-1020. Georgia Institute of Technology, Atlanta, Georgia.
Scharfetter, D. L., and Gummel, H. K. (1969). *IEEE Trans. Electron Devices* **16,** 64–77.
Scharfetter, D. L., Evans, W. J., and Johnston, R. L. (1970). *Proc. IEEE* **58,** 1131–1133.
Seidel, T. E., Niehaus, W. C., and Iglesias, D. E. (1974). *IEEE Trans. Electron Devices* **21,** 523–531.

CHAPTER 5

Dielectric Waveguide-Type Millimeter-Wave Integrated Circuits

Tatsuo Itoh

*Department of Electrical Engineering
and Electronics Research Center
The University of Texas at Austin
Austin, Texas*

I.	INTRODUCTION	199
II.	DIELECTRIC WAVEGUIDES FOR INTEGRATED CIRCUITS	201
	A. *Forms of Dielectric Waveguides*	201
	B. *Comparison of Millimeter-Wave and Optical Dielectric Waveguides*	205
III.	PROPAGATION CHARACTERISTICS OF DIELECTRIC WAVEGUIDES	207
	A. *Phase Constant and Field Distributions in Dielectric Waveguides*	207
	B. *Attenuation Characteristics*	227
	C. *Some Recent Developments in Theoretical Analysis*	236
IV.	PASSIVE COMPONENTS MADE OF DIELECTRIC WAVEGUIDES	243
	A. *Directional Couplers*	243
	B. *Resonators and Filters*	245
	C. *Nonreciprocal Devices*	250
V.	ACTIVE COMPONENTS	253
	A. *Phase Shifters*	254
	B. *Oscillators*	257
	C. *Mixers and Self-Oscillating Mixers*	260
VI.	ANTENNAS FOR DIELECTRIC MILLIMETER-WAVE INTEGRATED CIRCUITS	262
	A. *Surface-Wave Antennas and Arrays*	262
	B. *Leaky-Wave Antennas*	265
VII.	SUBSYSTEMS	269
VIII.	CONCLUSIONS	271
	REFERENCES	271

I. Introduction

Although dielectric waveguide structures have been in existence for some time (Chandler, 1949), only very recently extensive research interest and developmental efforts became widespread in connection with

ISBN 0-12-147704-5

fiber and integrated optical techniques. Low-loss fibers and functional devices in integrated forms are either available or being developed at a rapid pace. Numerous theoretical and experimental studies have been carried out for understanding the characteristics of various dielectric waveguide structures (Miller *et al.*, 1973). Once considered a dream, the optical communication is now a reality, due in large part to the availability of low-loss dielectric waveguides.

At much lower microwave frequencies, the past ten years have seen the development of integrated circuit approaches (Young and Sobol, 1974). Instead of relying on the traditional rectangular or circular waveguides, more and more microwave systems are made of microwave integrated circuit (MIC) components. They are more compact in size, reliable, and cost effective than the waveguide counterparts. Availability of better solid-state devices and planar technology accelerated the industry-wide trend to switch to MIC approaches where practical. In these MIC structures, several different types of printed transmission lines are used, including microstrip lines, coplanar waveguides, and slot lines. Characteristics of these transmission lines have been investigated by a large number of workers.

Let us now discuss the situation for the spectral range located between the two described above. In terms of frequency, this range corresponds to roughly 35 GHz through 30 THz and spans the millimeter wave to far infrared. This is the transition region of microwave and optical technologies. In the lower millimeter-wave spectrum 35–250 GHz, MIC technologies have been extended to apply, and an impressive list of successful component and system developments is ever growing. However, as the operating frequencies get higher, i.e., as we approach optical spectrum, the extension of conventional MIC technology is expected to meet increasing difficulties. Structural dimensions become too small, the ohmic loss increases as the metal is no longer a good conductor, and the tolerance for surface roughness gets extremely severe. Problems of dispersion and higher-order modes in printed transmission lines become pronounced.

Dielectric waveguide-type millimeter-wave integrated circuits have been introduced for circumventing the difficulties the conventional MIC technology is expected to encounter at higher frequencies. The basic idea is to replace microstrip and other printed transmission lines with dielectric waveguides appropriately designed for circuit integrations. The integrated circuit structures, therefore, resemble somewhat those used in integrated optics.

The cross-sectional dimensions of single-mode dielectric waveguides are typically of the order of the wavelength, whereas they are required to be about one tenth of the wavelength in the case of microstrip lines. The wave-propagation mechanism in dielectric waveguides does not depend on the existence of metallic conductors, but rather on the total internal reflection at the dielectric boundaries. Hence due to the absence of conductors the transmission loss in such structures can be made smaller than in printed lines. Instead of conductor losses, the main contribution to the wave attenuation comes from the dielectric material loss and radiation loss. The latter is caused by the surface irregularity of the waveguide, or generated at bends and junctions. Therefore, at the junction of a solid-state device and a dielectric waveguide, quite complicated wave interactions take place. The success of dielectric waveguide techniques largely depends on the correct understanding of such phenomena and subsequent suppression or prevention of them.

In what follows, we first review several methods for characterizing the wave-propagation phenomena in a number of dielectric waveguides developed for millimeter-wave integrated circuits in the past several years. Next, we describe various components developed to date, such as directional couplers, resonators, antennas, oscillators, and mixers. Several subsystems made of dielectric waveguide-type millimeter-wave integrated circuits are presented. Problems to be investigated will be indicated where applicable in the text.

II. Dielectric Waveguides for Integrated Circuits

A. Forms of Dielectric Waveguides

Several different types of dielectric waveguides have been investigated for millimeter-wave integrated circuit applications. The most frequently used ones are the rectangular rod waveguide and its variants (see Fig. 1). The original version [Fig. 1a] is made of a rectangular dielectric rod placed in free space (Jacobs and Chrepta, 1974). The structure in Fig. 1b is called the image guide (Toulios and Knox, 1970; Knox and Toulios, 1970) and consists of the dielectric rod placed on a large ground plane. The image of the rod due to the ground plane makes the propagation characteristics of certain guided modes identical to those of the rectangular rod waveguide in Fig. 1a; thus the wave "image guide" originated.

In many cases the image guide is more advantageous in circuit integration than the rod waveguide. First, it is placed on the ground plane that makes all the waveguides at the same vertical location. In addition, the presence of the ground plane is convenient as a heat sink and dc bias return when solid-state devices are mounted in the circuit. Another advantage of the image guide over the rod guide is that in the latter there are two

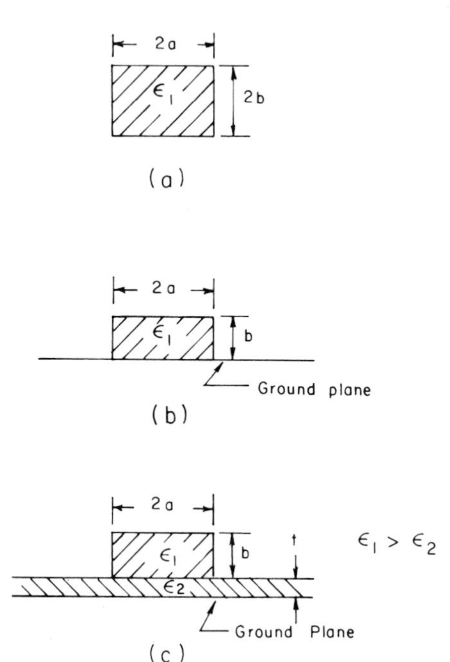

FIG. 1 Cross sections of typical dielectric waveguides for millimeter-wave integrated circuits. (a) Rod guide, (b) image guide, (c) insulated image guide.

almost degenerate lowest-order modes propagating whereas in the former such degeneracy is resolved.

There are, however, several practical problems associated with the introduction of the ground plane in the image guide. One is the loss caused by adhesive materials that bond the dielectric rod on the ground plane. Such materials are necessary to hold the rod in place and especially to fix the relative positions of more than one waveguide used in such structures as the directional coupler. However, most adhesive materials available have poor loss characteristics at microwave and millimeter-wave frequencies and contribute higher attenuations of the modal fields in the waveguide. In addition, the existence of the ground plane itself increases the wave attenuation due to the conductor loss.

Another problem is caused by the air gaps created between the dielectric rod and the ground plane. Because of surface irregularities caused in the manufacturing process, these gaps are often unavoidable. In the section of image guide where the gap is located, the propagation constant differs from the one where no gap is present. There, random gaps seriously affect the propagation characteristics of the waveguide. One way to circumvent the gap and adhesive problem is to deposit a conducting film directly on the underside of the image guide rod. However, such a process requires an extremely smooth surface finish on the dielectric rod. Obviously, such a process is relatively expensive.

It has been reported that the introduction of a thin dielectric layer with a lower dielectric constant between the rectangular rod and the ground plane can reduce the conductor loss. Such a structure [see Fig. 1c] may be called an insulated image guide or "insular" guide and has been extensively studied (Knox, 1976; McLevige *et al.*, 1975). In typical insular guide structures, alumina or other ceramic material is used as the rod, whereas a thin film of low-dielectric material such as Teflon or polyethylene is used as an insulating layer.

Let us now turn our attention to the dielectric waveguide structures in Fig. 2 in which the wave-guiding mechanism is completely different from that in Fig. 1. The structure in Fig. 2a is called the strip dielectric (SD) waveguide; whereas the one in Fig. 2b is the inverted strip dielectric (IS) waveguide (McLevige *et al.*, 1975; Itoh, 1976). In both structures, the electromagnetic energy is concentrated in the layered region, which has the highest dielectric constant ε_2. In the sideward direction, the presence of the dielectric strip ε_1 makes the field confined near the central region of the waveguide. Hence the actual wave propagation takes place in that portion of the ε_2 layer immediately below (SD guide case) or above (IS guide case) the dielectric strip.

These waveguides are, therefore, somewhat similar to optical strip waveguide (Furuta *et al.*, 1974). However, in the present structures, a large ground plane is provided for convenience of installing solid-state devices since it can be used as dc bias return and/or heat sink. The conductor loss in the ground plane is small since most of the energy propagates in the layer away from the ground plane.

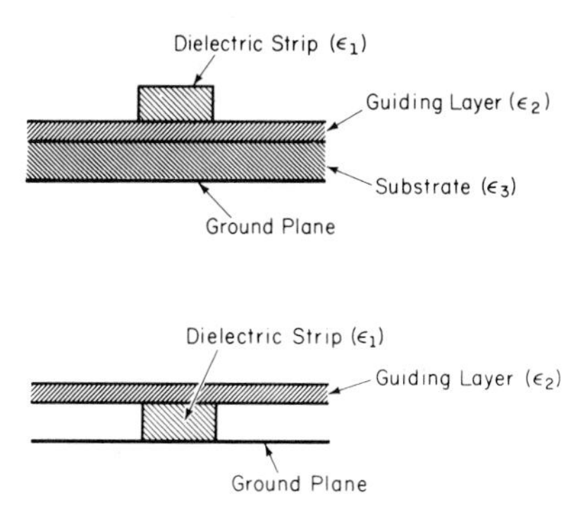

FIG 2 Strip-type dielectric waveguides. (a) Strip dielectric waveguide ($\varepsilon_2 > \varepsilon_1$), (b) inverted strip dielectric waveguide ($\varepsilon_2 > \varepsilon_1$). (From Itoh, 1976. Reprinted from *IEEE Transactions on Microwave Theory and Techniques* **24,** 821–827. © 1976 IEEE.)

These waveguides have certain advantages and disadvantages when compared with image-guide-type structures. In the SD and IS structures, radiation loss due to surface roughness is smaller because the dielectric strips are used only for concentrating the electromagnetic energy and the majority of the field lies in the layered medium ε_2. It is relatively easy to construct a layered structure with extremely smooth surfaces. On the other hand, in other dielectric waveguides the energy propagates in the rod. Hence the surface roughness of its side walls enhances radiation losses. The major disadvantage lies in the fact that the SD and IS waveguides are essentially so-called weakly guided structures. Compared with the image guide, the field is not tightly confined in the waveguide in the SD and IS structures. Hence the radiation loss at the bending section is larger in the weakly guided structures, such as SD and IS waveguides, than in the image guides. In any open waveguide, the radiation loss is inherently present at the bending section (Lewin *et al.*, 1977). Therefore, it is important in the circuit design to use waveguides that are less radiative or to choose the circuits containing fewer bends and make bending radii larger.

It has been proven that at lower millimeter-wave frequencies the IS guide is more practical in many respects than the SD guide. In the IS guide only two dielectric regions are involved as compared with three in the SD. This may reduce the dielectric loss caused in the substrate in Fig. 2a. Further, in the IS structure one can avoid bonding of two dielectric materials. Once the dielectric strip is securely placed on the ground plane, the guiding layer is placed on top of it without the use of adhesive materials at the interface of two dielectric media. Instead, it is held securely by the mechanical pressure between the guiding layer and the ground plane. The elimination of bonding material from the wave-guiding region is useful from the point of view of reducing the attenuation caused by the bonding material, which is usually quite lossy and further reduces the restriction on the choice of dielectric materials. Note that the strip can be bonded, if necessary, to the ground plane without significantly affecting the loss characteristics because the field near the ground plane is small due to the nature of the waveguide.

There are still other types of novel waveguide structures for millimeter waves. Two of them, the fence guide (Tischer, 1979) and the H-guide (Harris *et al.*, 1978), are given in Fig. 3. The former consists of two rows of periodically located pins perpendicularly on the ground plane. The wave propagates between the fencelike rows of pins with the electric field parallel to the pins. The H-guide consists of a thin dielectric film placed vertically between two parallel plates. The structure is closed in the vertical direction and open in the sideward direction. The film thickness is

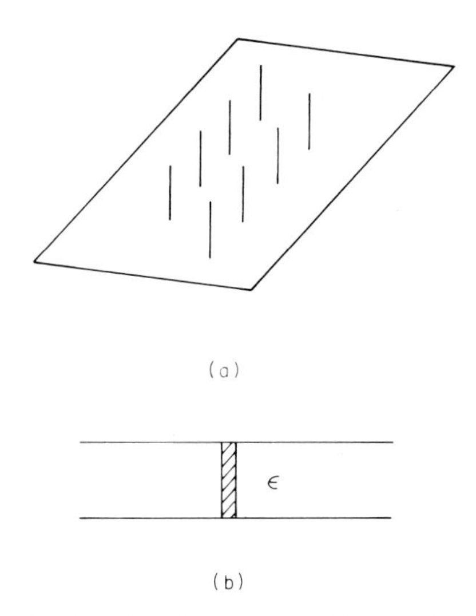

FIG. 3 Waveguides for millimeter waves. (a) Fence guide, (b) H guide.

made small so that most of the energy, though bounded near the film, propagates in the air, and hence the attenuation is kept extremely small.

B. Comparison of Millimeter-Wave and Optical Dielectric Waveguides

Although the idea of using certain dielectric waveguides for millimeter-wave integrated circuits is similar to the situation in integrated optics, there are a number of differences in structures and applications in these two spectral regions. A typical rectangular dielectric waveguide in integrated optical circuits is buried in a substrate having the dielectric constant ε_2 only incrementally lower than that in the core ε_1. The purpose of having ε_2 only slightly smaller than ε_1 is to allow for single-mode operation while the width of the waveguide is several wavelengths. If the ratio of ε_1 to ε_2 is large, the single-mode operation requires the waveguide dimension of the order of one-half wavelength. Such a structure is extremely difficult to fabricate as the dimension becomes much less than a micrometer. Optical waveguides are made of glass, semiconductors such as GaAs or crystals ($LiNbO_3$, etc.), and are formed using planar techniques such as sputtering, ion implantation, diffusion, ion beam, and etching.

In the millimeter-wave integrated circuits, the ratio of the core dielectric constant ε_1 to that of the surrounding medium is usually large. For instance, in a typical image guide the rod is made of high-permittivity material ($\varepsilon_1/\varepsilon_0 = 3.5 \sim 30$) and is surrounded by air. The materials used for the millimeter-wave structures include high-purity alumina, boron nitride, beryllium oxide, and high-resistivity semiconductors. The semiconductor materials such as silicon and GaAs have certain promising features that cannot be obtained in a passive dielectric material. For instance, some active devices may be built *in situ* (Jacobs and Chrepta, 1974) or some nonlinear interaction can be effected (Thompson and Coleman, 1974). The fabrication methods of dielectric waveguides include machining and injection molding.

The higher dielectric constant ratio $\varepsilon_1/\varepsilon_2$ between the guide and surrounding media is practical in millimeter-wave application because the guide wavelengths and the waveguide dimensions are not unreasonably small for economical manufacturing. The additional advantage is that the field is much more strongly confined and the use of much smaller radius of curvature (in wavelength) may be permitted without unacceptable level of radiation loss.

As will be discussed later, there are two families of guided modes in the rectangular dielectric waveguide. E^y_{mm} modes have the principal electric field in the vertical y direction; whereas in E^x_{mn} modes the principal electric field is in the sideward x direction. Here the combination of m and n specifies the mode order. The dominant E^y_{11} and E^x_{11} modes are in general degenerate, and hence at any imperfections and discontinuities, they may couple. In the image guide, however, the presence of the ground plane resolves the degeneracy because the strongest electric field component of the E^x_{11} mode that would have been existing at the center of the rod guide is shorted out, and the single-mode operation over a considerable frequency range can be attained. Of course, the introduction of the image

TABLE I

TYPICAL PROPERTIES OF MICROWAVE AND OPTICAL DIELECTRIC WAVEGUIDES[a]

	Microwave	Optical
Dielectric materials (ε_1)	Alumina, semiconductors	Glass, semiconductors
Dielectric materials (ε_2)	Plastic	Glass, semiconductors
Dielectric constant (ε_1)	10–15 or higher	2–4; 12
Dielectric constant (ε_2)	2.5	2–4; 12
Index ratio $\varepsilon_1/\varepsilon_2$	2	1.1–1.01
Waveguide width (in λg)	0.5	2–10
Degenerate E^x_{11} mode suppressed?	Yes	No
Radius of curvature (in λg)	2–5	30–1000

[a] ε_1 is for the core, and ε_2 is for the substrate. From Knox, 1976. Reprinted from *IEEE Transactions on Microwave Theory and Techniques* **24,** 806–814. © 1976 IEEE.

plane is practical only at the microwave and millimeter-wave frequencies. At optical frequencies, the use of metals must be avoided as they become lossy materials.

Table I (Knox, 1976) summarizes comparison between dielectric waveguides for optical and millimeter-wave integrated circuits. In the table, essentially the insular guide is compared with the optical rectangular waveguide. Some, but not all, of the features in Table I apply to the IS waveguide. In this waveguide, the presence of the ground plane and the fabrication process are similar to those for the insular guide, whereas the ratio of dielectric constants is usually lower and is more similar to optical structure.

III. Propagation Characteristics of Dielectric Waveguides

In the design of millimeter-wave integrated circuits, propagation characteristics of the transmission lines are of primary importance. The phase and attenuation constants and often the field distribution in the waveguide need to be known for design of passive and active components. Theoretical and experimental analyses for the waveguide structures will be reviewed in this section. Typical data useful for design are also reproduced.

A. Phase Constant and Field Distributions in Dielectric Waveguides

Dielectric waveguides shown in Figs. 1 and 2 belong to so-called nonseparable geometrics, and no closed form solutions are available for analysis of propagation characteristics. A number of approximate techniques have been developed for the past several years. All of these methods require numerical calculations to obtain the phase constants and field distributions. Some of them are relatively simple and others more complex in numerical processing. For dispersion characteristics of the rectangular or image guides, all of the methods provide accurate answers at higher frequencies where the field is well guided. At lower frequencies and near the cutoff of guided modes, some methods do not provide data as accurate as others do.

Guided modes in the structures in Figs. 1 and 2 are hybrid in nature, i.e., all six components of electromagnetic fields (three for electric and three more for magnetic fields) exist. These modes can be classified into E^y_{pq} and E^x_{pq} families. In the former, the principal electric field component is in the y direction (see Fig. 4 for coordinate orientation); whereas in the latter such a component is in the x direction. The subscripts p and q indicate the modal order with respect to the x and y directions. Note that for a

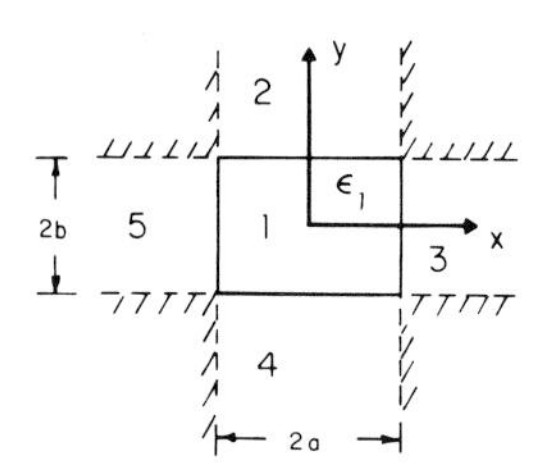

FIG. 4 Cross sections of a rectangular dielectric waveguide for analysis.

given frequency there exist only a finite number of guided modes. Therefore, when a guided mode is scattered by an object or a discontinuity in the waveguide, the scattering phenomena cannot be described by this finite number of guided modes only. The complete description requires the inclusion of the radiation mode having a continuous spectrum. This situation is in contrast to the closed metal waveguide in which an infinite number of guided modes, both propagating and evanescent, can describe the scattering phenomena completely. In the following, we only consider the guided modes and review several different techniques for analysis of these modes.

1. *Marcatili's Method*

This method has been developed for the "well-guided" modes in an optical rectangular rod waveguide (Marcatili, 1969). In the waveguide cross section, most of the guided field is strongly concentrated in the rod with the relative dielectric constant ε (see Fig. 4). Since the guided modes exist by virtue of the total internal reflection at each side wall, the field decays in an exponential fashion as you move away from the side wall in each region 2, 3, 4, or 5. The fields in the shaded four-corner regions must be even less. Consequently, only a small error should be introduced in the calculation of dispersion characteristic if one completely neglects the fields in these regions and does not match the fields at the boundaries between shaded regions and regions 2–5. Such an approximation drastically simplifies the boundary value problem associated with the structure. We now can assume simple field distributions in regions 1–5 and match the fields only along the periphery of the dielectric rod.

Hybrid field components in each region can be expressed in terms of two scalar potentials ϕ^e and ϕ^h as (McLevige *et al.*, 1975)

$$E_x = \frac{1}{\varepsilon_r(y)} \frac{\partial^2 \phi^e}{\partial y \partial x} + \omega\mu k_z \phi^h, \tag{1a}$$

$$E_y = \frac{1}{\varepsilon_r(y)} \left(k_z^2 - \frac{\partial^2}{\partial x^2} \right) \phi^e, \tag{1b}$$

$$E_z = \frac{-jk_z}{\varepsilon_r(y)} \frac{\partial \phi^e}{\partial y} - j\omega\mu \frac{\partial \phi^h}{\partial x}, \tag{1c}$$

$$H_x = -\omega\varepsilon k_z \phi^e + \frac{\partial^2 \phi^h}{\partial y \partial x}, \tag{1d}$$

$$H_y = \left(k_z^2 - \frac{\partial^2}{\partial x^2}\right) \phi^h, \tag{1e}$$

$$H_z = j\omega\varepsilon \frac{\partial \phi^e}{\partial x} - jk_z \frac{\partial \phi^h}{\partial y}, \tag{1f}$$

where ε_r is the relative dielectric constant in the respective region, k_z the phase constant of the guided wave, and ε and μ are free-space permittivity and permeability. The time factor $\exp(j\omega t)$ and the z dependence $\exp(-jk_z z)$ are suppressed in the expression. Since E_{pq}^y modes have principal E- and H-field components in the y and x directions, respectively, ϕ^e has the dominant contribution to the modal field. Similarly for E_{pq}^x modes, the principal field components are E_x and H_y, and ϕ^h has the dominant contribution to the modal field. Hence we can set $\phi^h = 0$ in (1) and write the solution of E_{pq}^y modes or set $\phi^e = 0$ and obtain the solution for E_{pq}^x modes. Since the solution processes of both families of modes are similar, we present the one for E_{pq}^y modes based on Marcatili's approximation. For E_{pq}^y mode, we let $\phi^h = 0$ and choose ϕ^e as

$$\begin{aligned}
\phi^e &= A_1 \cos(k_x x + \alpha) \cos(k_y y + \beta), && \text{region } 1 \\
&= A_2 \cos(k_x x + \alpha) \exp(-\eta_2 y), && 2 \\
&= A_3 \exp(-\xi_3 x) \cos(k_y y + \beta), && 3 \\
&= A_4 \cos(k_x x + \alpha) \exp(\eta_4 y), && 4 \\
&= A_5 \exp(\xi_5 x) \cos(k_y y + \beta), && 5
\end{aligned} \tag{2}$$

where α and β are some constants as yet unknown,

$$\varepsilon_1 k_0^2 - k_y^2 = \varepsilon_2 k_0^2 + \eta_2^2 = \varepsilon_4 k_0^2 + \eta_4^2 = k_z^2 + k_x^2, \tag{3a}$$

$$\varepsilon_1 k_0^2 - k_x^2 = \varepsilon_3 k_0^2 + \xi_3^2 = \varepsilon_5 k_0^2 + \xi_5^2 = k_z^2 + k_y^2, \tag{3b}$$

and k_0 is the free-space wave number given by $\omega\sqrt{\varepsilon\mu}$. After obtaining field components in each region, we match tangential field components along the interface and obtain two eigenvalue equations

$$2k_y b = q\pi - \tan^{-1}(\varepsilon_2 k_y/\varepsilon_1 \eta_2) - \tan^{-1}(\varepsilon_4 k_y/\varepsilon_1 \eta_4), \quad q = 1, 2, \ldots, \tag{4a}$$

$$2k_x a = p\pi - \tan^{-1}(k_x/\xi_3) - \tan^{-1}(k_x/\xi_5), \quad p = 1, 2, \ldots. \tag{4b}$$

The phase constant in the z (axial) direction is given by

$$k_z = (\varepsilon_1 k_0^2 - k_x^2 - k_y^2)^{1/2}, \tag{5}$$

where k_x and k_y are the solutions of Eqs. (4b) and (4a). Note that Eqs. (4a) and (4b) are independent of each other. In fact, they correspond to eigenvalue equations for the TM and TE modes of two-dimensional slab waveguides (Marcatili, 1969). This is a direct consequence of the assumption that the fields in the shaded area are neglected and Eq. (2) is used for the rest of the regions.

We note that the identical technique applies to the solution of image guide structures. However, we have to consider the boundary condition at the surface of the image (ground) plane. The dominant E_{11}^y mode is unaffected by the presence of the image plane. The E_{11}^x mode, on the other hand, cannot exist in the image guide, because the E_x field must be zero on the image plane and E_{11}^x mode has its E_x field maximum there. As we discussed before, E_{11}^y and E_{11}^x are degenerate, and such degeneracy is resolved by the introduction of the image plane. It is then possible to operate the image guide under the single-mode condition and to simplify the design of various waveguide components described later.

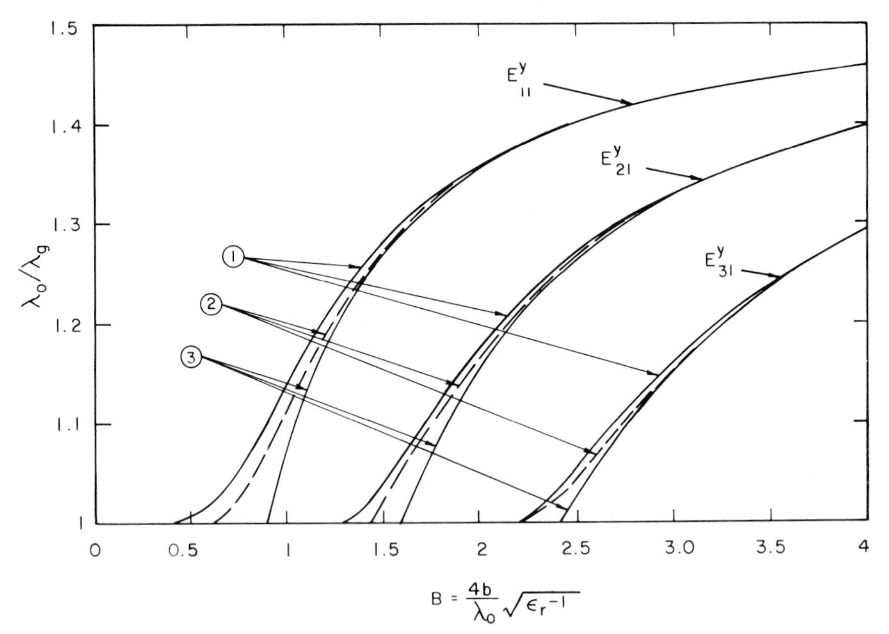

FIG. 5 Guided wavelength for various modes versus normalized height B of the guide ($\varepsilon_r = 2.25$, $a/b = 1$). 1, theory; 2, Goell's theory; 3, Marcatili's theory (Knox and Toulios, 1970).

Marcatili's method is used, and the dispersion characteristics of dielectric waveguides are computed from Eqs. (4) and (5). Typical results are shown in Fig. 5 (Knox and Toulios, 1970). The results are compared with those by Goell (1969), who used a point-matching technique. Both results agree well at higher frequencies where the field is well guided or tightly bound to the waveguide. Discrepancy occurs at lower frequencies or near the cutoff. Actually Marcatili's method cannot predict the fact that the dominant mode E_{11}^{y} has no cutoff. This inability is caused by Marcatili's basic assumption that the fields in the shaded areas in Fig. 4 are negligible because the mode is tightly bound to the waveguide. At low frequencies or near the cutoff, such an assumption does not hold and the fields in the shaded areas cannot be neglected any more.

In spite of the deficiency described above, the work by Marcatili should not be discredited. The method has been designed for high-frequency operation for which the mode is strongly bound to the waveguide. In such frequency regions, the method is one of the simplest to formulate and produces extremely accurate answers to the dispersion problem. Obtaining more accurate solutions near cutoff and at lower frequencies requires more elaborate and complicated analytical processing, as we shall see in what follows in this section.

2. *Effective Dielectric Constant Technique*

It has been shown that the method developed by Marcatili (1969) is not accurate enough at low frequencies and near the cutoff of a particular mode, because the waves in the shaded area in Fig. 4 are neglected in analysis. The effective dielectric constant (EDC) method can circumvent such deficiencies, though in an approximate fashion. The technique was first introduced by Knox and Toulios (1970) for the image guide solution and later by McLevige *et al.* (1975) and Itoh (1976) for more general waveguide structures. We shall first review the original version as applied to the image guide and then introduce more general application. By doing so, we can illustrate not only the difference from the Marcatili method, but also the fact that the EDC method is actually the first step of the more rigorous transverse resonance method discussed later.

Knox and Toulios (1970) recognized that Eqs. (4a) and (4b) are the eigenvalue equations of TM and TE modes in slab waveguides and are independent of each other. They devised a modification in which two eigenvalue equations are coupled. Their analysis starts with Eq. (4a) with the dispersion relation [Eq. (3a)]. Instead of solving Eqs. (4b) and (3b) independently, they first introduced the effective dielectric constant ε_{eff} defined by

$$\varepsilon_{\text{eff}} k_0^2 = \varepsilon_1 k_0^2 - k_y^2 = k_{z\text{slab}}^2. \tag{6}$$

The effective dielectric constant can be thought of as that of a hypothetical medium in which the phase velocity of the plane wave is identical to that of the surface wave in a slab in Fig. 6a. Having defined the effective dielectric constant, they solved Eq. (4b) under the condition

$$\varepsilon_{\text{eff}} k_0^2 - k_x^2 = \varepsilon_3 k_0^2 + \xi_3^2 = \varepsilon_5 k_0^2 + \xi_5^2, \tag{7}$$

in place of Eq. (3b). Equation (4b) used with Eq. (7) corresponds to the TE mode equation of a hypothetical slab waveguide shown in Fig. 6b. Since ε_{eff} in Fig. 6b contains the information obtained in solving the structure in Fig. 6a, two equations [(4a) and (4b)] are now coupled. The phase constant k_z of the original rod waveguide is given by

$$k_z = (\varepsilon_{\text{eff}} k_0^2 - k_x^2)^{1/2}. \tag{8}$$

Numerical data generated by the present method by Knox and Toulios (1970) are plotted in Fig. 5 for comparison with those computed by Marcatili's (1969) and Goell's (1969) methods. Results clearly indicate marked differences from Marcatili's (1969) method near the cutoff and at lower frequencies and are more realistic. In the EDC method, fields in the shaded area are not completely neglected, but are taken into account in the solution process, though in an approximate manner, via the use of the effective dielectric constant.

Knox (1976), McLevige *et al.* (1975), and Itoh (1976) extended the ap-

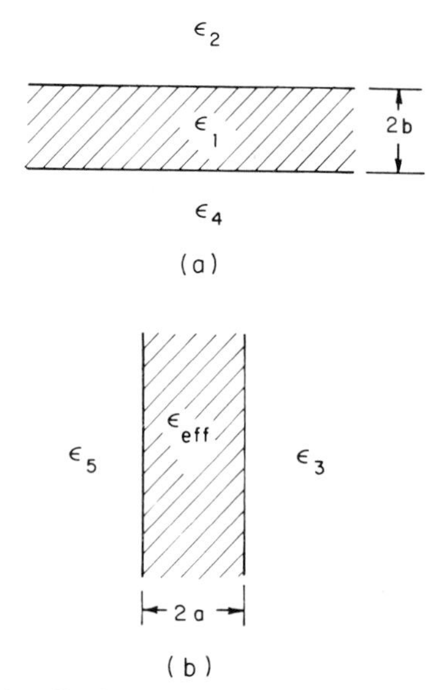

FIG. 6 Procedure for the effective constant method. (a) Vertical direction; (b) sideward direction with a slab having effective dielectric constant ε_{eff} ($\varepsilon_{\text{eff}} = \varepsilon_1$ in Marcatili's method).

plication of EDC methods to more general waveguide structures. Specifically, the phase constants of the "insular" and the inverted strip (IS) dielectric waveguides have been computed by the EDC method. We shall present the analysis for the IS waveguide because this structure is completely different from the rectangular rod or image guide, and hence Marcatili's approach cannot be applied even as the crudest approximation. On the other hand, the insular guide is a simple modification of the image guide and is amenable to Marcatili's method.

Figure 7 shows the cross section and the solution process of a coupled IS waveguide structure. An isolated IS guide can be recovered by letting the separation $x_3 - x_2$ infinity. In Fig. 7a, if each of regions I–V is infinitely long in the x direction, five slab waveguides are obtained. Regions I,III, and V have a single slab raised from the ground plane by a distance of h (Fig. 7b), whereas, regions II and IV are double-layered waveguides backed by a ground plane (Fig. 7c). The phase constants in these slab

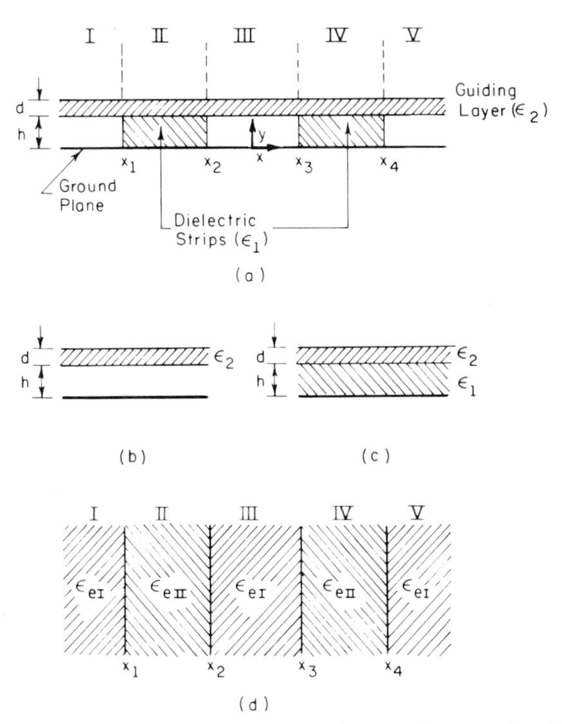

FIG. 7 Coupled inverted strip dielectric waveguide. (a) Cross section; (b) structures for analyzing y variations in I, III, and V regions; (c) structures for analyzing y variations in II and IV regions; (d) structure for analyzing x variations using concept of effective dielectric constants. (From Itoh, 1976. Reprinted from *IEEE Transactions on Microwave Theory and Techniques* **24,** 821–827. © 1976 IEEE.)

waveguides can be obtained by solving transcendental eigenvalue equations that are obtained by matching the tangential electric and magnetic fields at each interface. Both of these structures can then be replaced by hypothetical homogeneous regions in which the phase constants of the plane wave are identical to those of the surface waves in the slab structures. These hypothetical media are assumed to have "effective" dielectric constants ε_{eI} (for regions I, III, and V) and ε_{eII} (II and IV).

The effective dielectric contant ε_{eII} may be obtained from the scalar potential defined by

$$
\begin{aligned}
\phi^e(y) &= \cosh \eta_1 y \qquad (0 < y < h), \\
&= \left\{\cosh \eta_1 h \cos[k_y(y-h)] + \frac{\eta_1\varepsilon_2}{k_y\varepsilon_1} \sinh \eta_1 h \sin[k_y(y-h)]\right\} \\
&\qquad (h < y < h+d), \\
&= \left\{\cosh \eta_1 h \cos[k_y d] + \frac{\eta_1\varepsilon_2}{k_y\varepsilon_1} \sinh \eta_1 h \sin[k_y d]\right\} \\
&\qquad \times \exp[-\eta_3(y-h-d)] \qquad (y > h+d),
\end{aligned} \tag{9}
$$

where the normalization constant is omitted. Since Eq. (9) already incorporates continuity of H_x at each interface, the eigenvalue equation is obtained by matching E_z at each interface

$$
\begin{aligned}
&\eta_3\varepsilon_2 k_y\varepsilon_1 \cosh[\eta_1 h] \cos[k_y d] + \eta_3\varepsilon_2^2\eta_1 \sinh[\eta_1 h] \sin[k_y d] \\
&\quad - k_y^2\varepsilon_1 \cosh[\eta_1 h] \sin[k_y d] + k_y\varepsilon_2\eta_1 \sinh[\eta_1 h] \cos[k_y d] = 0.
\end{aligned} \tag{10}
$$

The transverse wave numbers k_y, η_3, and η_1 are related via

$$k_0^2 + \eta_3^2 = \varepsilon_2 k_0^2 - k_y^2 = \varepsilon_1 k_0^2 + \eta_1^2. \tag{11}$$

The effective dielectric constant becomes

$$\varepsilon_{eII} = \varepsilon_2 - (k_y^2/k_0^2). \tag{12}$$

The corresponding quantity ε_{eI} can be similarly obtained.

Having obtained ε_{eII} and ε_{eI}, we replace regions I, III, V and II, IV by hypothetical vertical slabs with the effective dielectric constants ε_{eI} and ε_{eII}, respectively (Fig. 7d). We now derive the eigenvalue equation for this hypothetical structure by matching the tangential fields E_y and H_z at each interface. These fields may be derived from the scalar potential

$$
\begin{aligned}
\phi^e(x) &= A \exp[\xi(x - x_1)], \qquad (x < x_1), \\
&= B^c \cos[k_x(x - x_1)] + B^s \sin[k_x(x - x_1)], \\
&\qquad (x_1 < x < x_2), \\
&= C^c \cosh[\xi(x - x_2)] + C^s \sinh[(x - x_2)], \\
&\qquad (x_2 < x < x_3),
\end{aligned}
$$

$$= D^c \cos[k_x(x - x_3)] + D^s \sin[k_x(x - x_3)], \quad (x_3 < x < x_4),$$
$$= E \exp[-\xi(x - x_4)], \quad (x > x_4), \tag{13}$$

where

$$k_z^2 = \varepsilon_{eII} k_0^2 - k_x^2 = \varepsilon_{eI} k_0^2 + \xi^2. \tag{14}$$

The eigenvalue equation is solved for k_x, and the phase constant k_z is obtained from Eq. (14). We assume that the value of k_z thus obtained is used as the phase constant of the original IS guide structures.

Figure 8 presents typical results of dispersion characteristics for both coupled and uncoupled (isolated) IS guide structures. Figure 9 shows the strength of E_y at $x = 3w$ and some y, e.g., y_0 and that of H_x at $x = 0$ on the ground plane in the uncoupled IS guide versus the ratio of the strip thickness h to the guiding layer thickness d for several values of d. The strength of E_y is normalized by that of E_y at $x = 0$ and $y = y_0$. These curves can be used as a measure of the guiding ability of the waveguide, because as the value decreases, more energy is concentrated toward the

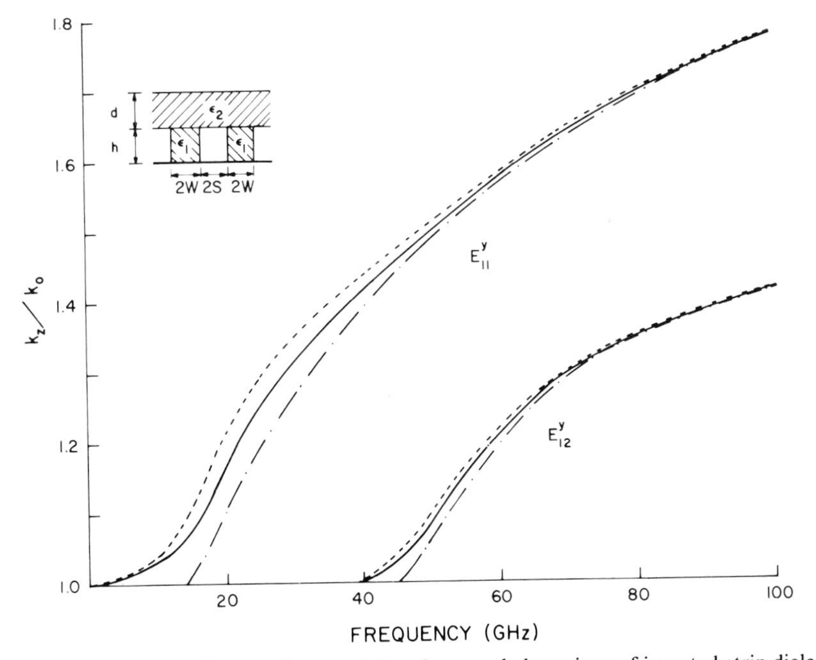

FIG 8 Dispersion diagrams for coupled and uncoupled versions of inverted strip dielectric waveguides. ---, even; –·–, odd; —— uncoupled guide; $\varepsilon_1 = 2.1$; $\varepsilon_2 = 3.8$; $s = 1$ mm; $w = 2$ mm; $h = d = 1.5875$ mm. (From Itoh, 1976. Reprinted from *IEEE Transactions on Microwave Theory and Technique* **24,** 821–827. © 1976 IEEE.)

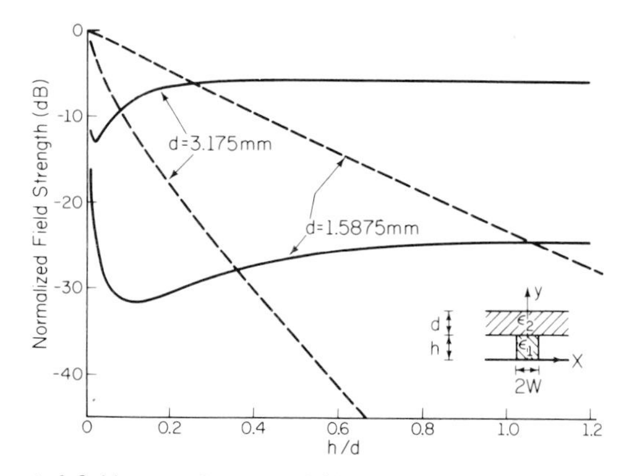

FIG. 9 Computed field strength versus h/d in an inverted strip dielectric waveguide. ——, $E_y(x = 3w)/E_y(x = 0)$; – –, $H_x(y = 0)/H_x$ max. $W = 2$ mm; $f = 81.7$ GHz; $\varepsilon_1 = 2.1$; $\varepsilon_2 = 3.8$. (From Itoh, 1976. Reprinted from *IEEE Transactions on Microwave Theory and Techniques* **24**, 821–827. © 1976 IEEE.)

center of the guide. The accuracy of the present method has been experimentally confirmed. Such experimental check will be summarized in the later part of this chapter.

3. *Variational Method*

This method developed by Pregla (1974) makes use of a variational formulation in conjunction with a field expression more rigorous than Marcatili's for solving rectangular dielectric waveguide structures. The problem is formulated for a coupled-waveguide structure shown in Fig. 10. Because of structural symmetries, either electric or magnetic walls can be inserted at the xz and yz planes to reduce the problem to that with one quarter of the cross section. The uncoupled waveguide can be obtained by taking an appropriate limit in the dimension.

A general variational expression for the phase constant k_z is given by

$$\begin{aligned} I = \iint \Bigg[&(k_z^2 - \varepsilon_r k_0^2)\left(|E_x|^2 + |E_y|^2 + \left|\frac{\partial E_z}{\partial x}\right|^2 + \left|\frac{\partial E_z}{\partial y}\right|^2 \right. \\ &+ \left|\frac{\partial E_y}{\partial x} - \frac{\partial E_x}{\partial y}\right|^2 - \varepsilon_r k_0^2 |E_z|^2 \\ &\left. - jk_z \left(E_x^* \frac{\partial E_z}{\partial x} - E_x \frac{\partial E_z^*}{\partial x} + E_y^* \frac{\partial E_z}{\partial y} - E_y \frac{\partial E_z^*}{\partial y} \right) \right] ds \\ = 0,& \end{aligned} \tag{15}$$

where the integration is over the entire cross section and the propagation factor $\exp(-jk_z z)$ is assumed. k_z^2 is stationary with respect to variations of

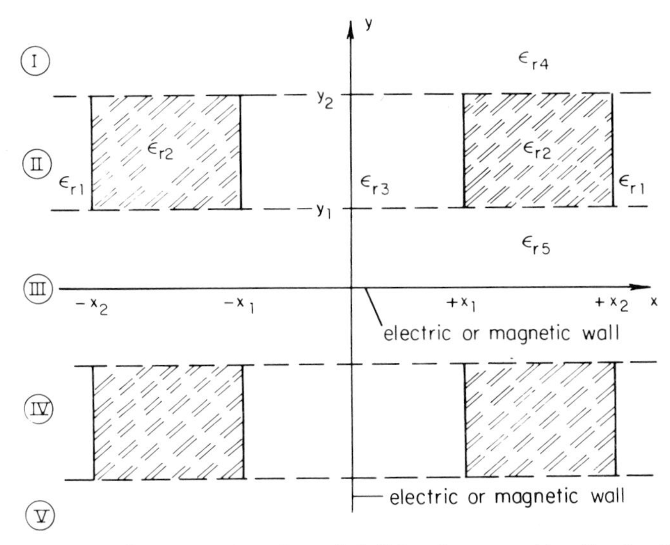

FIG. 10 General cross section of coupled-dielectric waveguides (Pregla, 1974).

the electric field. The next step is to find a suitable approximation for the electric field. k_z^2 obtained from Eq. (15) becomes more accurate if the approximate field is closer to the true-field distribution.

The waveguide cross section is now divided into five subregions (I–V in Fig. 10) and the x-directed scalar potentials $\phi_e(x, y)$ and $\phi_h(x, y)$ are used to describe the electromagnetic fields. In each region, ϕ_e and ϕ_h must satisfy

$$\frac{\partial^2 \phi_h}{\partial x^2} + \frac{\partial^2 \phi_h}{\partial y^2} + [\varepsilon_r(x) k_0^2 - k_z^2]\phi_h = 0, \tag{16a}$$

$$\frac{\partial^2 \phi_e}{\partial x^2} + \frac{\partial^2 \phi_e}{\partial y^2} + [\varepsilon_r(x) k_0^2 - k_z^2]\phi_e - \frac{\partial \varepsilon_r}{\partial x} \frac{1}{\varepsilon_r(x)} \frac{\partial \phi_e}{\partial x} = 0. \tag{16b}$$

Since each subregion is infinitely long in the x direction, we take Fourier transform via

$$\Phi_{e,h}(k_x, y) = \int_{-\infty}^{\infty} \phi_{e,h}(x, y) \exp(-jk_x x)\, dx. \tag{17}$$

Let us now consider all the regions in $y > 0$, beause the structure is either symmetric or antisymmetric. In region II, Eq. (16) becomes

$$\left(\frac{d^2}{dy^2} - \gamma_1^2\right) \Phi_h = -k_0^2 \int_{-x_2}^{x_2} \delta\varepsilon_r(x) \phi_h \exp(-jk_x x)\, dx, \tag{18a}$$

$$\left(\frac{d^2}{dy^2} - \gamma_1^2\right) \Phi_e = -k_0^2 \int_{-x_2}^{x_2} \delta\varepsilon_r(x)\phi_e \exp(-jk_x x)\, dx$$
$$+ \sum_{j=-2}^{2} d\varepsilon_r(x) \left[\frac{1}{\varepsilon_r(x)} \frac{\partial \phi_e}{\partial x}\right] \exp(-jk_x x_j)$$
$$= S(k_x, y), \qquad (x = x_j), \tag{18b}$$

where

$$\gamma_1^2 = k_z^2 + k_x^2 - \varepsilon_{r1} k_0^2, \tag{19}$$

$$\delta\varepsilon_r(x) = \varepsilon_r(x) - \varepsilon_{r1},$$
$$d\varepsilon_r(x_j) = \varepsilon_r(x_j + 0) - \varepsilon_r(x_j - 0). \tag{20}$$

In regions I and III, equations similar to Eq. (18) hold except that the right-hand sides are zero, as these regions are homogeneous in the x direction.

Instead of solving Eq. (18) exactly for ϕ_e and ϕ_h, we use these equations for obtaining better approximations from the assumed ϕ_e and ϕ_h substituted into the right-hand sides. For E_{pq}^x modes, we let $\phi_h = 0$, and the field given by Marcatili (1969) can be used for ϕ_e:

$$\phi_e = M \cos A(x - x_0) \cos B(y - y_0), \qquad (x_1 < |x| < x_2),$$
$$= M \frac{\cos A(x_0 - x_1)}{\begin{Bmatrix} \cosh \alpha_3 x_1 \\ \sinh \alpha_3 x_1 \end{Bmatrix}} \begin{Bmatrix} \cosh \alpha_3 x \\ \sinh \alpha_3 x \end{Bmatrix} \cos B(y - y_0), \tag{21}$$
$$(|x| < x_2 \quad \text{or} \quad |x| < x_1),$$

where $\cosh \alpha_3 x$ and $\sinh \alpha_3 x$ are for the even and odd modes in x, respectively.

After obtaining refined expressions for the scalar potentials, we match the fields at the boundary between regions I–III. The completed fields are now substituted in the stationary expression (15). The results obtained by this method give better approximation near the cutoff and at low frequencies than those by Marcatili (1969). However, at higher frequencies both results agree well.

4. *Mode-Matching Techniques*

The methods presented so far in this section directly deal with eigenvalue problems for the open structures. Alternative techniques presented in this and the next subsections consider the open dielectric-waveguide structures as limiting cases of closed inhomogeneously filled waveguides. For instance, the image guide is obtained by letting d equal infinity in the

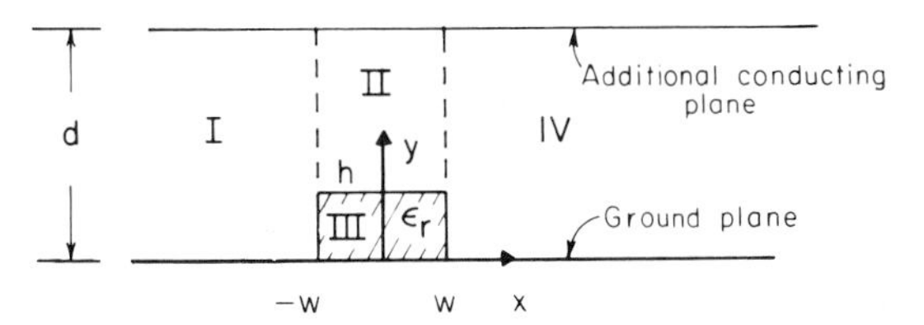

FIG. 11 Cross section of a dielectric waveguide with an additional conducting plane.

cross section shown in Fig. 11. Solbach and Wolff (1978) analyzed the image guide structure by formulating the closed cross section with a large d in terms of mode-matching technique. It is necessary to check how large d is needed so that the solution of the closed structure approximate the open image guide. A similar approach has been taken by Schlosser and Unger (1966).

The procedure for the solution of closed structures is as follows. First, a complete set of field expansions with unknown amplitude coefficients is derived for each subregion I–IV. It is assumed that the field dependence on the z coordinate in any region is described by $\exp(-jk_z z)$. These sets must be such that tangential electric field components on the metalic conductors vanish and all the field components decay exponentially for $|x| \to \infty$. Next, the tangential field components are matched at the interface between regions II and III. This process can be performed independently of the remaining interface conditions such as those between regions I and II. We now have complete sets of field descriptions in regions I, II, plus III and IV. The remaining task is to match the field along the two interfaces, $x = \pm w$, $0 < y < d$. When this is done, we obtain an infinite set of linear simultaneous equations for the unknown amplitude coefficients used to expand the fields in each region. The phase constant k_z is obtained by searching for the zero of the determinant of the truncated infinite set of equations. Once k_z is obtained, all the unknown amplitude coefficients can be computed and the field distributions of the mode corresponding to each k_z are derived.

The principal contribution by Solbach and Wolff (1978) is the intensive study of modal configurations. They compared E^y_{pq} and E^x_{pq} modes (in their paper referred to as EH_{pq} and HE_{pq} modes, respectively) with the modes in a circular dielectric rod or a dielectric slab depending on the aspect ratio w/h in Fig. 11. For instance, for $w \approx h$, the fundamental mode E^y_{11} corresponds to the dipole EH_{11} in the circular dielectric waveguide. Both modes converge into a plane wave for low frequencies and have no cutoff. The next two higher-order image guide modes with p, q = 2 and 1 correspond to the rotationally symmetric modes TE_{10} and TM_{10} in the circular dielectric waveguides.

For waveguides with a large w, the E^y_{11} mode is nearly transversely magnetic and resembles the TM_0 mode of the slab waveguide. Also, the image guide modes with $p = 1, 2, 3, \ldots$ and $q = 2, 3, 4, \ldots$ may be divided into two groups with TE or TM characteristics and called E^y_{pq} and E^x_{pq} modes.

In Marcatili's approximation, E_y and H_x are assumed much larger than the other field components in E^y_{pq} modes and E_x and H_y in E^x_{pq} modes. It is shown by Solbach and Wolff (1978) that this assumption is not valid in general, but is applicable only to the fundamental E^y_{11} mode of low-permittivity lines and arbitrary aspect ratio w/h or of high-permittivity lines of very low or very high aspect ratio.

5. *Generalized Telegrapher's Equation Method*

The technique introduced by Ogusu and Hongo (1977a) and Ogusu (1977) once again makes use of the closed boundary shown in Fig. 12. They approached the problem of inhomogeneously filled waveguide by means of the generalized telegrapher's equation developed by Schelkunoff (1952) and sought solutions for which the phase constant is $k_0 < k_2 < \varepsilon_{\max} k_0$, where $\varepsilon_{\max}$ is the relative dielectric constant of the highest-permittivity material. All the surface wave modes supported by the dielectric structure have phase constants in this range.

The transverse fields $\bar{E}_t$ and $\bar{H}_t$ may be expanded in terms of orthogonal mode functions $\bar{e}(x, y)$ and $\bar{h}(x, y)$ of the empty metallic waveguide as

$$\bar{E}_t = \sum_i^\infty V_{(i)}(z)\bar{e}_{(i)}(x, y) + \sum_j^\infty V_{[j]}(z)\bar{e}_{[j]}(x, y), \tag{22a}$$

$$\bar{H}_t = \sum_i^\infty I_{(i)}(z)\bar{h}_{(i)}(x, y) + \sum_j^\infty I_{[j]}(z)\bar{h}_{[j]}(x, y), \tag{22b}$$

where V and I are modal voltages and currents to be determined and the

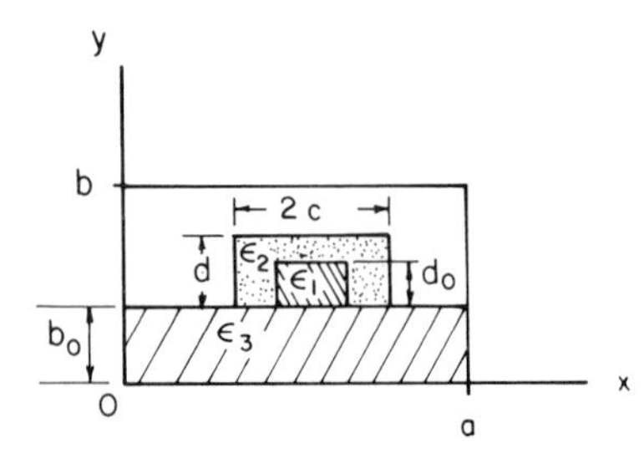

FIG. 12 Cross section of the generalized dielectric waveguide for analysis by the generalized telegrapher's equation.

subscript in parenthesis denotes the TM mode, whereas the one in bracket is for the TE mode. The indices i and j represent double index such as mn. Equations (22a) and (22b) are substituted in Maxwell's equation for the entire region in the waveguide $0 < x < a$ and $0 < y < b$ in Fig. 12. Next we take inner products of the resultant equations with the known mode functions $\bar{e}_{(i)}$, $\bar{h}_{(i)}$, $\bar{e}_{[j]}$, $\bar{h}_{[j]}$ for different values of i and j. The results are the following generalized telegrapher's equations:

$$\frac{dV_{(n)}(z)}{dz} = -\frac{1}{j\omega\varepsilon_0}\sum_i^{\infty} Z_1(n, i)\, I_{(i)}(z) - j\omega\mu I_{(n)}(z), \tag{23a}$$

$$\frac{dV_{[m]}(z)}{dz} = -j\omega\mu I_{[m]}(z), \tag{23b}$$

$$\frac{dI_{(n)}(z)}{dz} = -j\omega\varepsilon_0\left[\sum_i^{\infty} Y_1(n, i)V_{(i)}(z) + \sum_j^{\infty} Y_2(n, j)V_{[j]}(z)\right], \tag{23c}$$

$$\frac{dI_{[m]}(z)}{dz} = -j\omega\varepsilon_0\left[\sum_i^{\infty} Y_3(m, i)V_{(i)}(z) + \sum_j^{\infty} Y_4(m, j)V_{[j]}(z)\right] - \frac{k_{c[m]}^2}{j\omega\mu} V_{[m]}(z), \tag{23d}$$

where $k_{c(i)}$ or $k_{c[j]}$ is the cutoff wavenumber of the empty waveguide mode. The coefficients in Eq. (23) are given by

$$Z_1(n, i) = k_{c(i)}^2 k_{c(n)}^2 \iint \frac{1}{\varepsilon_r(x, y)} \phi_{(i)}(x, y)\phi_{(n)}(x, y)\, ds, \tag{24}$$

$$Y_1(n, i) = \iint \varepsilon_r(x, y)\bar{h}_{(i)}(x, y) \cdot \bar{h}_{(n)}(x, y)\, ds, \tag{25a}$$

$$Y_2(n, j) = \iint \varepsilon_r(x, y)\bar{h}_{[j]}(x, y) \cdot \bar{h}_{(n)}(x, y)\, ds, \tag{25b}$$

$$Y_3(m, i) = \iint \varepsilon_r(x, y)\bar{h}_{(i)}(x, y) \cdot \bar{h}_{[m]}(x, y)\, ds, \tag{25c}$$

$$Y_4(m, j) = \iint \varepsilon_r(x, y)\bar{h}_{[j]}(x, y) \cdot \bar{h}_{[m]}(x, y)\, ds, \tag{25d}$$

and $\phi_{(i)}$ satisfies

$$\left(\frac{\partial^2}{\partial x^2} + \frac{\partial^2}{\partial y^2} + k_{c(i)}^2\right) \phi_{(i)}(x, y) = 0. \tag{26}$$

We shall now consider the sinusoidal propagation such as $V_{(i)}(z) = V_{(i)} \exp(-jk_z z)$ and $V_{[j]}(z) = V_{[j]} \exp(-jk_z z)$. Substituting these expressions into the truncated version of Eq. (23), we obtain a set of linear algebraic homogeneous equations

$$\begin{bmatrix} a_{11} & \cdots & a_{1N} & b_{11} & \cdots & b_{1M} \\ \cdot & & \cdot & \cdot & & \cdot \\ \cdot & & \cdot & \cdot & & \cdot \\ \cdot & & \cdot & \cdot & & \cdot \\ a_{N1} & \cdots & a_{NN} & b_{N1} & \cdots & b_{NM} \\ c_{11} & \cdots & c_{1N} & d_{11} & \cdots & d_{1M} \\ \cdot & & \cdot & \cdot & & \cdot \\ \cdot & & \cdot & \cdot & & \cdot \\ \cdot & & \cdot & \cdot & & \cdot \\ c_{M1} & \cdots & c_{MN} & d_{M1} & \cdots & d_{MM} \end{bmatrix} \begin{bmatrix} V_{(1)} \\ \cdot \\ \cdot \\ \cdot \\ V_{(N)} \\ V_{[1]} \\ \cdot \\ \cdot \\ \cdot \\ V_{[M]} \end{bmatrix} = k_z^2 \begin{bmatrix} V_{(1)} \\ \cdot \\ \cdot \\ \cdot \\ V_{(N)} \\ V_{[1]} \\ \cdot \\ \cdot \\ \cdot \\ V_{[M]} \end{bmatrix}, \tag{27}$$

where the elements a_{ij}, b_{ij}, c_{ij}, and d_{ij} are expressed in terms of Z_1, Y_1, Y_2, Y_3, and Y_4 in Eqs. (24) and (25), and are known. It is a straightforward process to solve Eq. (27) for the eigenvalues k_z and the eigenvectors $V_{(1)}, \ldots, V_{(N)}, V_{[1]}, \ldots, V_{[M]}$. From these quantities we can compute all the field components of the guided waves.

The method presented in this subsection has two useful features. First, the profile of the dielectric waveguide cross section appears only in the computations of Z_1, Y_1, . . . , Y_4 in Eqs. (24) and (25). These quantities can be computed numerically even when the cross-sectional shape and the distribution of dielectric constant are arbitrary. Hence the method is adaptable to a wider class of dielectric waveguides with little modification of the computational algorithm. Second, since the process of deriving Eq. (27) is identified as Galerkin's method, the eigenvalue k_z derived in this method is stationary.

The accuracy of the present method was confirmed by comparison with other available data for the dispersion characteristics of the image guide as well as by experiments (Ogusu and Hongo, 1977b). The method has also been used for computing the field distributions in the cross sections of various types of dielectrical waveguides. Some examples are given in Fig. 13. Although this method cannot predict the deflection of electric field lines at the dielectric interfaces due to the nature of the formulation, general pictures as to where the field is concentrated and how it decays can be clearly identified. Such a knowledge of course is useful in the design of integrated circuits based on dielectric waveguides.

6. *Some Experimental Analyses*

The most widely used method for measuring the guide wavelength and the cross-sectional field distributions in the dielectric waveguide is by means of an electric probe (Knox and Toulios 1970; McLevige *et al.*, 1975; Ogusu and Hongo 1977b; Solbach, 1978b). An electric probe may be

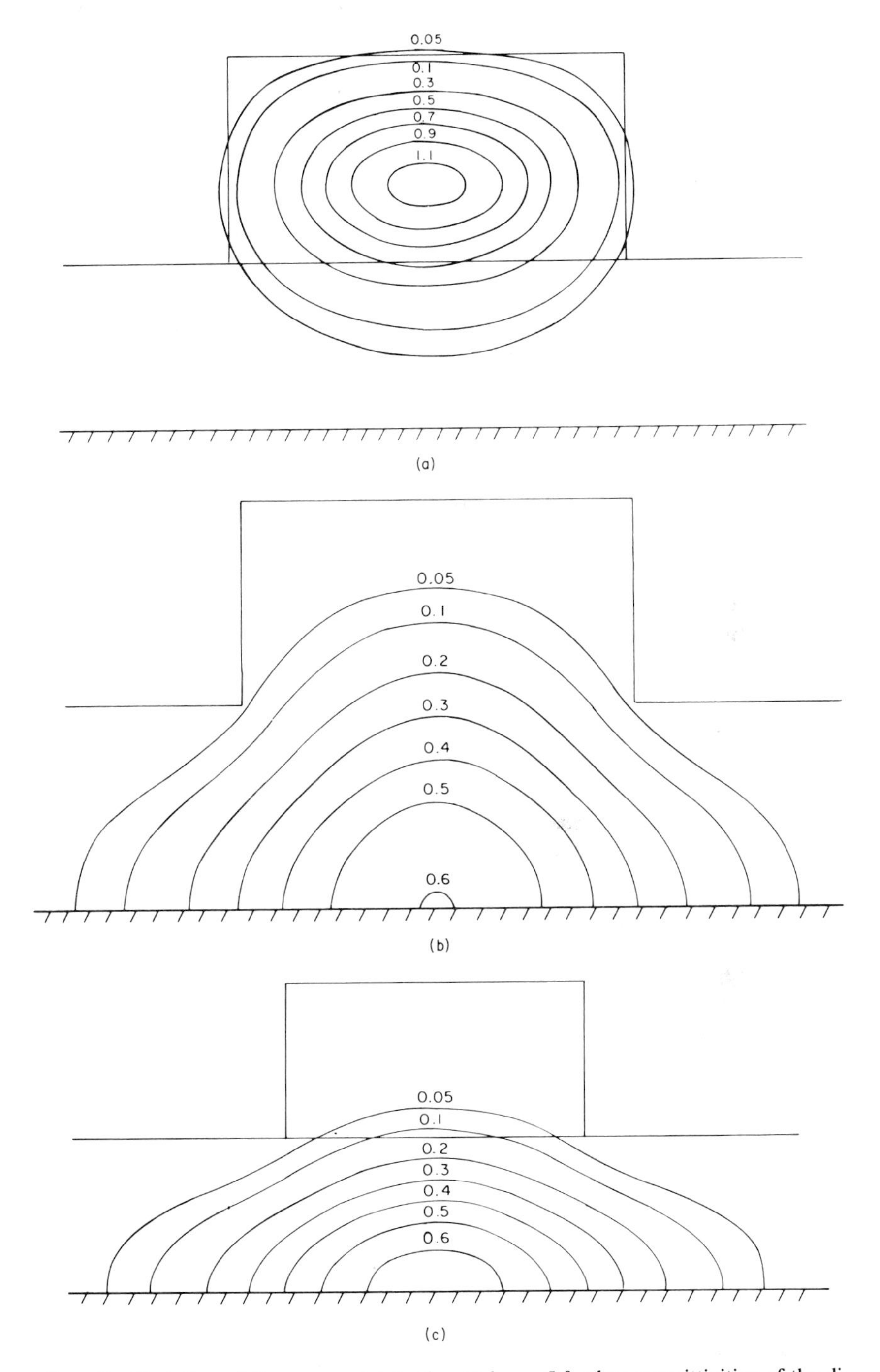

FIG. 13 Variation of the power distribution at $k_o c = 5.0$ when permittivities of the dielectric strip and substrate are changed. $\varepsilon_1 = \varepsilon_2$, $b_o/c = 1.0$, and $d/c = 1.0$. (a) $\varepsilon_1 = 3.0$, $\varepsilon_3 = 2.5$; (b) $\varepsilon_1 = \varepsilon_3 = 3.0$; (c) $\varepsilon_1 = 2.75$, $\varepsilon_3 = 3.0$. (From Ogusu, 1977. Reprinted from *IEEE Transactions on Microwave Theory and Techniques* **25,** 874–885. © 1977 IEEE.)

made of a thin semirigid coaxial cable by removing the short length of the outer conductor from its open end.

The guide wavelength may be obtained by measuring the standing wave pattern. This can be easily accomplished by moving the probe in the axial direction along the waveguide, provided the probe is lightly coupled to the waveguide field so that the latter is not heavily perturbed (see Fig. 14). From the distance of adjacent minima or adjacent maxima of the standing wave patterns, the guide wavelength can readily be obtained. Dispersion characteristics may be obtained by measuring the guide wavelengths for changing frequencies.

An alternative method used by Ogusu and Hongo (1977b) is to use a movable short to establish a standing wave pattern (see Fig. 15). In their method, a probe is inserted from the back side of the ground plane. The guide wavelength was derived from the distance of travel of the movable short and the number of peaks in readings of the standing wave meter.

Some resonant structures may also be used for determining the relation between the guide wavelength and the frequency. For instance, a ring resonator made of high-dielectric image guide resonates only at frequencies for which the total circumferential length is an integral multiple of the guide wavelength (Knox and Toulios, 1970). There are two drawbacks in this method. First, once a resonator is built, only at certain discrete resonant frequencies can we measure the guide wavelength. On the other hand, the frequency can be continuously varied in the probe method. Second, it is not clear what is the correct circumferential length of the ring. If the radius of curvature is relatively large, such a distance may be well approximated by 2π times the arithmetic mean of outer and inner radii of the ring. Also, the guide wavelength in the ring resonator is, strictly speaking, different from that in the straight section of the dielectric waveguide due to the radiation effect inherently associated with the curved section of the open dielectric waveguide. The curvature effect will be discussed later in this chapter.

A dielectric waveguide Gunn oscillator was used by Jacobs *et al.* (1976)

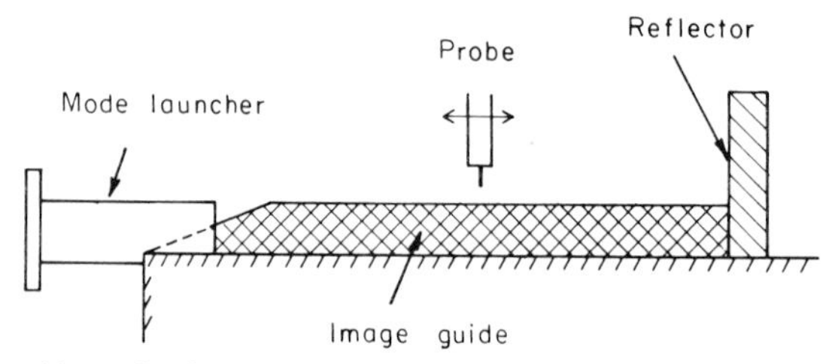

FIG. 14 Movable probe for detecting the standing wave patterns in the short-circuited image guide.

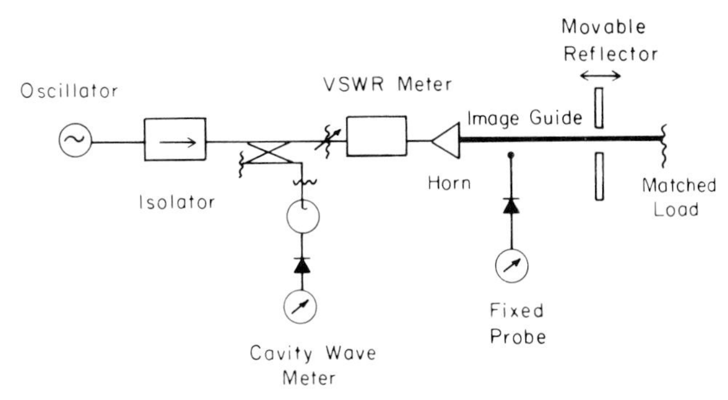

FIG. 15 Block diagram of setup for measuring dispersion characteristics (Ogusu and Hongo, 1977b).

for measurement of guide wavelength in a rectangular image guide. The Gunn oscillator is made of a section of rectangular image guide and a Gunn diode implanted in a hole drilled in a guide as shown in Fig. 16. The former creates a cavity resonator. By adding a small slice of image guide section at the end of the cavity, the length of the cavity is altered and the oscillation frequency and the output power change. By successively adding dielectric pieces, power and frequency variation versus the added length can be plotted as shown in Fig. 17. The frequency variation repeats itself at every added length equal to the guide wavelength. Therefore, from Fig. 17, a dispersion curve may be obtained.

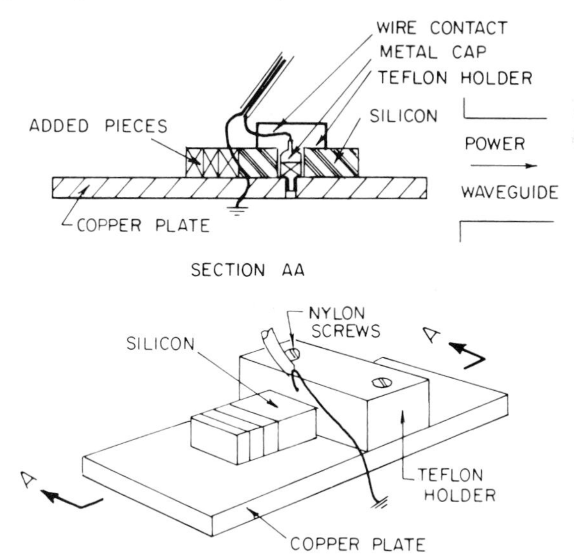

FIG. 16 Dielectric cavity structure, 18.0 mm long. (From Jacobs *et al.*, 1976. Reprinted from *IEEE Transactions on Microwave Theory and Techniques* **24,** 815–820. © 1976 IEEE.)

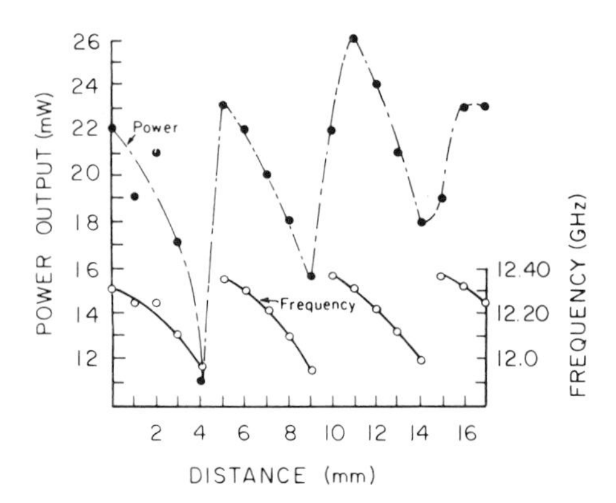

FIG. 17 Power and frequency as a function of added resonator length, with no tuning disk added: $\lambda g = 10.0$ mm. (From Jacobs *et al.*, 1976. Reprinted from *IEEE Transactions on Microwave Theory and Techniques* **24,** 815–820. © 1976 IEEE.)

The field distributions in the waveguide cross section can be measured with a movable coaxial probe used for guide wavelength measurements or with a pinhole-type probe (Itoh, 1976). By moving a coaxial probe across the waveguide, the field distributions can be measured in a straightforward manner. It is, however, not convenient to detect the field distribution inside the dielectric portion with this type of probe. For detecting such a field, the pinhole-type probe is more convenient as demonstrated by Itoh (1976). This probe is made of a pinhole created in a large conducting plane that is used as a short for the truncated waveguide (see Fig. 18). Under the single-mode operation, the shorting plane does not disturb

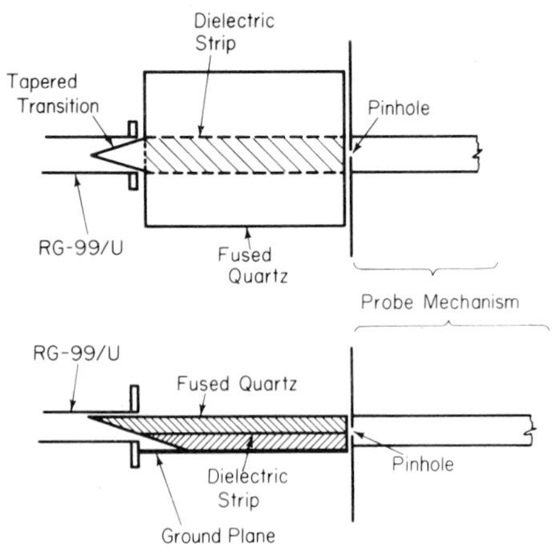

FIG. 18 Arrangement of the setup for measuring field distributions in a dielectric waveguide. (From Itoh, 1976. Reprinted from *IEEE Transactions on Microwave Theory and Techniques* **24,** 821–827. © 1976 IEEE.)

the field distribution in the cross section. The field leakage from the pinhole on the shorting plane can be made quite small so as not to perturb the field distribution. The strength of the leakage field is measured while the pinhole probe is moved in the transverse directions. This process results in the field distributions in the cross section. The fields inside the dielectric portion of the waveguide are as accessible as those outside.

An example of comparisons between the measured and computer dispersion relations is shown in Fig. 19 and can be used for checking the analytical method for the dispersion characteristics. Similar results have also been reported by a number of authors. Figures 20 and 21 are the measured field distributions by the coaxial and the pinhole probes. For comparison purposes, thoretical results are also plotted. Once again, the experimental results are useful for checking the accuracy of the analytical method. Once the accuracy of analytical methods is checked, these methods may be used for design of various components with confidence.

B. ATTENUATION CHARACTERISTICS

There are two kinds of loss mechanisms associated with dielectric waveguide structures. One is due to dissipation in the materials of which the waveguide is composed, and the other is by way of radiation due to mechanical imperfections of the waveguide structure. In addition, when a dielectric waveguide is bent, a portion of energy carried by the guided wave is lost by radiation. Although one can decrease the amount of this radiation loss by an appropriate choice of waveguide structures and mate-

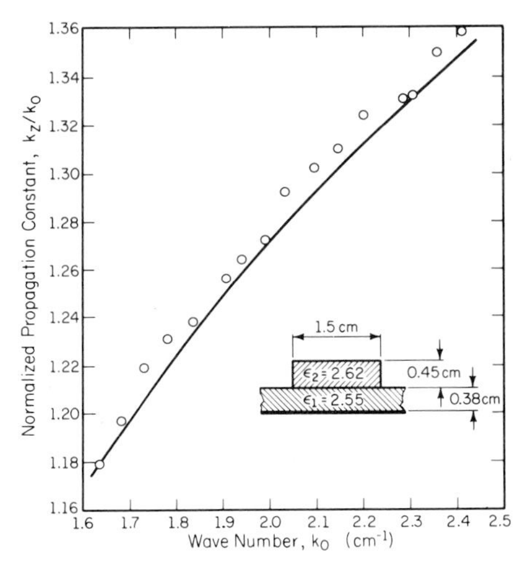

FIG. 19 Dispersion characteristics of the insulated image guide. ——, theory; ○, experiment. (From McLevige *et al.*, 1975. Reprinted from *IEEE Transactions on Microwave Theory and Techniques* **23**, 788–794. © 1975 IEEE.)

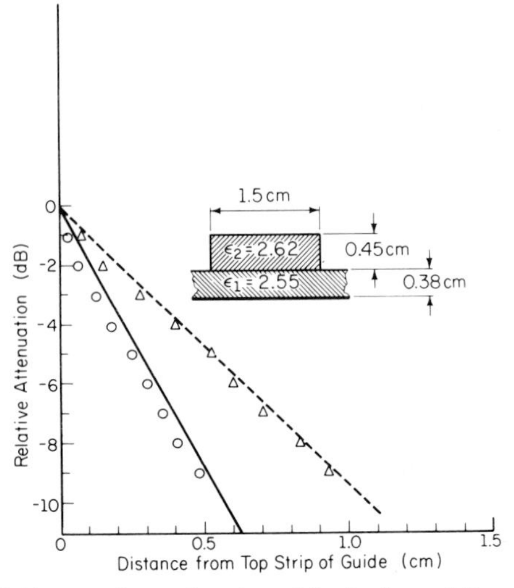

FIG. 20 Electric field strength as a function of the horizontal distance from the side wall of the insulated image guide. At 8 GHz: – –, theory; △, experiment. At 11 GHz: ——, theory; ○, experiment. (From McLevige *et al.*, 1975. Reprinted from *IEEE Transactions on Microwave Theory and Techniques* **23,** 788–794. © 1975 IEEE.)

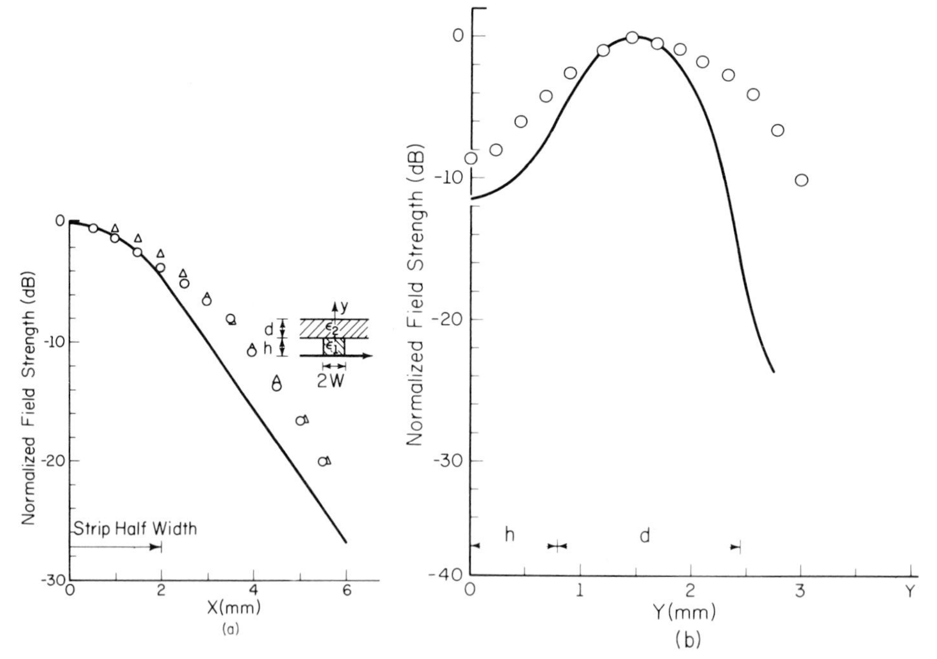

FIG. 21 Computed and measured field distributions at 81.7 GHz; $d = 1.5875$ mm, $h = 0.794$ mm, $w = 2$ mm, $\varepsilon_1 = 2.1$, and $\varepsilon_2 = 3.8$. (a) Sideward direction. (b) vertical direction; ——, computed; ○, measured $(+x)$; △, measured $(-x)$. (From Itoh, 1976. Reprinted from *IEEE Transactions on Microwave Theory and Techniques* **24,** 821–827. © 1976 IEEE.)

rials, it is not possible to eliminate completely this phenomenon, as it is inherent to the curved dielectric waveguide. In this section we pay our attention to the loss mechanism and the attenuation constant in the dielectric waveguides.

1. *Material Losses*

The dissipation in the dielectric material causes the guided wave to attenuate as it propagates in the waveguide. Since, in practice, good dielectric materials with small dissipation factors are used, the attenuation due to the dielectric material is small. This permits us to derive the attenuation constant from the same eigenvalue equations as those used in computing the phase constant in lossless waveguides, except that the relative dielectric constants in the equations are now complex quantities.

Toulios and Knox (1970) calculated attenuation constant due to dielectric losses in the image guide from the eigenvalue equations (4a) and (4b), which become

$$F(k_y, \varepsilon_r) = bk_y + \tan^{-1}(k_y/\varepsilon_r\eta) - \tfrac{1}{2}q\pi = 0, \tag{28a}$$

$$G(k_x) = ak_x + \tan^{-1}(k_x/\xi) - \tfrac{1}{2}p\pi = 0, \tag{28b}$$

if $\eta = \eta_2 = \eta_4$, $\xi = \xi_3 = \xi_5$, $\varepsilon_1 = \varepsilon_r$, and $\varepsilon_1 = \varepsilon_2 = \varepsilon_4 = \varepsilon_5 = 1$ are introduced. For a complex dielectric constant

$$\bar{\varepsilon}_r = \varepsilon_r(1 - j\tan\delta), \tag{29}$$

where $\tan\delta$ is the loss tangent of the dielectric material. The variables k_x, ξ, k_y, and η all become complex and are represented by $\bar{k}_x$, $\bar{\xi}$, $\bar{k}_y$, and $\bar{\eta}$. It is assumed that these quantities satisfy the eigenvalue equations

$$F(\bar{k}_y, \bar{\varepsilon}_r) = (\bar{k}_y - k_y)F_{k_y} + (\bar{\eta} - \eta)F_\eta + (\bar{\varepsilon}_r - \varepsilon_r)F_{\varepsilon_r} + \cdots = 0, \tag{30a}$$

$$G(\bar{k}_x) = G(k_x) + (\bar{k}_x - k_x)G_{k_x} + (\bar{\xi} - \xi)G_\xi + \cdots = 0, \tag{30b}$$

where F_i and G_i are partial derivatives with respect to i. It is reasonable to assume that to a first-order approximation the imaginary parts of $\bar{k}_x$, $\bar{k}_y$, $\bar{\xi}$, and $\bar{\eta}$ are sufficiently smaller than k_x, k_y, ξ, and η. These assumptions have been used to show that the complex propagation constant is

$$\bar{k}_z = k_z - j\alpha_d, \tag{31}$$

$$\alpha_d/\alpha = (\lambda_g/\lambda)\sqrt{\varepsilon_r}(1 - L_1 - L_2), \tag{32}$$

where α is the attenuation constant in infinitely extending medium and is defined by

$$\alpha = (\pi\sqrt{\varepsilon_r}\tan\delta)/\lambda_0. \tag{33}$$

and

$$L_1 = k_x^2(1 - L_2)/[(k_x^2 + \xi^2)(1 + k_x)], \tag{34a}$$

$$L_2 = [k_y^2(\varepsilon_r + 2\eta^2/k_0^2)/b]/[\eta(k_y^2 + \varepsilon_r^2\eta^2) + \varepsilon_r(k_y^2 + \eta^2)]. \tag{34b}$$

Solbach (1978a) used a similar approach though the eigenvalue equation used is the one described in Section III.A.4. Furthermore, instead of closed-form solutions by the Taylor expansion method used by Toulios and Knox (1970), the complex roots of the complex eigenvalue equation have been sought numerically. The results obtained by his method indicate that the presence of dielectric loss not only causes the attenuation, but changes values of the phase constant k_z from the lossless case. However, when $\tan \delta < 10^{-2}$, such changes in k_z are found negligible.

Let us now turn our attention to the attenuation caused by the conductor loss in the image guide. In works of both Toulios and Knox (1970) and Solbach (1978a), the attenuation due to conductor loss in the ground plane is obtained by the perturbation approach traditionally used for computing the wall loss in a closed rectangular waveguide. In this type of approach, the field distributions in the cross section of the lossless waveguide are used to find the surface-current distribution on the conductor as well as the total power transmitted by the guided wave. This method is valid whenever the conductivity of the metal conductors is reasonably high. The attenuation constant due to conductor loss in the dielectric waveguide with a ground plane is

$$\alpha_c = \frac{\text{power lost per unit length due to conductors } (W_c)}{2 \times \text{power transmitted } (2\ W_t)}, \tag{35}$$

where

$$W_c = \frac{1}{2} R_s \int \{|H_x|^2 + |H_z|^2\}\, dl, \tag{36}$$

and the integral is over the ground plane in the cross section. R_s is the surface resistance of the conductor and is given by

$$R_s = (\omega\mu_0/2\sigma)^{1/2} \quad (\sigma = \text{conductivity}); \tag{37}$$

also, for E^y modes

$$W_t = \frac{Z_0}{2} \int |H_x|^2\, da. \tag{38}$$

Z_0 is the wave impedance of the guided mode, and the integral is over the entire cross section of the waveguide.

Some numerical results are reproduced in Figs. 22 and 23 as functions

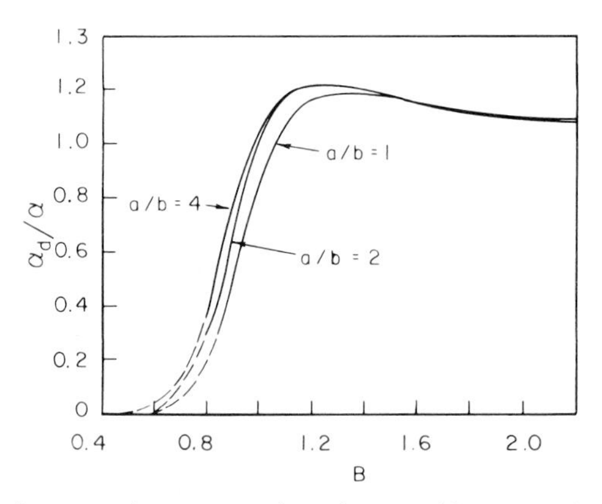

FIG. 22 Dielectric attenuation constant in an image guide (E^y_{11} mode, $\varepsilon_r = 9$): α = infinite medium loss, $\varepsilon_r = 9$ (Toulios and Knox, 1970).

of the normalized frequency $B = (4b/\lambda_0)(\varepsilon_r - 1)^{1/2}$ and the aspect ratio a/b. These results are for the dominant E^y_{11} mode in the image guide.

Experimental checks have been performed by measuring the loaded Q of an extremely lightly loaded image guide resonator (Toulios and Knox, 1970). The latter is made of a section of an image guide terminated with a large shorting flange. The excitation and measurement have been performed through extremely small holes on shorting flanges.

Since the resonator is extremely lightly loaded, the measured loaded Q is practically equal to the unloaded Q, say Q_0, which is related to the total attenuation constant $\alpha_t = \alpha_c + \alpha_d$ via

$$\alpha_t/\alpha = (\lambda_0/\lambda_g)/Q_0 \tan \delta\sqrt{\varepsilon_r} \tag{39}$$

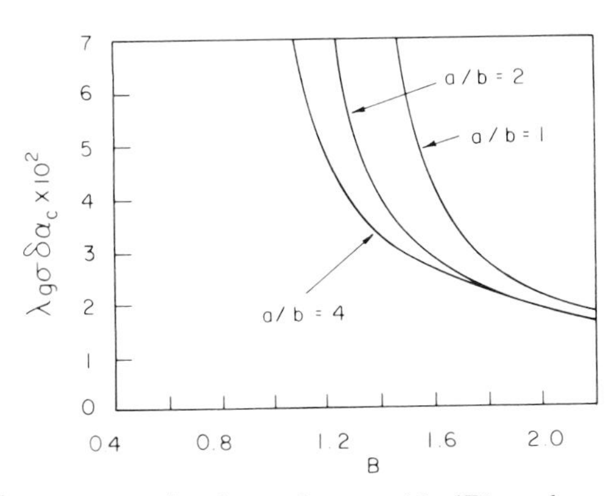

FIG. 23 Conductor attenuation for an image guide (E^y_{11} mode, $\varepsilon_r = 9$) (Toulios and Knox, 1970).

where α is the same as the one used in Eq. (32). Toulios and Knox (1970) reported that agreement of their theory and experiment was better than $\pm 10\%$.

Instead of measuring the unloaded Q of the cavity, the standing wave patterns can also be used for measuring the attenuation constant in the dielectric waveguide (Solbach, 1978b). The tapered end of the image guide is inserted into an open-ended metal waveguide, which is connected to a millimeter-wave source. The standing wave is established by the reflection from a large shorting flange placed at the truncated end of the image guide. The voltage standing wave ratio detected by a coaxial probe (see Fig. 24) is given by

$$\text{VSWR} = [1 + e^{-\tau} \exp(-2\alpha_t z)]/(1 - e^{-\tau} \exp(-2\alpha_t z)], \tag{40}$$

where z is the distance of the probe from the shorting flange, τ the ohmic loss in the shorting flange, and α_t the total attenuation constant. The two unknown quantities τ and α_t can be obtained by numerically processing the data of VSWR measured at several values of z.

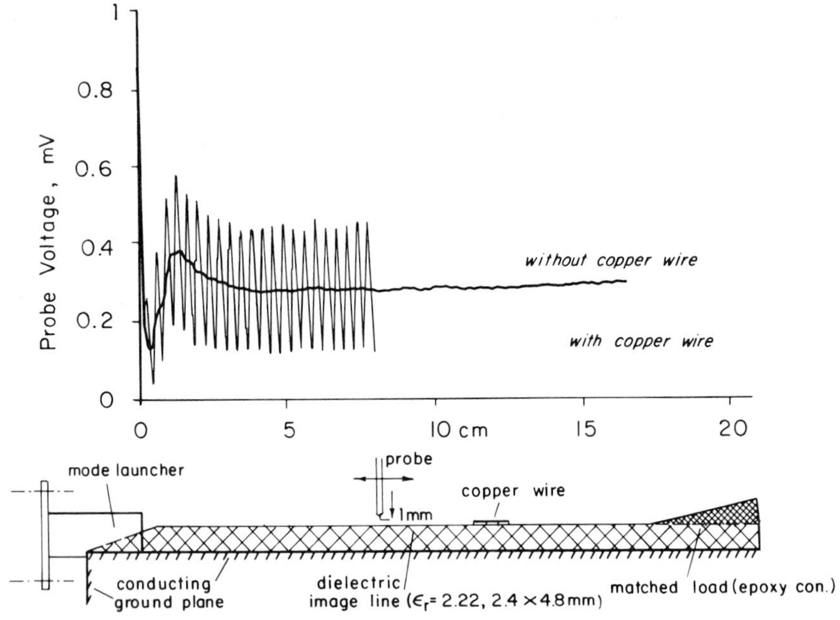

FIG. 24 Measurement of the voltage standing wave ratio on two short-circuited dielectric image guides and evaluation of the measurement points for the attenuation constants of the lines and the termination losses of the reflector walls. VSWR (matched load) = 1.02; VSWR (Cu wire) = 1.88; f = 35 GHz; radiated and dissipated power = 69%, detector operating in quadratic region. (From Solbach, 1978b. Reprinted from *IEEE Transactions on Microwave Theory and Techniques* **26,** 266–274. © 1978 IEEE.)

In this method, it is important to make sure that the probe does not perturb the guided field very strongly. Further, it must be ensured that the wave reflected from the shorting flange is totally transmitted past the mode launcher so as to prevent the launching of a secondary incident wave. Measured results were found to agree well with theoretical prediction as well as with the results from the resonator techniques (Solbach, 1978a).

2. *Radiation Loss*

In an open waveguide structure such as an image guide, any structural irregularities can cause a radiation loss. Imperfections in a dielectric material as well as the surface roughness at the dielectric–air and/or dielectric–dielectric interfaces are examples of these irregularities. When a guided mode is incident on such irregularities, the guided field is scattered. Scattered fields can be described in terms of a complete set of eigenmodes of the waveguide. It is well known that such a set for open waveguide structures consists of a finite number of surface wave modes and a continual spectrum of the radiation mode. A portion of the scattered wave energy can couple to the radiation mode and is no longer guided. Hence the energy of the incident guided wave is decreased by such a portion and contributes as the radiation loss.

Althoughh the radiation loss due to structural irregularities can be significant in optical integrated circuits, such loss seems to be relatively minor in most dielectric waveguide structures for millimeter-wave integrated circuits. The reason may be that any irregularities can be made much smaller than the wavelength, i.e., smooth surface finish of the dielectric rod and uniformity in material are more easily attainable. Also, the distance over which the guided wave travels is often much shorter in terms of wavelength in the millimeter-wave integrated circuit than in the integrated optical structures.

Radiation phenomena in millimeter-wave dielectric waveguide dielectric waveguide structures play a more important role in the following two situations. One is at the curved sections of waveguides that are often unavoidable in forming certain functional circuits or components. The other is at the junctions of two dielectric waveguides or between the waveguide and the active devices or even between the dielectric waveguide and other types of waveguides such as microstrip lines. Since some of the problems associated with the junctions will be treated in the later section, we shall here briefly summarize the radiation phenomena at curved sections of a dielectric waveguide. It is not intended to provide methods for computing radiation loss at curved sections for a specific dielectric waveguide. Rather, we shall describe why radiation occurs at the curved boundary between two dielectric media. Marcuse (1971) presented a simple heuristic explanation for radiation from a curved dielectric slab structure. We shall, however, follow the approach taken by Lewin *et al.* (1977).

Consider the curved slab waveguide dielectric materials as shown in Fig. 25, where R_0 and R_i are the outer and inner radii of curvature and the structure is assumed uniform and infinite in extent in the z direction perpendicular to the plane of the paper. Since the problem is a two-dimensional one, the field can be expressed in terms of a cylindrical coordinate (ρ, ϕ). The field in this structure may be derived from a z-directed scalar potential

$$\Phi(\rho, \phi) = X(\rho) \exp(-j\nu\phi). \tag{41}$$

The exponential term can be alternatively expressed as $\exp(-jk's)$, where $s = \phi R_0$ is the arc length along the boundary and $k' = \nu/R_0$ is the propagation constant of a wave along the curve structure. Accordingly, $X(\rho)$ satisfies the Bessel differential equation

$$\left[\frac{1}{\rho}\frac{\partial}{\partial\rho}\left(\rho\frac{\partial}{\partial\rho}\right) + \varepsilon_n k_0^2 - \left(\frac{k'\mathrm{R}_0}{\rho}\right)^2\right] X(\rho) = 0, \tag{42}$$

where

$$\varepsilon_n = \begin{cases} \varepsilon_r, & R_i < \rho < R_0, \\ 1, & \text{otherwise.} \end{cases}$$

Appropriate solutions satisfying (42) and required constraints such as the radiation condition at infinity are

$$X(\rho) = \begin{cases} AJ_\nu(k_0\rho), & \rho < R_i, \\ BJ_\nu(\sqrt{\varepsilon_r}k_0\rho) + \mathrm{C}H_\nu^{(2)}(\sqrt{\varepsilon_r}k_0\rho), & R_i < \rho < R_0, \\ DH_\nu^{(2)}(k_0\rho), & \rho > R_0, \end{cases} \tag{43}$$

where J_ν and $H_\nu^{(2)}$ are the Bessel and second-kind Hankel functions of

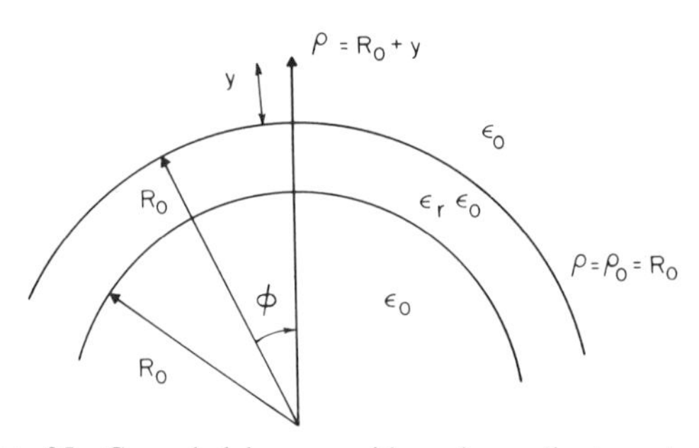

FIG. 25 Curved slab waveguide and coordinate system.

order ν, and A, B, C, and D are unknown coefficients. The guided wave solution can be derived by matching appropriate interface conditions at $\rho = R_i$ and R_o. However, we shall not proceed to obtain the guided mode, but rather investigate the origin of the radiation.

To this end, let us introduce a new coordinate system (u, s) given by

$$u = R_o \log \rho/R_o, \qquad s = \phi R_o. \tag{44}$$

We shall now concentrate our attention on the region $\rho > R_o$ into which the radiation occurs. Note that for the region close to the curved surface $\rho = R_o$, we have $(\rho - R_o) \ll R_o$ so that $u \simeq (\rho - R_o)$ is approximately the radial distance from the surface. Now, substituting Eq. (44) into Eq. (42) for $\rho > R_o$, we have

$$\{(d^2/du^2) + k_0^2[N^2(u) - \gamma 2]\}X(u) = 0, \tag{45}$$

where

$$N(u) = \exp(u/R_o) \tag{46}$$

and $\gamma = k'/k_o$ is the normalized propagation constant along the s axis. In the event $R_o \to \infty$, $N(u) = 1$ and Eq. (45) gives the transverse wave equation in rectangular coordinates, which with the radiation condition must admit the exponentially decaying field as the solution for a guided wave. For a finite R_o, i.e., for a finite curvature, $N(u)$ in Eq. (45) indicates that the homogeneous space $\rho > R_o$ is mapped into an inhomogeneous space $u > 0$ in the new rectangular coordinate system (u, s). $N(u)$ may be thought of as the inhomogeneous refractive index increasing exponentially as $\exp(u/R_o)$.

We now assume $\gamma = k'/k > 1$ and write down an approximate solution for the scalar potential in the (u, s) coordinate system as

$$\Phi(u, s) = \bar{D} \exp[-(\gamma^2 - N^2)^{1/2} k_o u] \exp(-jk_o \gamma s) \tag{47}$$

for the exterior region $u > 0$ $(\rho > R_o)$, with the assumption that $N(u)$ is slowly varying. If $R_o \to \infty$, $N(u) = 1$ and then Eq. (47) provides a field decaying in u and propagating in the s direction. However, in a curved structure R_o is finite and $N(u)$ has an ever-increasing profile. Hence the rate of decay for the evanescent field decreases continuously as it moves away from $u = 0$ $(\rho = R_o)$ surface. When the observation point reaches $u_t = R_o \log \gamma$, where N is equal to γ, the quantity under the square root in Eq. (47) becomes zero. For $u > u_t$, such quantity becomes negative and Eq. (47) now becomes no longer evanescent but provides an oscillatory (traveling) wave in the u (or equivalently ρ) direction.

Figure 26 illustrates how $X(u)$ changes as u is increased from $u = 0$ to the point past the turning point u_t. An evanescent field, say X^{inc}, pro-

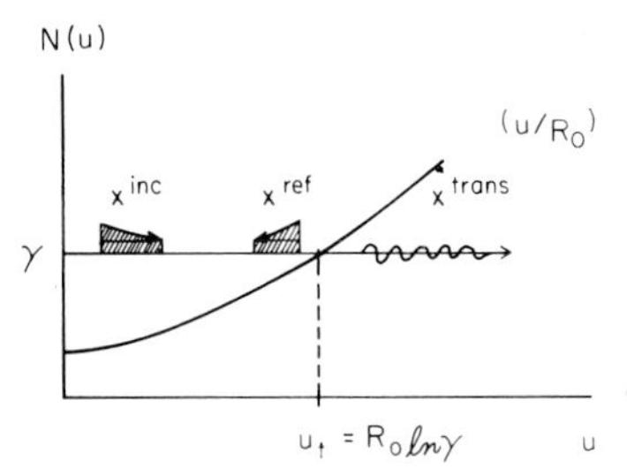

FIG. 26 Forward-progressing evanescent field x^{inc} partially reflected as evanescent field x^{ref} and partially transmitted as propagating wave x^{trans} (Lewin *et al.*, 1977).

gresses in the increasingly positive u and is incident on a boundary layer in the vicinity of the turning point. It is then partially transmitted and transformed into a forward-propagating wave X^{trans} while at the same time partially relfected by the boundary layer in the form of evanescent wave progressing toward the dielectric slab (toward $u = 0$). The value of this reflected field is usually small when it reaches the surface of the slab as the value of X^{inc} is already small at $u = u_t$. However, X^{trans} cannot be neglected since it carries some energy away from the guided wave in the dielectric slab. It is now clear that X^{trans} is solely caused by the curvature of the slab waveguide and is an inherent factor associated with the curved dielectric waveguide. This term X^{trans} is completely responsible for the radiation phenomena in the curved dielectric waveguide structure.

C. Some Recent Developments in Theoretical Analysis

As the dielectric waveguide-type millimeter-wave integrated circuits create increasing interest, some of the problems left unsolved in the process of component and system fabrications have come to the attention of a number of researchers. In addition, some problems have been investigated from a different point of view. In this section, we review some of the more recent developments in theoretical analysis on the dielectric waveguiding structures.

1. *Leaky Waves in Dielectric Waveguides*

In the dielectric waveguide, it is necessary that the field decay in any transverse direction as it moves away from the core area of the waveguide so that the guided electromagnetic wave is tightly bound to the structure. Recent studies indicate that some dielectric structures designed to guide the wave may produce some leakage of the wave as it propagates.

Oliner *et al.* (1978) investigated the modal structures in the inverted strip dielectric (IS) waveguide and reported that under some circum-

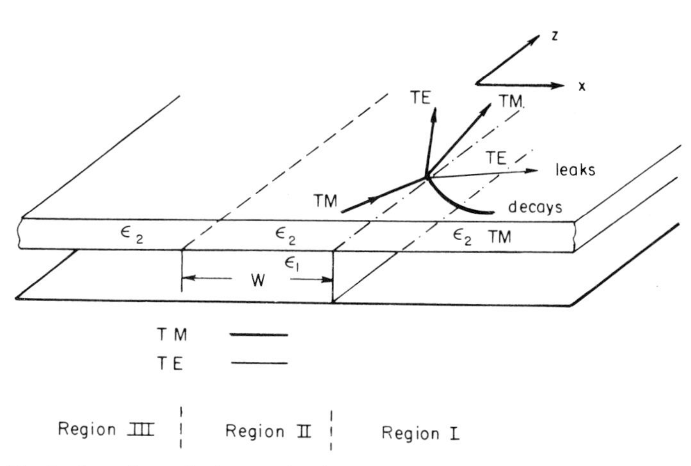

FIG. 27 Leakage from the inverted strip guide. (Courtesy of Professor A. A. Oliner.)

stances the guided wave loses its energy as it propagates, by leakage in the horizontal x direction (see Fig. 27).† It should be noted that there is no leakage in the vertical y direction, and the leakage from the IS waveguide travels in the guiding layer ε_2 as a surface wave bounded in the y direction and carries energy away from the central region. Such a phenomenon obviously causes a number of problems. For instance, the leakage significantly increases the attenuation of the guided wave even though the side walls of the dielectric strip are smooth so that no radiation takes place due to the wall irregularities. On the other hand, such a leakage may be used advantageously for design of certain components, because the leaky wave still propagates in the dielectric layer. For instance, such a leakage may be deliberately used for enhancing the coupling between one IS guide to another neighboring IS structure. Hence it is important to study under what circumstances this leakage phenomenon occurs and how it is controlled.

The first step toward this end is to recognize that the "effective dielectric constant" (EDC) method used by Itoh (1976) is essentially a simplified version of the transverse resonance technique. The latter, if rigorously applied, gives rise to a correct solution. In the EDC method, only one surface-wave mode, the lowest TM (to z), is assumed in each of the constituent transverse regions constituting the guide cross section (see Fig. 7 in Section III. A.2). Furthermore, the geometrical discontinuities at the sides of the dielectric strip are completely neglected since

† In addition to leakage, other interesting phenomena such as resonance, mode coupling, and modifications in transverse field distribution may occur. These problems, however, will not be treated here. Interested readers may refer to Oliner *et al.* (1978).

each constituent region is viewed as a uniform medium having an effective dielectric constant.

A more rigorous approach reveals that in each constituent region in the cross section of the guide, at least two surface wave modes, one TM and the other TE, must be simultaneously present under almost all conditions. In addition, if the thickness of the dielectric layer or dielectric strip were increased, additional surface wave modes would start propagating. These constituent modes are coupled to each other at the geometrical discontinuities corresponding to the sides of the strip. Hence in the rigorous transverse resonance method all of these modes and the coupling effects must be correctly accounted for, and the propagation constants of the IS guide so derived are expected to be different from those given by the simple EDC method. Some numerical studies, however, reveal that the simple EDC method gives rise to surprisingly accurate results for most IS waveguide structures developed to date.

We are now ready to investigate the leakage phenomenon. The principal TM wave components contained in the E^y mode will create TE components at the boundary between regions I and II or II and III in Fig. 27 because of the structural discontinuity. The nature of TE waves in each constituent region (I, II, or III) may be investigated by defining the effective dielectric constant for the TE waves. Note that the effective dielectric constant is a measure of the phase velocity of the surface wave in a constituent region. In a manner similar to the one for the TM wave in Section III. A.2, we write the field variations in region I in Fig. 27 in a sinusoidal form in the ε_2 layer and in exponential forms in the air regions above and below the ε_2 layer. Matching boundary conditions for TE waves, we obtain the transverse wave number $k_{y\mathrm{I}}^{\mathrm{TE}}$ in the ε_2 layer. The effective dielectric constant of the TE wave in region I is then

$$\varepsilon_{e\mathrm{I}}^{\mathrm{TE}} = \varepsilon_2 - (k_{y\mathrm{I}}^{\mathrm{TE}2}/k_0^2). \tag{48}$$

The propagation constant of the TE wave in region I is given by

$$k_{\mathrm{SI}}^{\mathrm{TE}} = (\varepsilon_{e\mathrm{I}}^{\mathrm{TE}})^{1/2} k_0. \tag{49}$$

If the propagation constant of the IS waveguide k_z is correctly calculated, the TE wave described above is a part of the solution and must have a propagation constant in the z direction equal to k_z. Hence the dispersion relation

$$(k_{\mathrm{SI}}^{\mathrm{TE}})^2 = k_z^2 + (k_{e\mathrm{I}}^{\mathrm{TE}})^2, \tag{50}$$

where $k_{x\mathrm{I}}^{\mathrm{TE}}$ is the transverse x phase constant so that the TE field varies as $\exp(-jk_{e\mathrm{I}}^{\mathrm{TE}}\, x)$ in the x direction. The guided wave in the IS waveguide requires that the field decay in the x direction, or equivalently that k_{xI}^{TE} be purely imaginary.

If, on the other hand, k_{xI}^{TE} is real, the TE wave propagates in region I as a surface wave away from the core region II and creates a leakage to E^y mode of the IS guide. Therefore, the condition for leakage is given by

$$(k_{xI}^{TE})^2 > 0, \tag{51}$$

or equivalently

$$\varepsilon_{eI}^{TE} > k_z/k_o. \tag{52}$$

Equation (52) requires the knowledge of k_z, which takes into account the effect of all of TE as well as TM waves in each constituent region. However, in practice, some approximate guidelines are often required. Such an approximation is readily obtained if we recognize that the EDC method discussed previously is quite accurate. Instead of Eq. (52), then, we can use the criterion

$$\varepsilon_{eI}^{TE} > k_z^{TM}/k_o, \tag{53}$$

where k_z^{TM} is the propagation constant of the IS guide as calculated by the EDC method described in Section III. A.2.

2. *Junction Problems*

In integrated circuits, a number of junctions and discontinuities appear so that appropriate circuit functions may be performed. Some examples include the junction between two dielectric waveguides with different cross-sectional dimensions (step discontinuity), the truncation of the waveguide (open-ended waveguide), and the discontinuity caused by the solid-state device implanted in a dielectric waveguide.

The theoretical analysis of this junction and discontinuity problems is still in its infancy and accordingly many approximations are employed in actual circuit designs. A number of studies on discontinuities have been made for optical dielectric waveguides. For example, Marcuse (1970) computed the radiation loss from the step discontinuity of two slab waveguides by approximating the step as a succession of infinitesimally small steps. For the latter, the radiation loss was computed by neglecting the backward radiation.

Because of the presence of the continuous radiation spectrum, the method of discrete mode matching developed for closed waveguide structures is not well suited in the present case. One of the better methods developed recently by Rozzi (1978) makes use of an integral equation with the Ritz–Galerkin (RG) variational technique for a quick convergence of

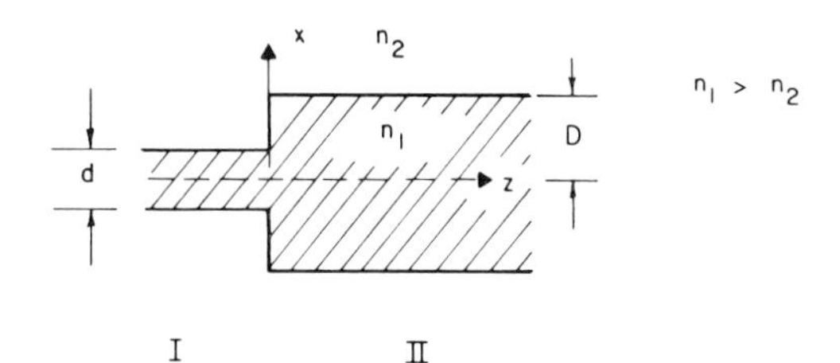

FIG. 28 Symmetric step junction between two slab dielectric waveguides: $n_1 = \varepsilon_1$, $n_2 = \varepsilon_2$. (From Rozzi, 1978. Reprinted from *IEEE Transactions on Microwave Theory and Techniques* **26,** 738–746. © 1978 IEEE.)

numerical solutions. We shall summarize his method in the following, as this method may be potentially applicable to other types of discontinuities often encountered in millimeter-wave integrated circuits.

Referring to Fig. 28, we study the scattering of the TE field that has transverse components E_y and H_x. We shall first describe both surface waves and radiation spectrum in the waveguide I as those for II will be obtained readily by replacing d by D. Assuming $\exp[j(\omega t - \beta z)]$, we have the expression of surface waves as

$$E_y(x) = u(x), \tag{54}$$

$$H_x(x) = -Y_0 u(x), \tag{55}$$

$$Y_0 = \beta/\omega\mu_0 = 1/Z_0. \tag{56}$$

We shall consider only symmetric modes for which

$$u(x) = \begin{cases} a \cos \kappa x, & 0 < x \le d, \\ a \cos \kappa d \exp[-\gamma(x - d)], & x \ge d, \end{cases} \tag{57}$$

where a is obtained from the normalization condition

$$\int_0^\infty u^2(x)\, dx = 1 \tag{58}$$

and κ and γ satisfy the eigenvalue equation

$$\kappa \tan \kappa d = \gamma \tag{59}$$

with

$$\kappa^2 + \gamma^2 = (\varepsilon_1 - \varepsilon_2)k_0^2 = v^2. \tag{60}$$

It is possible to have more than one solution to Eq. (59) for which β, given by

$$\beta^2 = \varepsilon_1 k_0^2 - \kappa^2 = \varepsilon_2 k_0^2 + \gamma^2, \tag{61}$$

takes discrete values in the range

$$\sqrt{\varepsilon_2}k_o < \beta < \sqrt{\varepsilon_1}k_o.$$

These solutions are labeled as $u_k(x)$, $1 \leq k \leq n_l$.

The radiation spectrum is described by

$$E_y = \phi(x, \rho) = \begin{cases} \left(\frac{2}{\pi}\right)^{1/2} \frac{\cos \sigma x}{[1 + (v/\rho)^2 \sin^2 \sigma d]^{1/2}}, & 0 < x \leq d, \\ (2/\pi)^{1/2} \cos[\rho(x - d) + \alpha], & x \geq d, \end{cases} \tag{62a}$$

$$H_x = -(1/Z_o(\rho))\phi(x, \rho), \tag{62b}$$

$$Z_o(\rho) = \omega\mu_o/(\varepsilon_2 k_o^2 - \rho^2)^{1/2}, \tag{62c}$$

where $0 < \rho < \infty$, $\sigma^2 = v^2 + \rho^2$, and $\alpha = \tan^{-1}(\sigma/\rho \tan \sigma d)$. The propagation constant is

$$\beta^2 = \varepsilon_1 k_o^2 - \sigma^2 = \varepsilon_2 k_o^2 - \rho^2. \tag{63}$$

For region II, ϕ is replaced by ψ.

It is of importance to note that the radiation spectrum, the portion belonging to

$$0 < \beta^2 < \varepsilon_2 k_o^2,$$

is called radiative, whereas the one belonging to

$$\beta^2 < 0,$$

is called reactive.

Let us consider the situation when there are n_l guided modes in guide I and n_r in guide II. The problem is to describe the discontinuity as an n_i $(= n_l + n_r)$ port by means of its scattering matrix. The effect of radiation spectrum must be included completely in the scattering matrix. It is noted that the reactive part of the radiation spectrum ($\beta^2 < 0$) attenuates in the direction of propagation ($\pm z$ directions) and represents energy stored in the neighborhood of the discontinuity. This may be thought of as the counterpart of the evanescent modes in a closed waveguide. On the other hand, the radiative part propagates in the axial direction and represents loss or coupling to the surroundings.

We express the electric field amplitudes of the incoming and reflected waves at $z = 0$ with the vectors $\bar{\mathbf{A}}$ and $\bar{\mathbf{B}}$ of sizes n_i. Let $b(\rho)$ and $d(\rho)$ be the amplitudes of the scattered radiation spectra in regions I and II, respectively. From the continuities of E_y and H_x at $z = 0$, an integral equation may be obtained:

$$s_l u_l(x) = \int_0^\infty dx'\, Z(x, x') h_l(x'), \tag{64}$$

$$Z(x, x') = \frac{1}{2} \sum_{k=1}^{n_l} Z_{ok} u_k(x) u_k(x')$$
$$+ \int_0^\infty d\rho Z_o(\rho)[\phi(x, \rho)\phi(x', \rho) + \psi(x, \rho)\psi(x', \rho)],$$

where $s_l = \pm 1$ according to $l \lesseqgtr n_l$ and h_l is the magnetic field H_x when only one mode is incident ($A_l = 1$, $A_{k \neq l} = 0$). The scattering matrix element becomes

$$S_{k,l} = \delta_{k,l} - s_k Z_{ok} \int_0^\infty u_k(x) h_l(x)\, dx. \tag{65}$$

Rozzi (1978) solved the integral equation (64) by the use of the Ritz–Galerkin (RG) variational method. To this end, both u_x and $\phi(x, \rho)$ are expanded in terms of a truncated complete set of functions $L_n(x)$ and converted Eq. (64) into a matrix equation. After this matrix equation is solved, one can obtain a variational expression for the scattering matrix element S_{kl}. Since the matrix from Eq. (64) is complex in nature, the solution is neither a maximum nor a minimum. However, by making an appropriate choice, such as the Lagurre functions for $L_n(x)$, a rapid convergence of the solution is found to be accomplished.

A number of interesting results have been obtained from numerical cal-

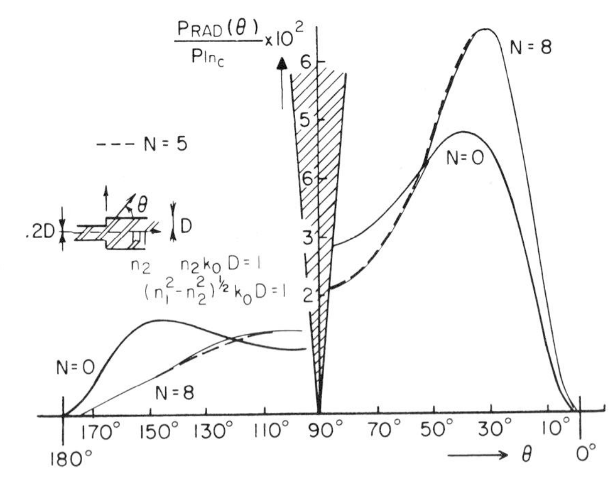

FIG. 29 Convergence of the radiation pattern (incidence from left) for increasing order of the variational solution $n_2 k_o D = 1$, $\nu D = 2$, $n_1 = \varepsilon_1$, $n_2 = \varepsilon_2$. (From Rozzi, 1978. Reprinted from *IEEE Transactions on Microwave Theory and Techniques* **26,** 738–746. © 1978 IEEE.)

culation based on the method presented. For instance, it is clear that the backward radiation cannot be neglected from a sizable step as illustrated in Fig. 29.

Since in millimeter-wave integrated circuit a large step is often encountered, unlike in optical circuit, a rigorous analysis like the one summarized here should be extended to various discontinuity problems so as to obtain more accurate design data.

IV. Passive Components Made of Dielectric Waveguides

In this section we review some of the works carried out for development of passive components made of dielectric waveguides such as couplers, resonators, and filters. In addition, some nonreciprocal devices involving gyrotropic material will be included. These components are essential in forming millimeter-wave integrated subsystems and systems in conjunction with active devices described in a later section of this chapter.

A. Directional Couplers

Directional couplers are used for a number of functions such as power splitting and signal combining. It is an important component in a balanced mixer. Most of the dielectric waveguide couplers are of distributed type in which the coupling effect takes place in a coupled waveguide that consists of two dielectric waveguides placed in a close vicinity of each other. Figure 30 shows a distributed type 3-dB directional coupler designed for 75–80-GHz operation made of a coupled inverted strip dielectric (IS) waveguide (Rudokas and Itoh, 1976).

The length L needed for complete power transfer from one waveguide to another in the coupled-waveguide structures is given by

$$L = \pi/(k_{ze} - k_{zo}), \tag{66}$$

where k_{ze} and k_{zo} are the propagation constants of two orthogonal modes, even and odd, in the coupled structure. They are functions of the separation S of the coupled guide, and the difference $k_{ze} - k_{zo}$ becomes larger for a smaller S. Given some length of the coupled guide l, the ratio of the power at ports 2 and 3 is

$$P_3/P_2 = \tan^2(\pi l/2L). \tag{67}$$

In the above analysis the coupling between the connecting guides as well as the junction effect between the connecting guides and the coupled guides are completely ignored. In many designs, the latter is considered small. The former may be taken into account by replacing l in Eq.

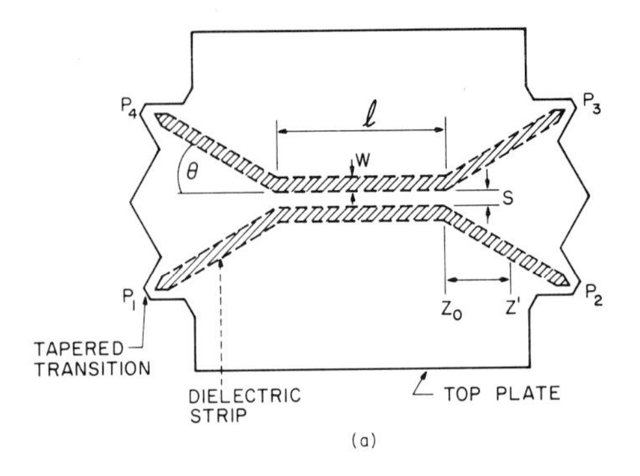

FIG. 30 Distributed directional coupler. (a) Top view; (b) 3-dB coupler. (From Rudokas and Itoh, 1976. Reprinted from *IEEE Transactions on Microwave Theory and Techniques* **24,** 978–981. © 1976 IEEE.)

(67) with an effective length $\bar{l}$ defined by

$$\bar{l} = l + (\Delta\phi/\pi)L, \tag{68a}$$

$$\Delta\phi = 2 \int_{z_0}^{z'} [k_{ze}(z) - k_{zo}(z)]\, dz. \tag{68b}$$

In Eq. (68b) the integration is carried out along the axial z direction of the directional coupler; z_0 corresponds to the junction between the coupled guide and the connecting arm, while z' is chosen to be some value of z

beyond which the coupling between two arms is negligible. The factor of 2 occurs because of the symmetry of the coupler structure. The theoretical results based on the effective length agree much better with experimental data obtained in 75–80-GHz region.

Directional couplers have also been made and tested by a number of workers. Knox (1976) reported the performance of a typical 3-dB quadrature hybrid made of "insular" waveguide. Figure 31 describes coupling characteristics of a 3-dB image guide hybrid coupler made of boron nitride (BN). The data were reported by Paul and Chang (1978).

A different type of coupler was reported by Rudokas and Itoh (1976) and is based on the principle similar to optical beam splitter. The coupler consists of two perpendicularly interesecting IS waveguides, in which a gap S, oriented at 45° to each guide, is made in the dielectric strip portion of the IS waveguide (see Fig. 32). The propagation constant corresponding to the gap portion is smaller than the one in the waveguide with dielectric strip. Therefore, the coupling structure may be replaced by the sandwiched structure (Fig. 32). A part of the wave entering from port 1 to this gap portion is, therefore, reflected into port 3 and the remaining energy is transmitted to port 2. If all of the ports are terminated appropriately, no output will appear in port 4.

Theoretical results based on a simple plane-wave approximation have been found to agree reasonably well with experimental data of a 20-dB coupler measured at 75–80-GHz range.

B. Resonators and Filters

Bandpass and band-reject filters may be made of one or more ring resonators. The circumferential length of ring resonators made of dielectric waveguides is usually several wavelengths long so that the radiation loss

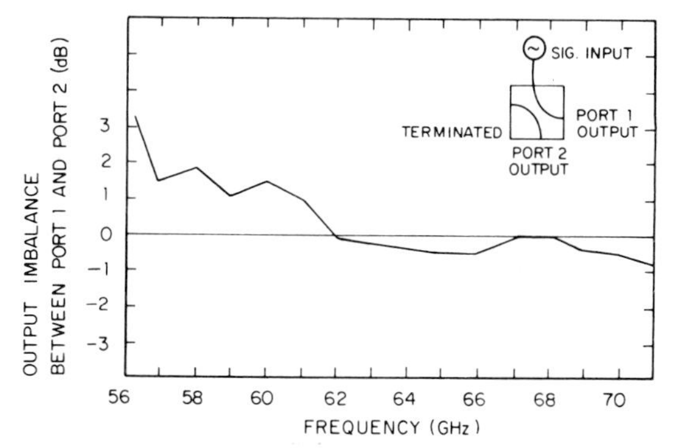

Fig. 31 Coupling characteristics of a BN image-guide 3-dB hybrid coupler. Cross section = 0.070 × 0.035 in.; coupling length = 0.151 in.; measured. (From Paul and Chang, 1978. Reprinted from *IEEE Transactions on Microwave Theory and Techniques* **26,** 751–734, © 1978 IEEE.)

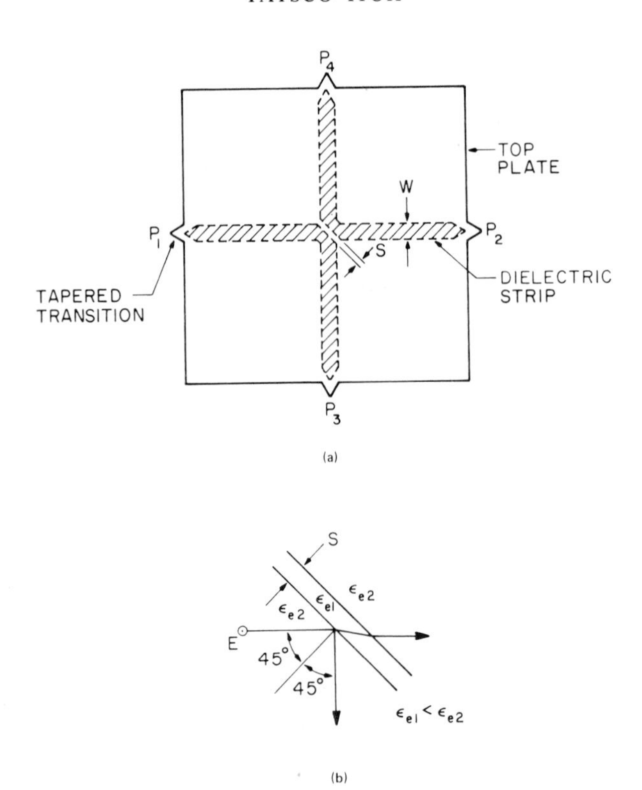

FIG. 32 Beam splitter-type coupler. (a) Top view; (b) equivalent structure. (From Rudokas and Itoh, 1976. Reprinted from *IEEE Transactions on Microwave Theory and Techniques* **24,** 978–981. © 1976 IEEE.)

due to curvature is kept at less than an acceptable level. In such structures, the multiple resonance phenomena occur. When applied to band-reject or bandpass filters, spurious responses at frequencies other than the specified one are not desirable. Resonance conditions for the ring resonator are given by

$$n\lambda_g = 2\pi\bar{r}, \tag{69}$$

where λ_g is the guide wavelength in the ring and $\bar{r}$ is the effective radius of the ring. If the width of the ring is small compared to the radius, $\bar{r} \approx (ab)^{1/2}$, where a and b are the outer and inner radii. When the radius of curvature is large and the radiation loss is small, λ_g may be approximated with the guide wavelength in a straight section of the dielectric waveguide. Since the dispersion characteristic of the straight waveguide gives a relation between the frequency and the guide wavelength, we can calculate the resonant frequency from such a relation and Eq. (69).

Rudokas and Itoh (1976) tested a ring resonator made of an inverted strip dielectric (IS) waveguide at 75–80-GHz range, and the measured resonant frequencies were found to agree favorably with theoretical calculations. In their structure, the radiation loss due to curvature was relatively high because the field in an IS waveguide made of a Teflon strip and a fused silica plate is rather weakly guided as compared to the image guide made of a high-permittivity material such as alumina.

Toulios and Knox (1970) have fabricated a ring resonator and subsequently developed a preselector filter having one or more resonators for V-band operation (Knox, 1976). The level of radiation due to curvature was maintained at the value below the acceptable level, and the high Q resonators were made.

Aylward and Williams (1978) reported bandpass filters consisting of a number of rings each coupled to the adjacent rings or the input and output lines in an insular guide setup. Since the rings are multiple numbers of the guide wavelengths in circumference, it is necessary to use rings of different sizes to reduce the spurious response at the next higher or lower resonant frequency of each ring. In their synthesis, coupling effects between adjacent rings are taken into account.

It is pointed out that due to manufacturing tolerances, adjustment of the resonant frequency of each ring is often needed for correct alignment of the filter. The tuning element devised by Aylward and Williams (1978) consists of a metalic disk located concentrically within the resonant ring. The electric field tangential to the wall of the disk is forced to zero, and the distortion in the field distribution results in a modification of the propagation constant in the ring, which can be completed from a consideration of the appropriate boundary value problem. As a secondary effect, the shift in effective path length of the ring tends to reduce the change in resonant frequency. If a variable height mechanism is incorporated, a linear frequency versus height characteristic may be obtained.

Two-, three-, and four-pole filters with bandwidths between 150 and 200 MHz and 0.25-dB ripple Chebyshev response have been considered as design examples. The ring sizes have been chosen on the basis of achieving a maximum rejection level at the spurious resonances. Figure 33 shows the top view of one of two-pole filters developed and its theoretical and experimental responses. Correlation between theoretical and experimental results is seen to be very good. The slightly lower level of the spurious responses may be accounted for by the marginal differences in the bandwidth of the filter and the change of coupling values with frequency.

Let us now turn our attention to a different type of filter structure made of a grating structure (Itoh, 1977). We shall present a band-reject filter

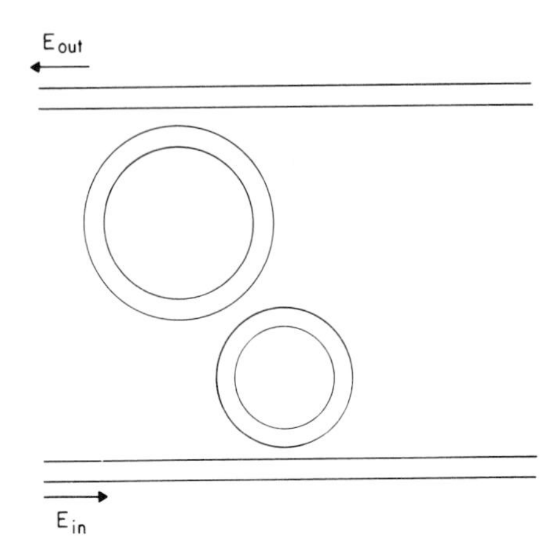

FIG. 33 Layout of an image guide two-pole filter.

based on the Bragg reflection in a periodic grating in a dielectric waveguide. The band-rejection due to the Bragg reflection occurs at certain frequencies determined by the electric length of the grating period, because of wave-coupling phenomena.

Gratings may be created by periodically modulating the geometrical or material nature of the dielectric waveguide. It is well known that electromagnetic waves in a grating structure can be represented in terms of space harmonics whose phase constants are

$$\beta_m = \beta_0 + (2m\pi/d), \qquad m = 0, \pm 1, \pm 2, \ldots, \tag{70}$$

where d is the grating period and β_0 the phase constant of the dominant ($m = 0$) space harmonic determined by the excitation of the grating. If perturbation due to unit cell of the grating is small, β_0 is very close to the propagation constant k_z in the unperturbed dielectric waveguide, except in the coupling regions of the frequency spectrum.

Figure 34 shows the k–β diagram for the $m = 0(\beta_0)$, $m = -1(\beta_{-1})$, and $m = -2(\beta_{-2})$ forward-traveling space harmonics and the $m = -1(-\beta_{-1})$ backward-traveling harmonic of the inverted strip dielectric (IS) waveguide. The curve for β_0 is obtained from the dispersion characteristic of k_z in the IS waveguide (Itoh, 1976). Curves for higher-order space harmonics may be obtained by shifting β_0. The curves in Fig. 34a correspond to a grating with vanishingly small-periodic perturbations. In fact, mode-coupling phenomena occur at synchronous points such as A, B, C, and D. The mode-coupling phenomena can be used for deriving the dispersion relations for the grating structure such as the one shown in Fig. 34b. Note that if for a given frequency the period d is chosen such that $\beta_0 d$ is

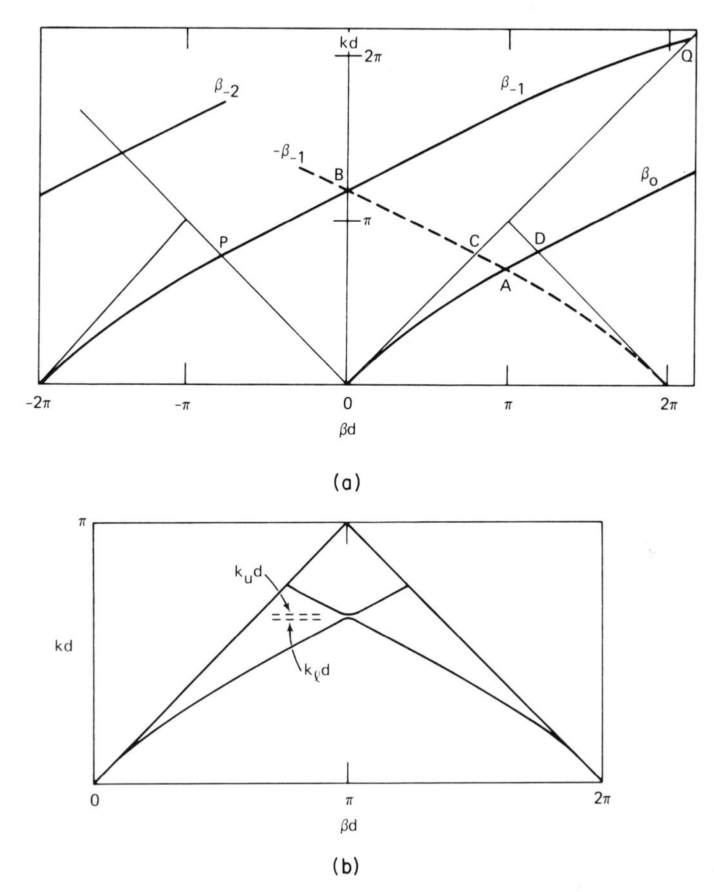

FIG. 34 Dispersion characteristics of a grating structure in the inverted strip dielectric waveguide: $\varepsilon_1 = 2.1$, $\varepsilon_2 = 3.75$, period d, thicknesses of guiding layer and strip are 1.58 mm. Strip width, 4 mm. (a) Wave coupling neglected; (b) surface-wave stop band. (From Itoh, 1977. Reprinted from *IEEE Transactions on Microwave Theory and Techniques* **25**, 1134–1138. © 1977 IEEE.)

less than the value at D, no higher space harmonic radiates and the grating supports a surface wave.

As an example, each unit period of the grating structure in Fig. 35 in the IS waveguide may be modeled by two cascaded transmission lines of lengths a and $d - a$. If we neglect the junction effect such as the one analyzed by Rozzi (1978), we obtain the following dispersion relation:

$$\cosh \gamma d = \cos(k_{zg}a) \cos[k_{zf}(d - a)] + \frac{1}{2} [(k_{zg}/k_{zf}) + (k_{zf}/k_{zg})] \sin(k_{zg}a) \sin[k_{zf}(d - a)], \tag{71}$$

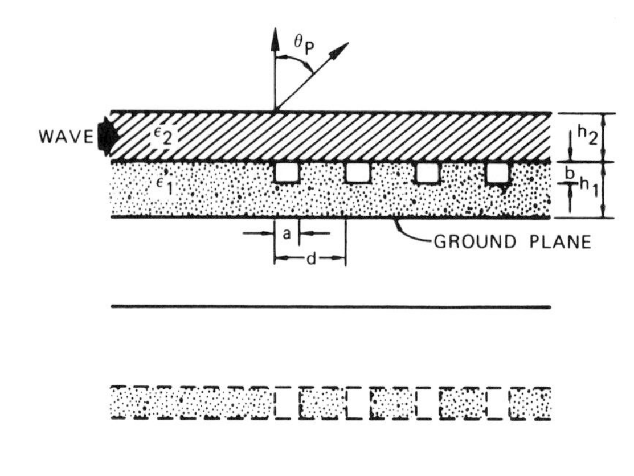

FIG. 35 Grating structure in an inverted strip dielectric waveguide. Under the surface-wave condition, no beam radiates into the direction identified by θ_p. (From Itoh, 1977. Reprinted from *IEEE Transactions on Microwave Theory and Techniques* **25**, 1134–1138. © 1977 IEEE.)

where k_{zg} and k_{zf} are the propagation constants in the grooved and non-grooved sections of IS guide. At the point A in Fig. 34a, a stop band appears as shown in Fig. 34b due to mode coupling and then $\gamma = \alpha + j\pi/d$, where α is a real quantity determining the attenuation. Any frequency component in the stop-band region is reflected and cannot pass through the grating section.

An example of the stop-band phenomena is shown in Fig. 36. A strong reflection occurs from a grating section created in an IS waveguide at the stop band around 15 GHz. The bandwidth and Q of the stop band may be adjusted by the profile of the grating structure. As demonstrated by successful use of such structures at integrated optical circuits, filters based on the grating structure may become increasingly more attractive at higher frequencies, because the grating may be created by the use of planar fabrication technology.

C. NONRECIPROCAL DEVICES

In traditional microwave circuits, ferrite materials are extensively used for realizing nonreciprocal devices such as isolators and circulators. One of the problems associated with the use of ferrite at higher frequencies is the increased loss. Recently, however, a considerable attention has been paid for reducing the loss in the ferrite devices. For instance, Babitt *et al.* (1978) developed high-performance phase shifters by the use of plasma-spray techniques.

To date, however, not much has been reported on the nonreciprocal devices compatible with dielectric waveguide-type millimeter-wave inte-

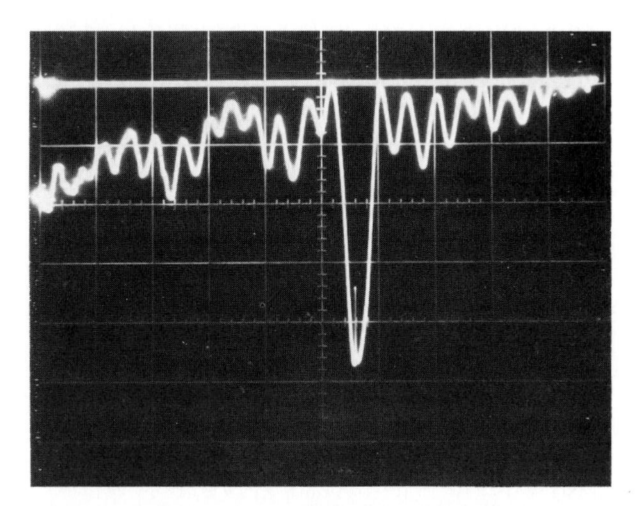

FIG. 36 Reflection property of a grating filter: d = 6.25 mm, a = 1.59 mm, b = 3.16 mm, $h_1 = h_2$ = 6.35 mm, ε_1 = 2.1, ε_2 = 3.75, number of grating elements = 30, 20 mV/div, 12.4–18 GHz. (From Itoh, 1977. Reprinted from *IEEE Transactions on Microwave Theory and Techniques* **25**, 1134–1138. © 1977 IEEE.)

grated circuits. We shall summarize the structure and operation of one of a few available, a field displacement-type ferrite isolator for V-band (50–75 GHz) developed by Nanda (1976).

The isolator is in the form of a rectangular image guide and made of a dielectric and a ferrite material as shown in Fig. 37. The principle of operation can be best understood with the help of Fig. 37. When the image guide operating in the dominant E_{11}^y mode is made of a low-loss dielectric material, the E_y or H_x field has a transverse distribution as shown in Fig. 37a. The distribution may be expressed as in Eq. (2) when Marcatili's approach is followed.

In the isolator arrangement, a portion of the image guide is replaced by a ferrite material and a dc magnetic bias field is applied in the y direction that is normal to the ground plane. Since the principal magnetic field components are H_x and H_z, the real part of permeability tensor components of an appropriately biased ferrite section takes positive and negative values of different magnitudes for two opposite directions of propagation in the structure. Hence the amplitude A_i and decay coefficients ξ_i and η_i in Eq. (2) are different for the two opposite directions of propagation. For instance, for the forward transmission case of Fig. 37b, smaller field amplitudes exist in the x_1 and x planes by an appropriate magnetic bias. On the other hand, for the reverse transmission case, the field amplitudes are made to be much larger in the x_1 and x planes. If one now places a resistive film in the vicinity of the ferrite loaded section of the image guide, the

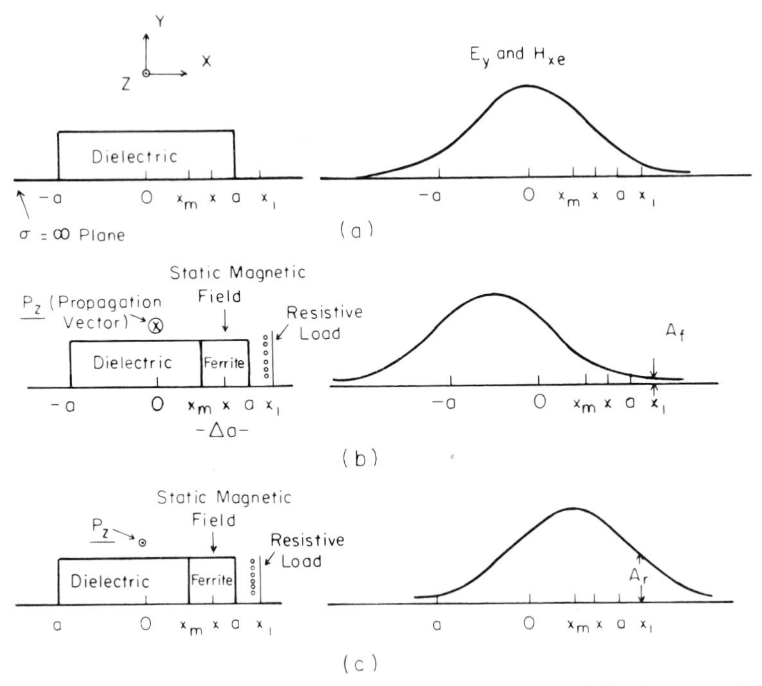

FIG. 37 Transverse field representation in the edge loaded dielectric waveguide. (a) Transverse field for a typical dielectric guide; (b) transverse field for the forward-transmission case of ferrite loaded guide; (c) reverse-transmission case. (From Nanda, 1976. Reprinted from *IEEE Transactions on Microwave Theory and Techniques* **24,** 876–879. © 1976 IEEE.)

film has much stronger interaction with the reverse-transmission case. Therefore, the insertion loss of the structure for the forward transmission can be made negligibly small, whereas it can be made quite large for the reverse-transmission case. Such phenomena are exactly those of an isolator. The figure-of-merit of the isolator may be approximately given by the ratio of amplitudes A_r and A_f in the x_1 plane.

The design of an isolator requires knowledge of several parameters, such as type of ferrite material, size of image guide, filling factor of ferrite slab, strength of biasing magnetic field, and position of resistive film.

Several image guide ferrite isolators of the type described above have been fabricated and tested. Characteristics of a V-band isolator are shown in Fig. 38. The maximum isolation of 11 dB was obtained at 61.25 GHz by using a nickel–ferrite of 0.8 × 0.040 × 0.018 in. loaded in an image guide of overall size 0.8 × 0.040 × 0.046 in. The forward-insertion loss was found to be 1 dB.

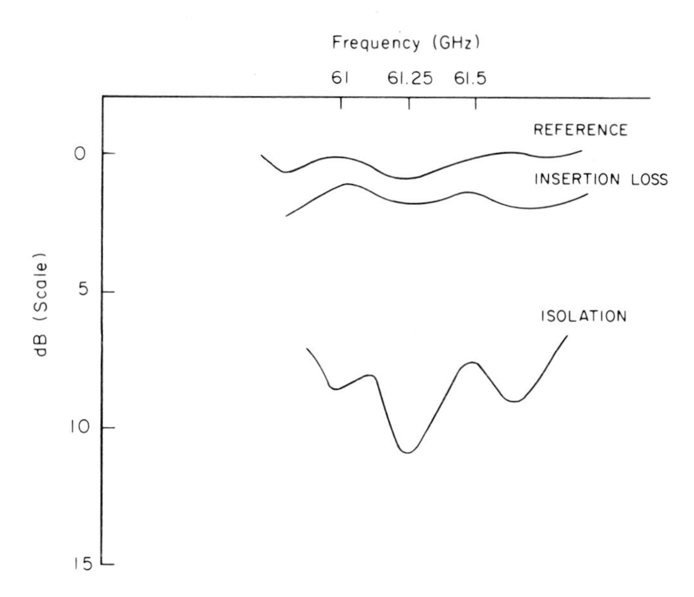

FIG. 38 *V*-band isolator characteristics. (From Nanda, 1976. Reprinted from *IEEE Transactions on Microwave Theory and Techniques* **24,** 876–879. © 1976 IEEE.)

Although much more is desired for performances, the concepts in the development of this isolator can lead to evolution of various other ferrite devices, such as a three-port circulator, a nonreciprocal phase shifter, and a differential-phase-shift circulator.

V. Active Components

Components made of active devices in conjunction with dielectric waveguides are another class of essential elements in millimeter-wave integrated circuits. Two approaches to device mounting have been attempted. In the first approach, several components were fabricated using metal split-block mounts in which the active device is mounted in a rectangular air-filled waveguide that incorporates a mode converter from the E_{11}^{y} mode dielectric waveguide to the TE_{10} metal waveguide. In the second approach, however, the active device is directly mounted in the dielectric waveguide. The second approach is more difficult, because one needs to understand the mode conversion to higher-order and radiation modes and to control these effects whenever necessary. However, future of dielectric millimeter-wave integrated-circuit technology is believed to depend on the successful implementation of the second approach, especially at higher frequencies. The original philosophy of introducing the integrated circuit techniques is to stay away from the use of conventional metal waveguides and to use more planar-oriented fabrication techniques as much as possible. Circuits at higher frequencies such as in the near-millimeter-wave regions can be cost effective and have satisfactory performance only if direct implementation of active devices is successfully worked out. This is because the situation becomes more similar to one we encounter at optical frequencies.

For this reason, we discuss mainly the direct implementation approaches in what follows. In many instances, the performances of the components fabricated in this manner do not outperform those of the ones developed by the state-of-the-art microstrip-type technology at frequencies as high as 250 GHz. However, efforts along this line are hoped to be rewarding in the future.

A. Phase Shifters

When a perturbation is applied to the modal field in a dielectric waveguide, the guide wavelength of the mode changes accordingly. For a given length of the waveguide, change of the guide wavelength results in a phase shift. Such a phase-shift mechanism is important in a number of applications such as the beam steering of the array antennas. Mechanical perturbation is well suited for dielectric waveguide structures as a method for studying the fundamentals of the phase-shift mechanism. Because of the open structure of the dielectric waveguides, it is relatively easy to incorporate a mechanical structure exterior to the dielectric material and to perturb the exponentially decaying field there. Jacobs and Chrepta (1974) studied mechanical phase shifter for a dielectric waveguide made of high-resistivity silicon.

The principle of operation of mechanical phase shifters may be understood by considering two extreme cases shown in Fig. 39. When a metal plate is placed on one of the surfaces ($y = b$) of the dielectric waveguide, the field distribution of the dominant E^y_{11} mode is the same as the one in the dielectric waveguide with its cross section $2a \times 2b$. When the metal plate is removed to $y = +\infty$, the E^y_{11} mode is the same as the one in the dielectric waveguide with the cross section $2a \times b$. Since the field distributions are different in the two situations (with or without a metal plate), the propagation constants and hence the electrical lengths of a given axial length of the structure also change. When the metal plate is at a finite distance away from $y = b$ surface, the situation is between the two extremes described above. It is now recognized that the phase shift of the guided wave after passing through a finite length of the structure described here is dependent on, and controlled by, the distance of the metal plate from the top surface $y = b$ of the waveguide. It is also possible to use a dielectric plate instead of a metal plate to control the phase shift. Such a mecha-

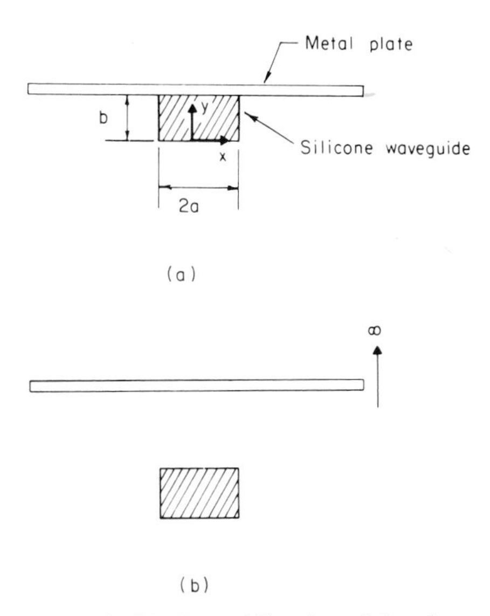

FIG. 39 Operating principle of PIN phase shifter in a dielectric waveguide. (a) Metal plate at contact with the waveguide, (b) metal plate removed to infinity.

nism has been used in a leaky wave antenna structure to control the direction of the main beam (Itoh and Hebert, 1978).

Use of a semiconductor PIN structure instead of the mechanical arrangement leads to an electronic phase shifter as demonstrated by Jacobs and Chrepta (1974). As shown in Fig. 40, a distributed PIN diode is attached to the upper surface of a rectangular dielectric waveguide. The application of dc bias on the PIN diode is equivalent to adjusting the location of the metal plate in the mechanical phase shifter.

Vanier and Mindock (1975) improved the bias current requirement of the PIN diode phase shifter by using a series stacked sandwich structure that reduces the thickness of the depletion layer. A typical PIN diode constructed in this manner has the width of 0.15 cm, height of 0.127 cm, and length of 0.762 cm. For this device, the forward voltage drop of 2.0 V at 10 mA and excess carrier life time of 30 μsec were measured. It was found that an applied current level of 20 mA or greater will drive the intrinsic region of the diode to high-conductivity levels at 70-GHz operation, yielding a phase-shift characteristic of the attached metal boundary.

This PIN diode is mounted on a dielectric waveguide made of a high-resistivity silicon having a cross section of 0.2×0.1 cm. Measured phase shift and dynamic insertion loss at 70 GHz are shown in Fig. 41. It is seen that a rapid increase in phase shift taking place for small current values is followed by a saturation region where any further increase of current has no appreciable change in phase shift. The transition point is around 20 mA

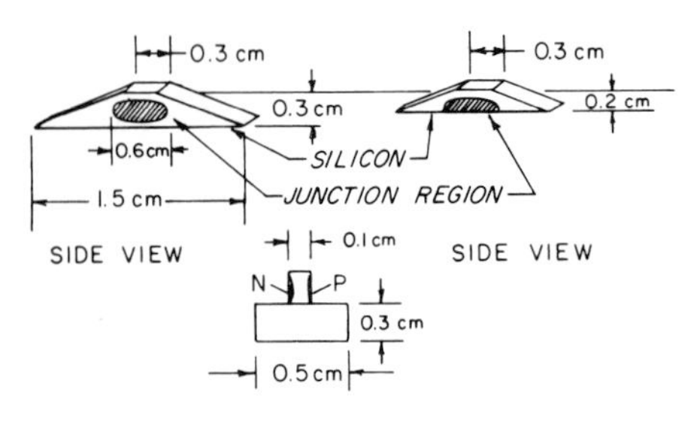

FIG. 40 Geometry of a typical electronic phase shifter. (From Jacobs and Chrepta, 1974. Reprinted from *IEEE Transactions on Microwave Theory and Techniques* **22,** 411–417. © 1974 IEEE.)

beyond which the diode enters high-conductivity regions. At this current level, the boundary condition at the top wall of the waveguide is that of a perfect conductor. Hence any further increase of the current does not change the boundary condition, and no further increase in phase shift may thus be explained.

From the loss characteristic in Fig. 41, it is apparent that this phase shifter is not appropriate to use at a small current bias. It is more useful for digital implementation, because the loss is reduced at higher current level.

In addition to the dynamic loss, there is an inherent insertion loss caused by the mounting of the PIN structure to the waveguide. Since the PIN structure is piggy-backed on the smooth surface of the dielectric

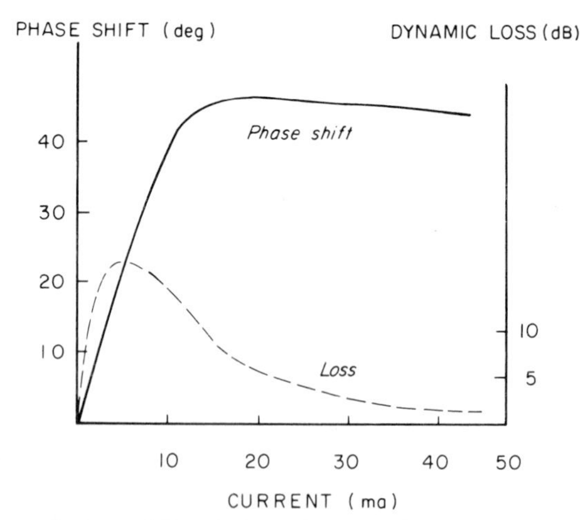

FIG. 41 Measured phase shift and dynamic insertion loss at 70 GHz. (From Vanier and Mindock, 1975. Reprinted from *IEEE MTT-S International Microwave Symposium*, pp. 173–175. © 1975 IEEE.)

waveguide, radiation losses due to the junction effect may be substantial. In the measurement, the insertion loss due to radiation was found to be as high as 3.5 dB. Although this device requires further improvement in efficiency and loss characteristics, there are possible applications to a number of components and subsystems, such as oscillators and phased arrays in the forms of dielectric integrated circuits.

B. Oscillators

Oscillators for dielectric millimeter-wave integrated circuits may be made of an active device such as a Gunn or IMPATT diode mounted in a cavity resonator. The latter is typically made of a finitely long section of a dielectric waveguide. The resonant frequency in such a cavity may be approximately obtained by assuming that the open ends of the cavity satisfy the magnetic wall (open-circuited) condition, if the dielectric constant of the material comprising the cavity is sufficiently high.

A number of works have been reported on the fabrication of Gunn or IMPATT oscillators. In this chapter only a few of them reported in open literature or symposia will be included. Chrepta and Jacobs (1974) used high-resistivity semiconductor materials such as 10,000 Ω-cm silicon or GaAs, as a cavity for Gunn or IMPATT oscillators in the K_u-band and V-band. The diode is mounted on a heat-sinking slab and connected by a post and disk through a hole formed in the dielectric cavity. The location of the hole is approximately at a multiple of the half-guide wavelength from one open end. The disk and post also work as an impedance tuner, and the frequency of oscillation is dependent on the disk diameter. An rf choke is built in this tuning and holder mechanism.

Dependence of the oscillation frequency and the power output on the cavity length have been studied by Jacobs *et al.* (1976). This oscillator is described in Section III of this chapter.

The equivalent circuit of these oscillators may be given as in Fig. 42 if the waveguide comprising the cavity supports only one guided mode. Y_0 and β are the characteristic admittance and propagation constants of such a mode, and Y_L is the load admittance seen from the cavity toward the load. If no load is connected on the left-hand side, the open-end boundary may be considered as an open circuit. B_D and $-G_D$ are the susceptance and the negative conductance attributed to the diode as seen from the cav-

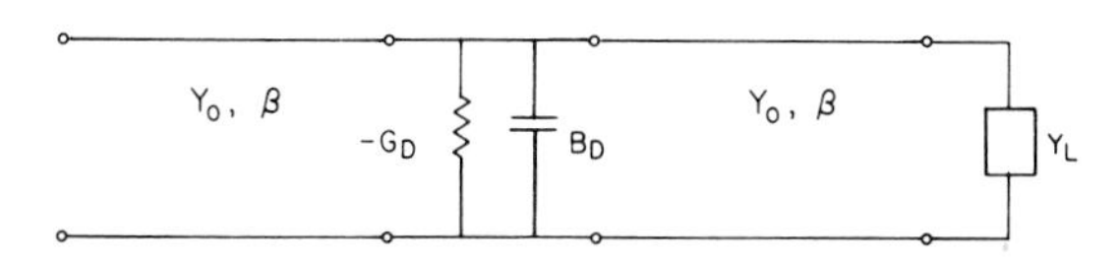

FIG. 42 Equivalent circuit of a dielectric waveguide oscillator.

ity. Since a diode is mounted in an open waveguide structure, the guided wave is scattered by the diode and only a fraction of guided wave energy couples with the diode. The rest of the energy will couple to the radiation mode and may be expressed as a radiation resistance (radiation loss) and a susceptance (stored energy due to reactive part of the radiation spectrum). The expressions of G_D and B_D must include these spurious components.

In practice, radiation problems associated with the diode structure need to be minimized for development of efficient oscillators. Use of a circular or rectangular disk on the upper surface above the diode of the resonator employed by Jacobs *et al.* (1976) is one of the schemes for such a purpose. It is also necessary to implement an rf choke or an equivalent in the bias circuit.

The oscillators described above have cavities made of open-ended waveguide. The Q of the resonator depends on the Fresnel reflection at the open end. Hence, if the dielectric constant of the waveguide material is not sufficiently high, a high-Q cavity may not be formed. Also, in some cases, one desires to make an oscillator without cutting a waveguide to form a finitely long cavity. As the operating frequency gets higher, fabrication of such a cavity is expected to become increasingly expensive.

To circumvent problems described in the preceding paragraph, Itoh and Hsu (1979) and subsequently Song and Itoh (1979) proposed oscillators made of cavities with two grating sections facing each other. As shown in Fig. 43, the oscillator consists of a diode implanted in a dielectric waveguide such as an image guide and grating structures created on the top surface of the waveguide on both sides of the diode. As demonstrated in Section IV, the surface-wave stop band of a periodic open-waveguide structure can be used for a band-reject filter. If the guided wave is incident on the section of waveguide with a grating, it is reflected only when the frequency lies in the stop band. Therefore, the grating can be thought of as a frequency-selective reflector.

The length of the cavity in Fig. 43 is l_1 and l_2. The cavity is terminated with two frequency-selective reflectors. The equivalent circuit of the oscillator is similar to the one in Fig. 42, except that the transmission lines l_1 and l_2 are now terminated with frequency-selective load admittances. If the grating is semiinfinitely long, the input admittance becomes purely reactive in the stop band and purely real at other frequencies. For a finitely long grating, it has also a small real part in the stop band. In any case, a high-Q cavity is realized only when the electrical lengths of l_1 and l_2 are such that the resonance condition is satisfied at a frequency in the stop band of the gratings.

This oscillator has potentially the following advantages over conven-

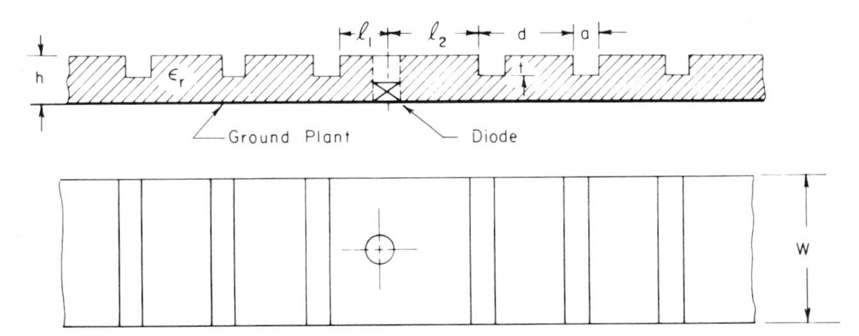

FIG. 43 Configuration of a distributed Bragg reflector oscillator. (From Itoh and Hsu, 1979. Reprinted from *IEEE Transactions on Microwave Theory and Techniques* **27,** 514–518. © 1979 IEEE.)

tional types: (1) since the grating may be created by the use of planar techniques, the structure becomes more cost effective at higher frequencies; (2) by changing the profile of grating structure, the Q of the resonator may be controlled because the width of the stop band depends on such profiles; and (3) high-Q cavities may be obtained in a waveguide made of relatively low-permittivity materials. This feature is also attractive at higher frequencies when choice of low-loss materials is quite limited.

In the oscillator, as in Fig. 43, the output is extracted through the finitely long grating section by means of a small resistive part in the equivalent admittance. The oscillator structure proposed by Song and Itoh (1979), on the other hand, has an added feature that the output can be extracted in the direction transverse to the waveguide axis.

As will be described later, a grating section created in the open dielectric waveguide can be used as a leaky-wave antenna if the period d of the grating is appropriately long. In such a leaky-wave antenna, the direction of the main beam is a function of the frequency. When such a beam is in the broadside (normal to the axis of the waveguide with gratings), a high VSWR results due to the wave coupling as the frequency is now in the so-called leak-wave stop band (Collin and Zucker, 1969). In ordinary applications, such an operation would be avoided as the efficiency of the antenna is not good. However, in the proposed oscillator (Song and Itoh, 1979), this high-input VSWR has been deliberately used for providing a frequency-selective feedback to the diode.

The proposed structure has a construction similar to the one in Fig. 43 except that one of the gratings has much longer period d so that at the desired oscillation frequency this grating works as a broadside firing leaky-wave antenna. The high-input VSWR is used to create a high-Q cavity in conjunction with the surface-wave stop-band phenomena of another grating on the other side of the diode. Therefore, one of the gratings in the oscillator performs two functions, an antenna and a frequency-selective reflector.

These proposed structures have been tested in the X-band and the K_u-band to demonstrate their operation. The results reported indicate that these components would be useful at higher frequencies.

C. Mixers and Self-Oscillating Mixers

A Gunn oscillator made with dielectric waveguide cavity may be operated as a self-oscillating mixer in which the local oscillator and mixer functions are realized by a single Gunn diode (Chrepta and Jacobs, 1974, 1977). The basic structure shown in Fig. 44 is quite similar to the standard dielectric waveguide Gunn oscillator. The incoming signal is mixed with the local oscillator output within the identical Gunn diode, and the if output is picked up through an rf choke.

Fabrication and measurements of self-oscillating mixers were performed both at K_u-band and V-band (Chrepta and Jacobs, 1977). At 60 GHz, conversion gains of V-band mixers were measured up to 15 dB. Typically, the local power output from the Gunn diode was 0.1 mW at 60.78 GHz when the conversion gain was set at 5 dB. At this operating condition, the minimum detectable signal was found to be in the range of -80 to -84 dBm over about 100-MHz if band. This yielded a total receiver noise figure of 4 dB. This is a quite low value in the 60-GHz range.

Balanced mixers have also been fabricated by a number of authors for V-band operations (Knox, 1976; Paul and Chang, 1978). Paul and Chang (1978) used image guides made of boron nitride (BN) for constructing a V-band (60 and 70 GHz) and W-band (94 GHz) balanced mixers. As shown in Fig. 45, the mixer consists of GaAs Schottky beam-lead diodes

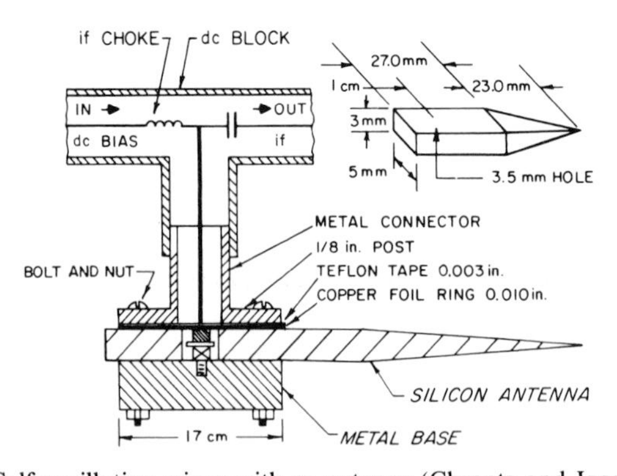

Fig. 44 Self-oscillating mixer with an antenna (Chrepta and Jacobs, 1974).

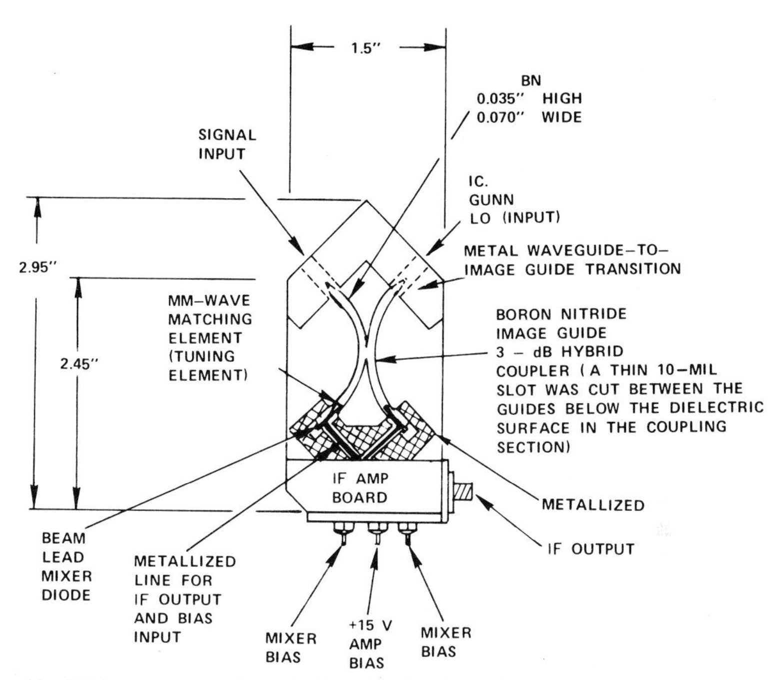

FIG. 45 BN image guide balanced mixer circuit. (From Paul and Chang, 1978. Reprinted from *IEEE Transactions on Microwave Theory and Techniques* **26,** 751–754. © 1978 IEEE.)

mounted at the two output ends of a 3-dB hybrid coupler. Intermediate frequency (IF) output lines are printed on the BN substrate and connected to an if amplifier. Behind each diode, an air-filled cavity is provided to create a $\frac{1}{4}$-wavelength termination. Although metal top plates were placed over the cavity section, they only improved the mixer noise figure by

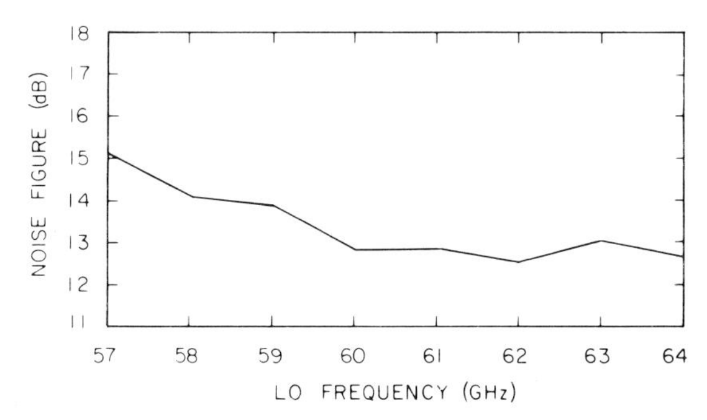

FIG. 46 Noise figures as a function of LO frequencies of the 60-GHz BN image guide balanced mixer. Tuning elements epoxied in place; IF passband: 600 NHz–2.2 GHz; LO power ≈ + 10 dBm; IF noise figure = 3.5 dB. (From Paul and Chang, 1978. Reprinted from *IEEE Transactions on Microwave Theory and Techniques* **26,** 751–754. © 1978 IEEE.)

about 1 dB. This means that the signal coupling to the diodes was quite adequate.

Two other ends of the coupler are tapered and connected to rf signals and a local oscillator (LO). Integrated image guide Gunn oscillator was used as a LO. Double sideband (DSB) noise figures of a 60-GHz BN-balanced mixer are shown in Fig. 46. Generally, the BN mixers exhibited bandwidths in excess of 5 GHz and the LO–rf isolation was found to be in the range of 15–20 dB.

The structure of the 94-GHz balanced mixer is very similar to that of *V*-band counterparts. The best DSB noise figure obtained is 14 dB. Lower noise-figure performances are expected by the use of better beam-lead diodes. Also, wider bandwidth capability of this type mixers is anticipated.

VI. Antennas for Dielectric Millimeter-Wave Integrated Circuits

Since dielectric waveguides are open structures, guided electromagnetic energy can relatively easily be transformed to radiating waves. Electromagnetic fields in an open structure can be completely described in terms of finitely many-guided modes and a continuous spectrum of radiation mode. When some discontinuities appear along the dielectric waveguide, a part or all of energy in a guided mode can be radiated. When this radiation is well controlled, an efficient antenna can be made from a dielectric waveguide. Hence it is possible to integrate antennas with the rf front end and other remaining dielectric integrated circuit structures. This is an attractive feature of dielectric waveguide techniques when designed for higher-frequency operation.

There are two types of antennas developed and studied to date.† They are surface-wave antennas and leaky-wave antennas. Since theoretical descriptions of these antennas are detailed in Collin and Zucker (1969), we shall emphasize in what follows particular features of these antennas developed for millimeter-wave dielectric integrated circuit technology.

A. Surface-Wave Antennas and Arrays

Surface-wave antennas are relatively simple to construct from dielectric waveguides if the operating frequency is not very high so that precision machining is not too expensive. The main beam of this antenna radiates usually in the endfire direction. The radiation occurs due to the trun-

† Other types such as dielectric lens antennas will not be discussed here, as they are not directly related to dielectric waveguide structures.

cation of surface-wave structures. Therefore, if the dielectric waveguide is suddenly terminated, an endfire surface-wave antenna may be created. However, in most cases, instead of a sudden termination, a gradual tapered transition from a dielectric waveguide to free space is used. The taper can reduce the reflection and hence the antenna may be made more efficient. In addition, the sidelobe level and pattern bandwidth may be somewhat improved.

One of the simplest surface-wave antennas used in dielectric millimeter-wave integrated circuits is the rod antenna used in the *V*-band communications receiver developed at IIT Research Institute (Knox, 1976). In this case the antenna is protected by a sleevelike radome made of a low-dielectric material. Another example is the silicon antenna implemented in the self-oscillating mixer (see Fig. 44).

Surface-wave antennas have been used as radiating elements of a frequency-scanned array (Williams *et al.*, 1977). An eight-element prototype array section is shown in Fig. 47. Alternate surface-wave radiating elements are fed from the main insular guide transmission line via proximity coupler. The array circuit is repeated on the reverse side of the ground plane, and each array half is excited from a waveguide input with an E-plane power splitter. This arrangement permits the use of couplers whose radii of curvature are limited to prevent excessive radiation from the curve sections and yet maintains close element spacing necessary for wide-angle scanning. The curved sections are identical and the required amplitude aperture distribution is established by adjusting the line spacing at each coupler location. The waveguide and antenna elements are made of Custom Material's High-K 707L, which is a machinable styrene-based ceramic and has a dielectric constant of around ten. Dielectric rods made

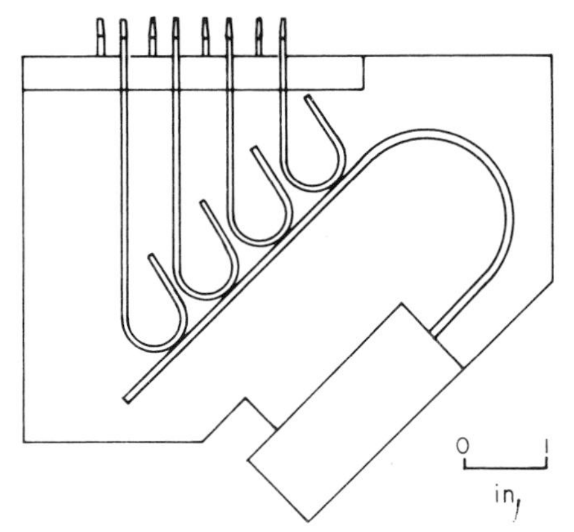

FIG. 47 Eight-element prototype array section. (From Williams *et al.*, 1977. Reprinted from *IEEE MTT-S International Microwave Symposium*, pp. 542–544. © 1977 IEEE.)

of this material are positioned and bonded on a thin polyethylene film to constitute insular guides.

In Fig. 48, radiation patterns of the eight-element array are presented. The higher sidelobe levels in experimental data are attributed to small distortions in the styrene-based material. These distortions arose during fabrication and result in different phase tapers in the two halves of the array.

It is seen that sidelobe levels increase as the main beam is steered away from broadside. The performance may be improved by tapering the aperture distribution and decreasing the element spacing. It is found, however, that a uniform scan performance more than ±20° is possible with a 10%-frequency bandwidth at the center frequency around 30 GHz.

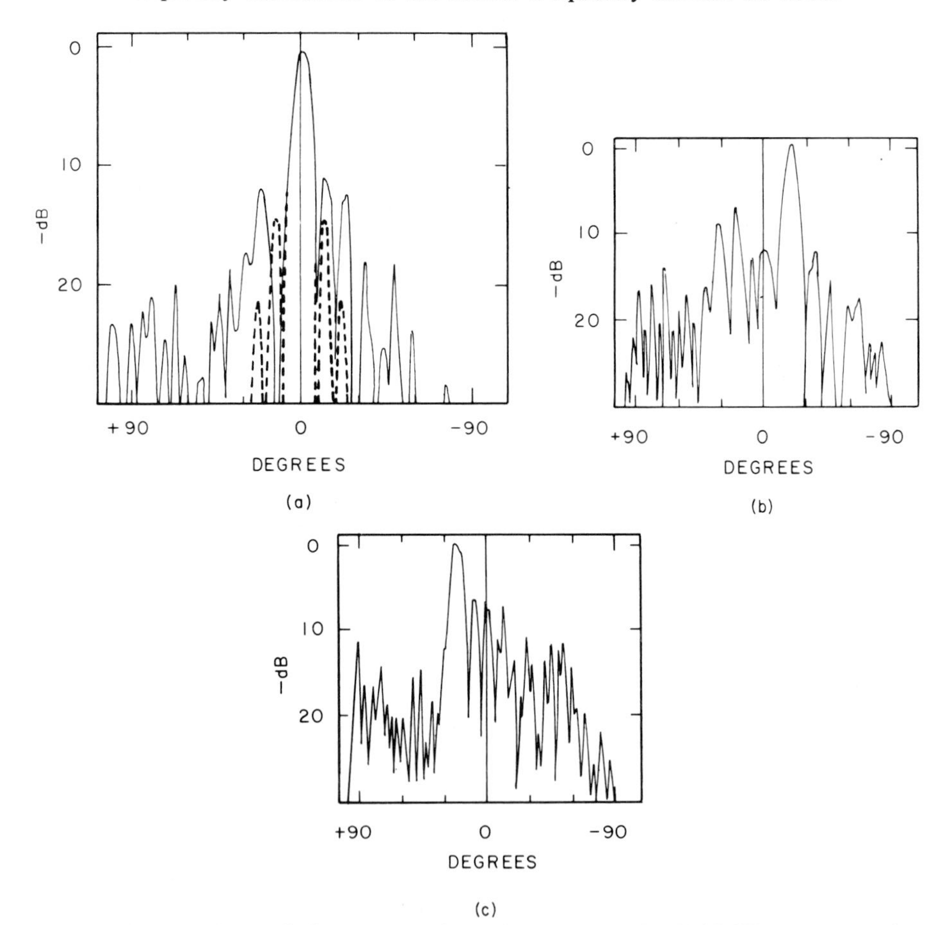

FIG. 48 H-plane radiation patterns of complete array. (a) f = 29.25 GHz; ——, experiment; --, theory; (b) f = 27.75 GHz, (c) f = 30.75 GHz. (From Williams *et al.*, 1977. Reprinted from *IEEE MTT-S International Microwave Symposium*, pp. 542–544. © 1977 IEEE.)

B. Leaky-Wave Antennas

In integrated optical circuits, the so-called grating couplers are frequently used for coupling a laser beam from free space to an optical waveguide or vice versa. In electromagnetic and microwave terminology, these grating couplers are nothing but leaky-wave antennas. Hence leaky-wave structures may be realized in dielectric waveguide-type millimeter-wave integrated circuit in a similar manner, viz., by the use of gratings. Since a number of planar technologies may be used for fabrication of gratings, leaky-wave antennas are potentially more cost effective at higher frequencies than surface-wave antennas.

A leaky-wave antenna generally radiates its main beam not in the endfire, but in a sector of sideward direction. Usually one avoids the broadside radiation since the input VSWR becomes excessive under such an operating condition due to wave-coupling mechanism.

The principles of operation of leaky-wave antennas are detailed in Collin and Zucker (1969). We shall here illustrate a typical example of a leaky-wave antenna integrated in a dielectric waveguide. The structural view of a leaky-wave antenna created in an inverted strip dielectric (IS) waveguide is the same as the one illustrated in Fig. 35 (Itoh, 1977). Gratings can be created by periodically modulating the geometrical or material nature of the dielectric strip. For instance, in Fig. 35 periodic grooves are created in the dielectric strip. These grooves periodically perturb the transmission characteristics of the IS guide, which then functions as a grating.

Electromagnetic waves in a grating region can be represented in terms of Floquet space harmonics with phase constants given by

$$\beta_p = \beta_0 + (2p\pi/d), \qquad p = 0, \pm 1, \pm 2, \ldots, \tag{72}$$

where d is the grating period and β_0 is the phase constant of the dominant ($p = 0$) space harmonic and is determined by the excitation of the grating. β_0 is a complex quantity in the grating operated as a leaky-wave antenna to account for the energy leakage. However, if perturbation at each grating element is small, β_0 is very nearly real and close to the phase constant, say k_z, of the dominant guided mode in the unperturbed IS waveguide except in the wave-coupling regions. In this discussion, effects of higher-order guided modes in the IS guide will not be considered.

Figure 34a shows the k–β diagram of a grating structure in an IS waveguide when perturbation is vanishingly small. Curves corresponding to each space harmonic are explained in Section IV.B. Once again, wave-coupling phenomena associated with synchronous points such as A, B, C, and D are not shown.

Let us choose the product of the free-space wave number k and the grating period d, i.e., the electrical length of this period, such that the phase constant of the pth space harmonic, say $p = -1$, satisfies

$$|\beta_p/k| < 1. \tag{73}$$

Then the grating supports a leaky wave, and the wave traveling in the grating leaks into free space along the direction given by

$$\theta_p = \sin^{-1}(\mathrm{R_e}\beta_p/k), \tag{74}$$

measured from the broadside of the grating, where Re β_p is the real part of β_p. From Eqs. (72) and (74) it is seen that θ_p is a function of the operating frequency. Hence the beam can be steered by changing the frequency. In Fig. 34a, $\beta_p(p = -1)$ lies on the curve from P and Q except near coupling points such as B, where a wave coupling occurs between different space harmonics. Since these points lie in the radiation region of k–β diagram, the wave coupling is different from the surface-wave stop band discussed earlier.

Since actual leaky-wave antennas have finite lengths, there are sidelobes in addition to the main beam in the direction given by Eq. (74). An approximate method for deriving such sidelobe characteristics will now be given. If the perturbation due to each of N grating elements in the antenna is infinitesimal and if the operating point is not in one of the wave-coupling regions, the grating may be viewed as an N-element linear array of spacing d, excited uniformly with a linear-phase taper corresponding to Re β_0. Since the energy loss at each grating element is assumed small, the amplitude taper is also small and may be neglected. Under such approximations the power radiation pattern may be expressed as

$$|f(\theta)|^2 = 1/N^2|\sin(N\Psi/2)/\sin(\Psi/2)|^2, \tag{75}$$

where $\Psi = kd \sin\theta - \mathrm{Re}(\beta_\gamma d)$.

A leaky-wave IS guide antenna designed for 60 GHz and scaled to 15-GHz operations is shown in Fig. 49. The smaller tip on one end of the guiding layer (made of Stycast HiK, $\varepsilon_r = 3.75$) is the transition for launching whereas the larger tip on the other end is intended to avoid the reflection of energy left unradiated at the grating. Figure 50 shows measured and computed radiation patterns of the model antenna. At 15 GHz, the main beam is directed $-26°$ (26° toward the backward direction) form the broadside. At 17 GHz, the main beam direction changes to $-10°$ from the broadside. It is noticed that computed and measured patterns differ considerably near the endfire direction. The major reason is that the antenna is not long enough, and considerable electromagnetic energy is left

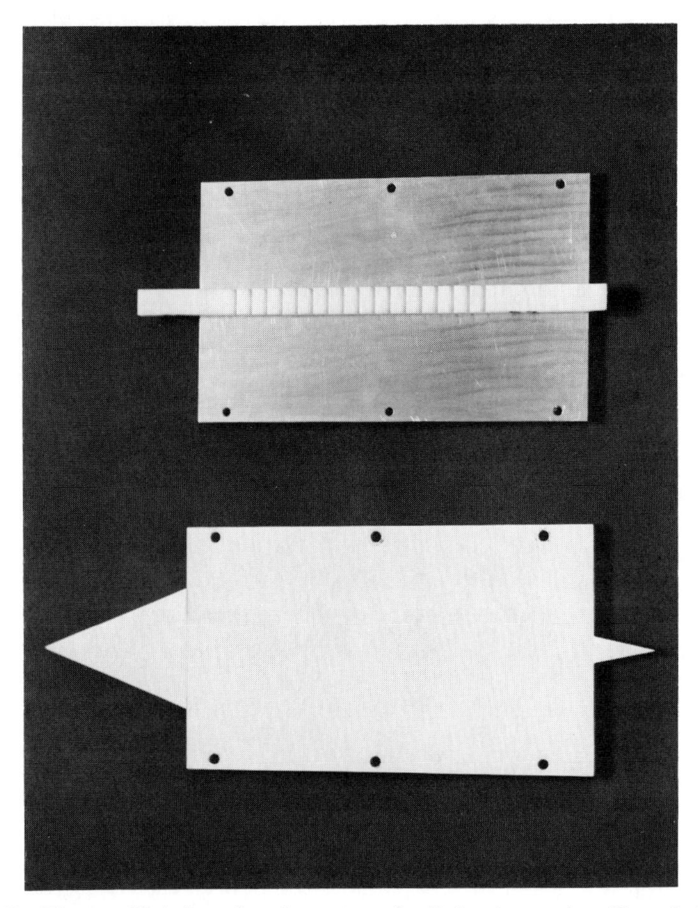

FIG. 49 Disassembled view of grating antenna for K_u-band operation. (From Itoh, 1977. Reprinted from *IEEE Transactions on Microwave Theory and Techniques* **25,** 1134–1138. © 1977 IEEE.)

unradiated beyond the last grating element. To avoid reflection, the forward end of the antenna has a tapered section that unintentionally works as a surface-wave antenna. It should be relatively straightforward to reduce this unwanted radiation by increasing the length of the antenna and creating more grating elements.

A frequency-scanned leaky-wave antenna was also created from rectangular silicon waveguide by Klohn *et al.* (1978). By changing the operating frequency from 57 to 65 GHz, the main beam direction changed from $-68°$ to $-7°$, resulting in a change of 7.5°/GHz. Experimental plot of radiation angle versus frequency is given in Fig. 51.

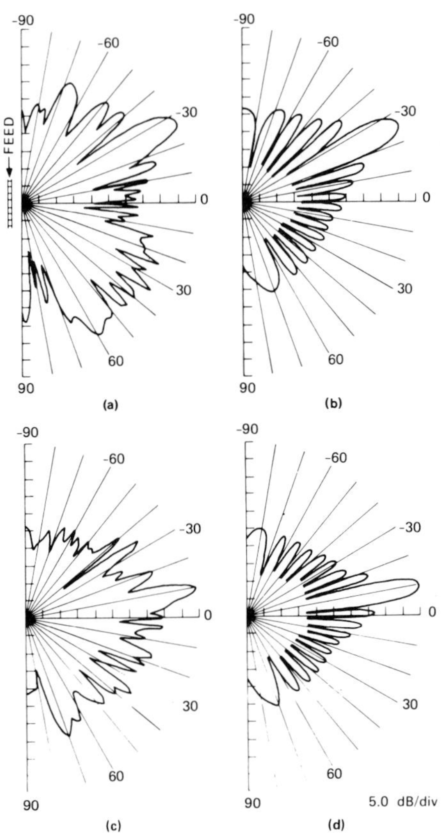

FIG. 50 Radiation patterns of grating antenna: d = 10.16 mm, a = 3.18 mm, b = 4.76 mm, $h_1 = h_2$ = 6.35 mm, ε_1 = 2.1, ε_2 = 3.75, number of grating elements = 17. (a) Measured at 15 GHz; (b) computed at 15 GHz; (c) measured at 17 GHz; (d) computed at 17 GHz. (From Itoh, 1977. Reprinted from *IEEE Transactions on Microwave Theory and Techniques* **25,** 1134–1138. © 1977 IEEE.)

Instead of frequency steering, the main beam direction may be scanned by creating phase shift in a waveguide either electrically or mechanically. If PIN phase shifters described in Section V.A are implemented in the leaky-wave antenna, the beam direction can be changed by adjusting the bias voltage of the PIN device. If speed is not the main concern, a phase shift may be created by mechanical means. If an additional layer of dielectric plate is placed on the grating section of dielectric waveguide, the phase constant may be varied by adjusting the vertical location of such a layer and thus the main beam direction changes. Such a model antenna was reported by Itoh and Hebert (1978). As the frequency becomes higher, however, the mechanical structure is not very attractive.

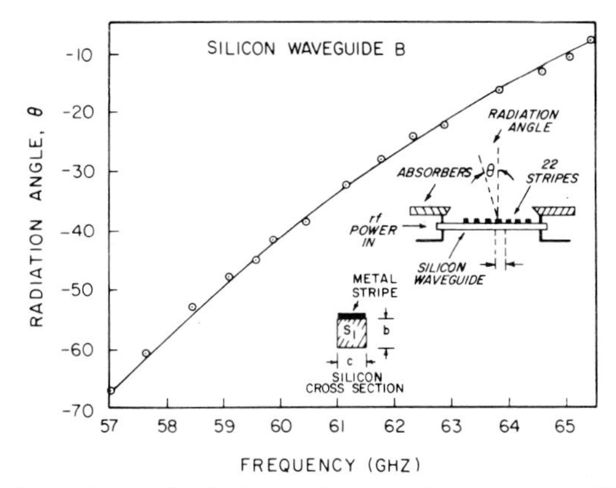

FIG. 51 Experimental plot of radiation angle versus frequency: $n = 22$ stripes, $d = 0.18$ cm, $\bar{Q} = 0.969$ mm, $b = 1.071$ mm. (From Klohn *et al.*, 1978. Reprinted from *IEEE Transactions on Microwave Theory and Techniques* **26,** 764–773. © 1978 IEEE.)

VII. Subsystems

Following description of individual components, we shall illustrate some of the subsystems that are defined here as combinations of more than one component developed in accordance with dielectric millimeter-wave integrated circuit techniques. We shall include only those works appearing in open literature.

FIG. 52 Oscillator, modulator detector module (Chrepta and Jacobs, 1974). (Courtesy of Dr. H. J. Kuno.)

A novel oscillator was created by integrating a leaky-wave antenna as a part of the oscillator cavity (Song and Itoh, 1979). In this structure, one of the grating reflectors in Fig. 43 is replaced with a leaky-wave antenna that radiates in the broadside direction at the designed oscillation frequency. Usually the broadside operation is avoided in a leaky-wave antenna because the input VSWR becomes higher under such an operation. The operating point on the k–β diagram is then at B in Fig. 34a at which the wave coupling takes place and a leaky-wave stop band is created (Collin and Zucker, 1969). The resulting high-input VSWR is deliberately used for providing the feedback for the solid-state device, a Gunn diode. Hence this grating structure has double function, one as a leaky-wave antenna and the other as a frequency-dependent reflector.

V-band communication receivers and transmitters have been developed at IIT Research Institute using insular guides, although not all of the microwave components, particularly solid-state devices, are integrated in dielectric waveguide structures (Knox, 1976; Chrepta and Jacobs, 1974). In the receiver, the incoming rf signal is coupled to a ring preselector filter ($Q \approx 2500$) and then to the 3-dB quadrature hybrid, both of which are made of insular guides. The if signal is generated in the balanced mixer. The receiver characteristics include conversion loss of 6.0 dB and noise figure 7.5 dB at the input frequency of 58.8 GHz.

An oscillator–modulator–detector module developed at Hughes Air-

FIG. 53 Photograph of a BN image guide balance mixer with a BIV image guide Gunn LO attached. (From Paul and Chang, 1978. Reprinted from *IEEE Transactions on Microwave Theory and Techniques* **26,** 751–754. © 1978 IEEE.)

craft Company using silicon waveguide structures is shown in Fig. 52. A 60-GHz balanced mixer-local oscillator module was developed using boron nitride image guide (Paul and Chang, 1978). Figure 53 shows the module. The balanced mixer of this module is described in Section V of this chapter. This module was used in the compact 60-GHz communication interconnect system.

VIII. Conclusions

In this article, we reviewed the latest development of dielectric waveguide-integrated circuit techniques. In the past few years, considerable interest has developed both in theoretical and developmental aspects of these structures. Although a number of impressive works have appeared, there remains much to be worked out. Much of the problem stems from the fact that the dielectric waveguide-type millimeter-wave integrated circuits are based on the hybrid technology. Most of the passive devices are based on the optical techniques, viz., dielectric waveguide, whereas solid-state devices available to date are of lumped element type originally developed for microstrip type millimeter-wave integrated circuits. Interfacing these two classes of configuration is naturally difficult, but challenging.

Most of the developmental works to date are for the frequency up to 100 GHz. In future, works at even higher frequencies will be attempted. It is then more important to recognize the advantageous features of dielectric waveguide techniques and to find a technique to circumvent difficulty in interfacing. It is desirable that new devices more appropriate for dielectric waveguide techniques be developed. When these are done, a wide-open frequency spectrum above 100 GHz will be much better utilized.

Acknowledgments

This work was supported by a U.S. Army Research Office Grant DAAG29-78-G-0145. The author acknowledges assistance and encouragement from a number of people. They include Dr. H. Jacobs, U.S. Army Electronics Technology and Device Laboratory, Dr. H. J. Kuno of Hughes Aircraft, Dr. Y. W. Chang of TRW, Dr. R. M. Knox of Epsilon Lambda, and Professor A. A. Oliner of Polytechnic Institute of New York.

References

Aylward, M. J., and Williams, N. (1978). *Eur. Microwave Conf., 8th, Paris* 319–323.

Babitt, R. W., Stern, R. A., Whicker, L. R., and Young, C. W., Jr. (1978). *Conf. Proc. Millimeter Submillimeter Wave Propag. Circuits, NATO/AGARD,* Munich, pp. 34-1–34-15.

Chandler, C. H. (1949). *J. Appl. Phys.* **20,** 1189–1192.

Chrepta, M. M., and Jacobs, H. (1974). *Microwave J.* **17**(11), 45–47.

Chrepta, M. M., and Jacobs, H. (1977). "Self-Oscillating Mixers in Dielectric Waveguide," ECOM Rep. Army Electronics Command, Fort Monmouth, New Jersey.
Collin, R. E., and Zucker, F. J. (1969). "Antenna Theory," Part II. McGraw-Hill, New York.
Furuta, H., Noda, H., and Ihaya, A. (1974). Novel optical waveguide for integrated optics. *Applied Optics* **13,** 322–326.
Goell, J. E. (1969). *Bell Syst. Tech. J.* **48,** 2133–2160.
Harris, D. J., Lee, K. W., and Reeves, J. M. (1978). "Groove- and H-Waveguide Design and Characteristics at short Millimeter Wavelengths," *IEEE Trans. Microwave Theory Tech.* **26,** 998–1001.
Itoh, T. (1976). *IEEE Trans. Microwave Theory Tech.* **24,** 821–827.
Itoh, T. (1977). *IEEE Trans. Microwave Theory Tech.* **25,** 1134–1138.
Itoh, T., and Hebert, A. S. (1978). *IEEE Trans. Microwave Theory Tech.* **26,** 987–991.
Itoh, T., and Hsu, F. J. (1979). *IEEE Trans. Microwave Theory Tech.* **27,** 514–518.
Jacobs, H., and Chrepta, M. M. (1974). *IEEE Trans. Microwave Theory Tech.* **22,** 411–417.
Jacobs, H., Novick, G., LoCascio, C. M., and Chrepta, M. M. (1976). *IEEE Trans. Microwave Theory Tech.* **24,** 815–820.
Klohn, K. L., Horn, R. E., Jacobs, H., and Freibergs, E. (1978). *IEEE Trans. Microwave Theory Tech.* **26,** 764–773.
Knox, R. M. (1976). *IEEE Trans. Microwave Theory Tech.* **24,** 806–814.
Knox, R. M., and Toulios, P. P. (1970). "Integrated Circuits for the Millimeter through Optical Frequency Range." Proc. Symp. on Submillimeter Waves Polytechnic Inst. of Brooklyn, New York.
Lewin, L., Chang, D. C., and Kuester, E. F. (1977). "Electromagnetic Waves and Curved Structures." Peter Peregrinus, Southgate House, Stevenage, Herts. SGI 1HQ, England.
McLevige, W. V., Itoh, T., and Mittra, R. (1975). *IEEE Trans. Microwave Theory Tech.* **23,** 788–794.
Marcatili, E. A. J. (1969). *Bell Syst. Tech. J.* **48,** 2071–2102.
Marcuse, D. (1970). *Bell Syst. Tech. J.* **49,** 273–290.
Marcuse, D. (1971). *Bell Syst. Tech. J.* **50,** 2551–2565.
Miller, S. E., Marcatili, E. A. J., and Li, T. (1973). *Proc. IEEE* **61,** 1703–1701.
Nanda, V. P. (1976). *IEEE Trans. Microwave Theory Tech.* **24,** 876–879.
Ogusu, K. (1977). *IEEE Trans. Microwave Theory Tech.* **25,** 874–885.
Ogusu, K., and Hongo, K. (1977a). *Trans. Inst. Electron. Commun. Eng. Jpn.* **J60-B,** No. 1, 9–16.
Ogusu, K., and Hongo, K. (1977b). *Trans. Inst. Electron. Commun. Eng. Jpn.* **J60-B,** No. 5, 358–359.
Oliner, A. A., Peng, S. T., and Hsu, J. P. (1978). *IEEE MTT-S Int. Microwave Symp. Dig.*, Ottawa, Canada, pp. 408–410.
Paul, J. A., and Chang, Y. W. (1978). *IEEE Trans. Microwave Theory Tech.* **26,** 751–754.
Pregla, R. (1974). *Arch. Elektron. Uebertragungstechn.* **28,** 349–357.
Rozzi, T. E. (1978). *IEEE Trans. Microwave Theory Tech.* **26,** 738–746.
Rudokas, R., and Itoh, T. (1976). *IEEE Trans. Microwave Theory Tech.* **24,** 978–981.
Schelkunoff, S. A. (1952). *Bell Syst. Tech. J.* **31,** 784–801.
Schlosser, W., and Unger, H. G. (1966). *Adv. Microwaves* **1,** 319–387.
Solbach, L. (1978a). *Arch. Elektron. Uebertragungstech.* **32,** 321–328.
Solbach, L. (1978b). *IEEE Trans. Microwave Theory Tech.* **26,** 755–758.
Solbach, K., and Wolff, I. (1978). *IEEE Trans. Microwave Theory Tech.* **26,** 266–274.
Song, B. S., and Itoh, T. (1979). *Int. Microwave Symp., Orlando, Fla.* pp. 217–219.

Thompson, D. E., and Coleman, P. D. (1974). *IEEE Trans. Microwave Theory Tech.* **22,** 995–1000.

Tischer, F. J. (1979). "Transmission Media for Millimeter-Wave Integrated Circuits," *IEEE MTT-S Int. Microwave Symp., Orlando, Fla.* pp. 203–207.

Toulios, P. P., and Knox, R. M. (1970). *Rectangular Dielectr. Image Lines Millimeter Integrated Circuits. 1970 Wescon Conf., Los Angeles, Calif.* pp. 1–10.

Vanier, G., and Mindock, R. M. (1975). *IEEE MTT-S Int. Microwave Symp., Palo Alto, Calif.* pp. 173–175.

Williams, N., Rudge, A. W., and Gibbs, S. E. (1977), *IEEE MTT-S Int. Microwave Symp. Dig., San Diego, Calif.* pp. 542–544.

Young, L., and Sobol, H., eds. (1974). *Adv. Microwaves* **8.**

CHAPTER 6

Submillimeter Guided Wave Experiments with Dielectric Waveguides

M. Tsuji, H. Shigesawa, and K. Takiyama

Department of Electronics
Doshisha University
Kyoto, Japan

I.	Introduction	275
II.	Dielectric Slab Waveguides	277
	A. *Analysis*	278
	B. *Experiments*	283
III.	Dielectric Rib Waveguides	297
	A. *Analysis*	298
	B. *Experiments*	302
IV.	Dielectric Cylindrical Waveguides	310
	A. *Analysis*	310
	B. *Experiments*	316
V.	Conclusion	319
	Appendix I	322
	Appendix II	324
	Appendix III	325
	References	325

I. Introduction

Recent developments of both power sources and detectors in the submillimeter-wave region require and encourage further research and improvement of the wave-guiding structures suitable for this spectral range. The microwave and optical techniques have become more sophisticated, and it is easy to expect that the technical progress in the submillimeter-wave region will be affected from both spectral ends of the optical and the millimeter wave regions. Moreover, it is safe to say that strong efforts will be made to develop dielectric waveguides for the integrated circuit uses in the submillimeter-wave region because of their mechanical stability and precision of construction.

However, a major difficulty now is to find low-loss dielectric materials

ISBN 0-12-147704-5

to be used in this spectral range. We consider that at this stage, the most important and urgent problem is to investigate what transmission characteristics are practically realized in typical waveguides made of already existing low-loss dielectrics.

From this point of view, this chapter deals with our recent experimental investigations of the transmission characteristics of three kinds of dielectric waveguides, viz., slab waveguides (Tsuji *et al.*, 1979c), rib waveguides (Shigesawa *et al.*, 1979a), and dielectric cylindrical waveguides (Tsuji *et al.*, 1979a).

In Section II, the slab waveguides are investigated as a basic structure common to all dielectric waveguides, although this waveguide is a two-dimensional structure that confines the fields only in one direction, say, vertically. The minute examinations of this waveguide, including experiments as well as analysis, will give us significant information on how to devise more practical waveguides in the submillimeter-wave region.

The exact analysis of the slab waveguide is indeed possible, but it is quite difficult as yet to discuss its characteristics obtained experimentally with sufficient accuracy and reliability corresponding to the theoretical analysis.

Most of our endeavors to develop the advanced experimental techniques in the submillimeter-wave region have proved to be successful. This, including theoretical discussions, will be described in detail in Section II.

Section III will treat the transmission characteristics of rib waveguides described by Goell (1973). Practical integrated circuits will properly require the waveguides having a three-dimensional structure, which confines the fields horizontally as well as vertically. Such circuits will likely take the form of dielectric strip waveguides placed on or embedded in a dielectric substrate.

A typical structure of this type of waveguides will be a dielectric rib waveguide where the material of the strip portion is the same as that of the film of the substrate. Experimental work on this waveguide has been successfully performed by means of the same techniques as those developed in Section II.B. However the theoretical considerations encounter many difficulties because of the irregular and unbounded structure of its boundaries, and we propose a new approximate analytical method, which satisfies the boundary conditions in the least-squares sense.

Finally, some discussions about dielectric cylindrical waveguides will be presented in Section IV. Such waveguides will be utilized as a transmission line to connect a number of isolated integrated circuits to one another.

A typical dielectric cylindrical waveguide has the same structure as an optical fiber (Kao and Hockham, 1966), which usually consists of a high-index core surrounded by a cladding of a lower-index material. Such a structure in which most of power is concentrated in and near its core is not utilized at present in the submillimeter-wave region because of the nonexistence of low-loss core materials, as we have mentioned above. Then the application of an O-type guide (Sugi and Nakahara, 1959), which consists of a thin film dielectric tube, is experimentally examined in Section IV.B. The low-loss property of this guide arises from the transmission of only a fraction of power through the dielectric. But this waveguide has a practical difficulty in wiring or supporting without excess losses, so that an improvement of O-type guides is discussed. In a proposed waveguide, the metal wire in its center is coated with a low-index material and isolated from an outer cylindrical layer with a higher index. It is then expected that this guide will again keep the low-loss characteristics like an O-type guide, even if a lossy metal wire is included as the tension member of a guide. This waveguide will be discussed in Section IV.A.2.

II. Dielectric Slab Waveguides

In this section, we study the transmission characteristics of slab waveguides as the basic guide form of a dielectric strip waveguide described in Section III.

The attenuation constant is one of the most important parameters for estimating the applicability of a guide for the submillimeter integrated circuits. Tacke and Ulrich (1973) have measured the attenuation constant of a polyethylene slab waveguide by varying the distance between the two prism couplers, and Danielwitz and Coleman (1977) have performed the same experiment by using a crystal quartz as the slab. However, in such a type of coupler based on the frustrated total reflection, the power coupling to a slab waveguide is so sensitive to a gap between a slab and a prism that it is practically impossible to keep up the constant coupling of a movable prism over the sufficient length for loss measurements. It is evident that the accuracy of loss measurements is perfectly due to this constant coupling, and then the movable metal-grating coupler has been proposed to perform such measurements with sufficient accuracy and reliability (Tsuji *et al.*, 1979c). First, we measured both the phase and the attenuation constants of a number of symmetric polyethylene slab waveguides using this grating coupler and obtained the typical attenuation constant of 1.3 dB/m for the TM_0 mode with 10-μm thickness.

Next, since a slab waveguide provides confinement of the field in only one dimension, the transverse broadening of a guided wave was measured to show the accuracy of our loss measurements, and it has been concluded that this broadening has negligible effects on our measured attenuation constants.

Finally, the bending losses of a slab waveguide were measured (Tsuji *et al.*, 1979b). Dielectric waveguides cannot guide electromagnetic energy around their bends without the radiation loss, but our measurements at the wavelength of 337 μm show that the guides with slab thickness greater than 50 μm are little affected by bends with radius of curvature larger than 0.5 m.

A. Analysis

A dielectric slab waveguide of thickness $2a$ is shown schematically in Fig. 1. The slab is assumed to have refractive index n_1 and is embedded in an infinite medium with refractive index n_2. By considering the infinite extent of the slab in both y and z directions, the field can be decomposed into TE and TM modes and each of them should independently satisfy the boundary conditions.

As mentioned before, the present slab waveguide in the submillimeter-wave region will always be constructed with a lossy dielectric material, and all of the constituent variables of the eigenvalue equation become complex quantities. To solve such a complex eigenvalue equation, Burke (1970) has shown an interesting method in which an eigenvalue equation is transformed into two simultaneous equations. Our theoretical treatments indeed follow his method, but the more general expressions applicable to all modes, including TM modes, are derived.

Now, the wave-guiding structure shown in Fig. 1 has a plane of symmetry in the y–z plane, and the propagating modes may then be divided roughly into two groups: symmetric and antisymmetric. Following the notations employed by Kapany and Burke (1972), the well-known eigenvalue equations are shown as follows:

$$q = \begin{cases} \eta u \tan u & \text{for symmetric modes,} \quad (1) \\ -\eta u \cot u & \text{for antisymmetric modes,} \quad (2) \end{cases}$$

where u and q are the normalized wave numbers in the x direction of mediums 1 and 2, respectively, and

$$\eta = \begin{cases} 1 & \text{for TE mode,} \\ n_2^2/n_1^2 & \text{for TM mode.} \end{cases} \tag{3}$$

The properties of eigen modes propagating along the z axis can then be derived from these equations, in conjunction with the conservation relation of wave numbers:

$$u^2 + q^2 = (n_1^2 - n_2^2)k_0^2a^2 = R^2, \tag{4}$$

where $k_0(=2\pi/\lambda_0)$ is the free-space wave number and the propagation

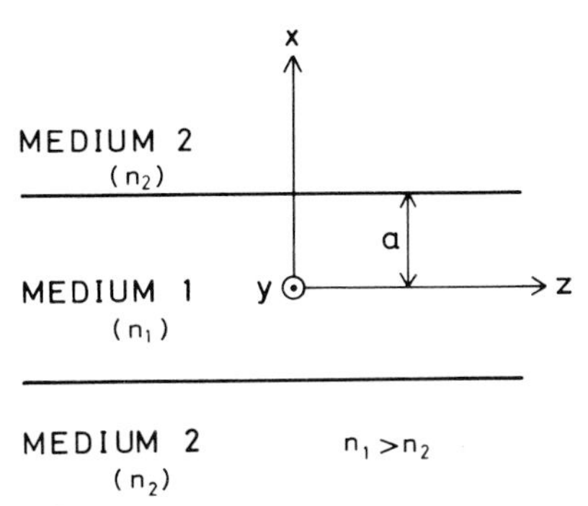

FIG. 1 Cross section of a symmetric slab waveguide. (From Tsuji *et al.*, 1979c. Reprinted from *IEEE Transactions on Microwave Theory and Techniques* **27**, 873–878. © 1979 IEEE.)

constant h in the z direction is given by

$$(ha)^2 = (n_1 k_0 a)^2 - u^2 = (n_2 k_0 a)^2 + q^2. \tag{5}$$

Substituting Eq. (4) into Eqs. (1)–(2), we then have

$$u^2(1 + \eta^2 \tan^2 u) = R^2 \text{ for symmetric modes} \tag{6}$$

and

$$u^2(1 + \eta^2 \cot^2 u) = R^2 \text{ for antisymmetric modes.} \tag{7}$$

For the lossy dielectric slab, the constituent variables of Eqs. (6)–(7), i.e., u, η, and R^2 become complex, say, $u = u' + ju''$. In this case, it is convenient to indicate R^2 in terms of its magnitude M and argument Φ as follows:

$$R^2 = M \exp(j\Phi). \tag{8}$$

Substituting Eq. (8) into Eqs. (6)–(7), each of those eigenvalue equations are transformed into two simultaneous equations for u' and u'', and these equations independent of modes can be shown as follows:

$$M = (u'^2 + u''^2)(m_r^2 + m_i^2)^{1/2}, \tag{9}$$

$$\Phi = \tan^{-1}\{2u'u''/(u'^2 + u''^2)\} + \tan^{-1}(m_i/m_r), \tag{10}$$

where m_r and m_i are given as follows:

$$\begin{aligned} m_r = \{&(\cos 2u' \pm \cosh 2u'')^2 + (\eta'^2 - \eta''^2) \\ &(\sin^2 2u' - \sinh^2 2u'') \\ &\mp 4\eta'\eta'' \sin 2u' \sinh 2u''\}/(\cos 2u' \pm \cosh 2u'')^2, \end{aligned} \tag{11}$$

$$\begin{aligned} m_i = 2\{&\eta'\eta''(\sin^2 2u' - \sinh^2 2u'') \\ &\pm(\eta'^2 - \eta''^2) \sin 2u' \sinh 2u''\}/ \\ &(\cos 2u' \pm \cosh 2u'')^2. \end{aligned} \tag{12}$$

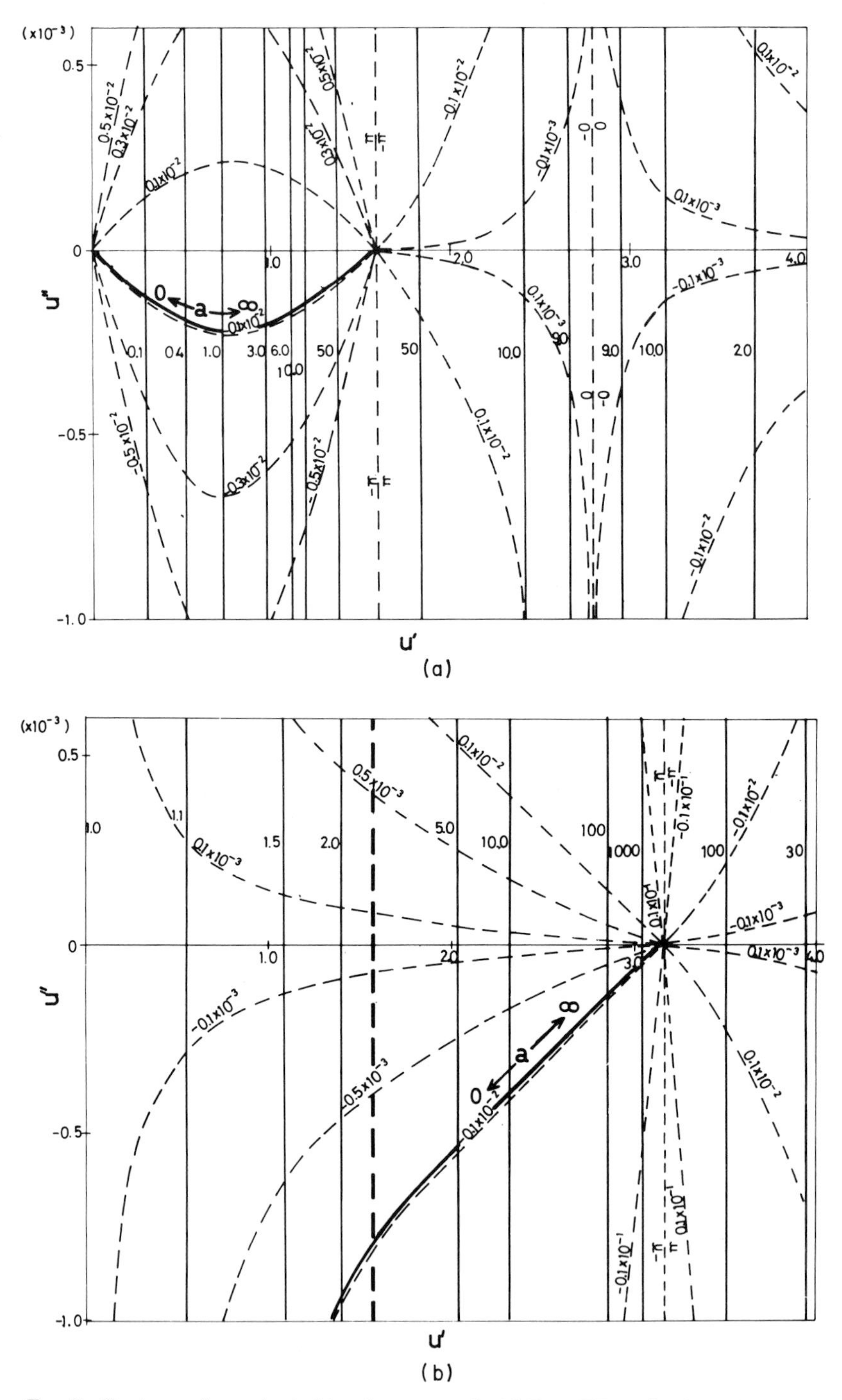

FIG. 2 Contours of magnitude M and argument Φ. (a) Even TE mode, (b) odd TE mode.

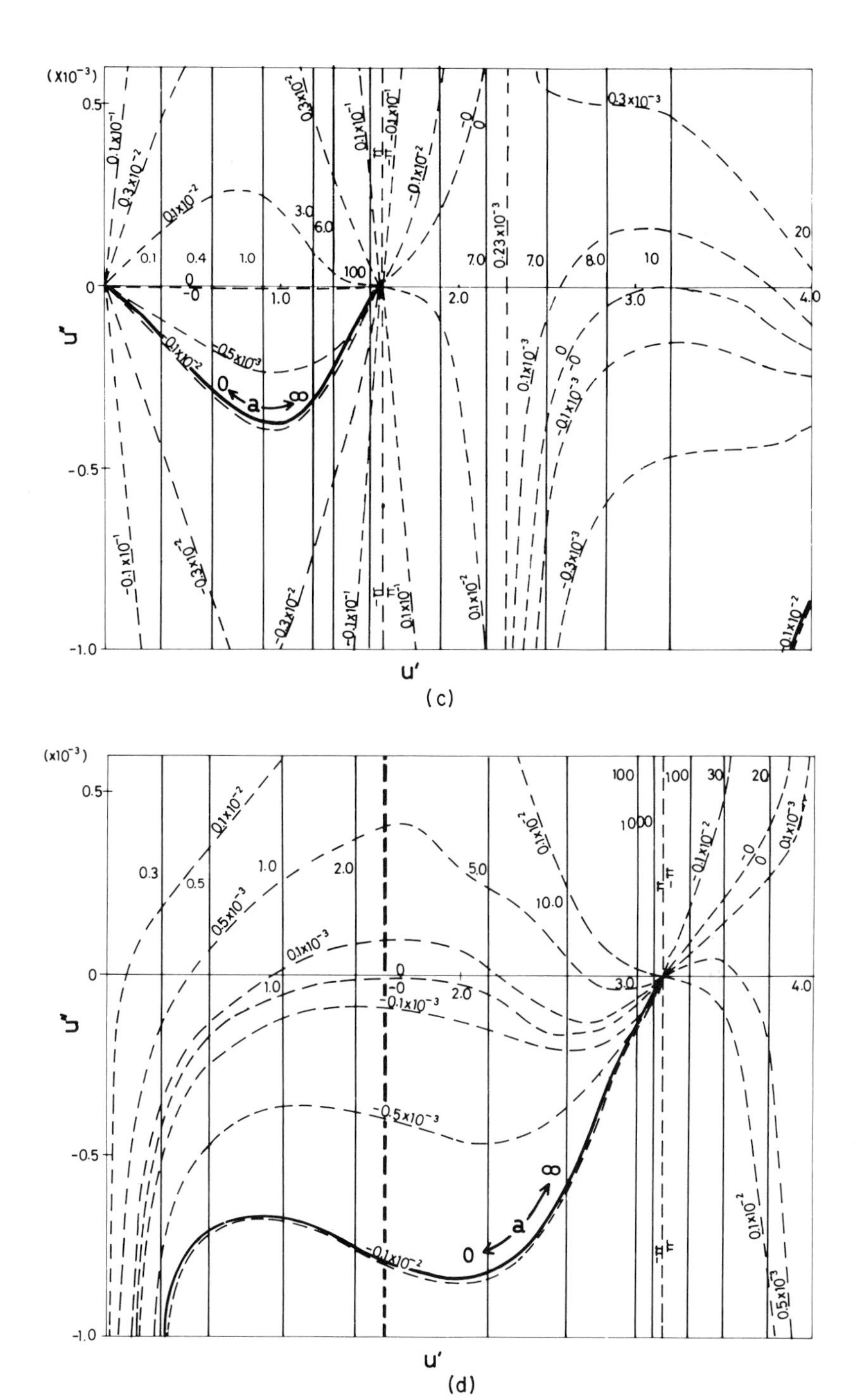

FIG. 2 (*Continued.*) (c) Even TM mode [$\eta = (1.461 - j0.000387)^{-2}$], (d) odd TM mode [$\eta = (1.461 - j0.000387)^{-2}$].

The upper sign holds for the symmetric modes, while the lower sign holds for the antisymmetric modes. Equations (9) and (10) may be solved graphically by plotting both contours of equal magnitude and argument on the complex u plane as shown in Figs. 2a–2d.

In these plots, the contours of equal magnitude (M = const) are shown by the solid curves, while those of equal phase are shown by the dashed ones. Then, if the complex number R^2 of Eq. (4) is given, the root of Eqs. (9)–(10) will be found from the point (u', u''), which is determined by the contours of both M and Φ calculated through Eq. (8). It should be noted that the scale on the u'' axis is enlarged against that on the u' axis for indicating precisely the eigenvalues in the small-loss region of a waveguide. These contour plots show the typical examples in which the refractive index n_1 is assumed to be real and unity (air), while the dielectric slab is assumed to be a polyethylene film with $n_1 = 1.461 - j0.000387$ at $\lambda_0 = 337\ \mu m$ (Chantry *et al.*, 1971). Such a constitution of slab waveguide will appear in the succeeding experiments. In this case, the root loci of Eqs. (9)–(10) are given by the bold solid curves, as the slab thickness is varied from zero to infinity. In these figures, the root loci are shown only for the dominant mode in each mode group and the bold dotted curves in Figs. 2b and 2d indicate the cut-off curves where Re $\cdot\ q = 0$.

As a result, the calculated values of the phase constant β and the atten-

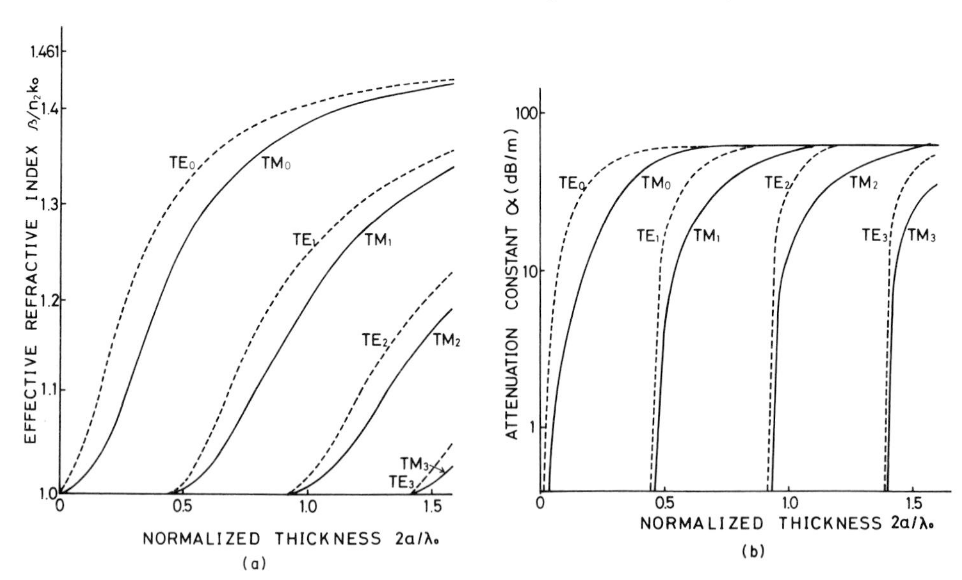

FIG. 3 Complex propagation constant h for various modes as a function of the normalized thickness $2a/\lambda_0$. (a) Phase constant, (b) attenuation constant. (From Tsuji *et al.*, 1979c. Reprinted from *IEEE Transactions on Microwave Theory and Techniques* **27,** 873–878. © 1979 IEEE.)

uation constant α for each propagating mode are plotted as a function of the normalized slab thickness $2a/\lambda_0$ in Figs. 3a–3b, where the effective refractive index β/n_2k_0 is employed instead of β. These results are available only for $\lambda_0 = 337\ \mu m$ because of the presence of the dispersion in the loss tangent of a polyethylene. Such a dispersion can be expressed in the spectral range of $200\ \mu m < \lambda_0 < 1$ mm by the following equation (Chantry *et al.*, 1971):

$$\tan \delta = (0.87 + 0.0015/\lambda_0) \times 10^{-4}, \tag{13}$$

where λ_0 should be given in meters.

Noting little dispersion in the real part of the refractive index of a polyethylene (Chantry, 1971) in that spectral range, the frequency dependence of the attenuation constant can be calculated in the same manner as above, and typical results are shown in Figs. 4a–4c.

In this case, the frequency dependence of the phase constant will be found to be almost the same as that obtained in Fig. 3a.

In the following sections, these results will be used successfully to explain the experimental results.

B. Experiments

1. *Movable Metal Grating Coupler*

In the first step of the experiments, the selective launch of each propagating mode has been studied. It is well known to use a prism (Tien *et al.*, 1969) or a grating coupler (Ogawa *et al.*, 1973) for this purpose. Especially the former can be applied to measure the attenuation constant of a slab waveguide by moving it along a slab. However, in such a type of coupler based on the frustrated total reflection, it is practically impossible to keep up the constant coupling of a movable prism over the sufficient length for loss measurements. The structure of our grating coupler shown in Fig. 5 is different from the conventional grating coupler fixed on a slab, and the fine metal wires are fixed periodically on the rigid frame. Unlike a prism coupler, our grating coupler provides the desirable behavior when the grating is mounted with a slight pressure to keep a close contact with the slab. In the grating coupler, the period of the grating L is related to the angle of incidence θ of the laser beam onto the metal grating through the following phase-matching condition (see Fig. 6):

$$\beta_m = k_0 \sin \theta + 2n\pi/L, \tag{14}$$

where β_m is the propagation constant of mth guided mode in a dielectric slab waveguide and n is the order number of diffraction. Considering both the use of an HCN laser ($\lambda_0 = 337\ \mu m$) and the backward-wave coupling

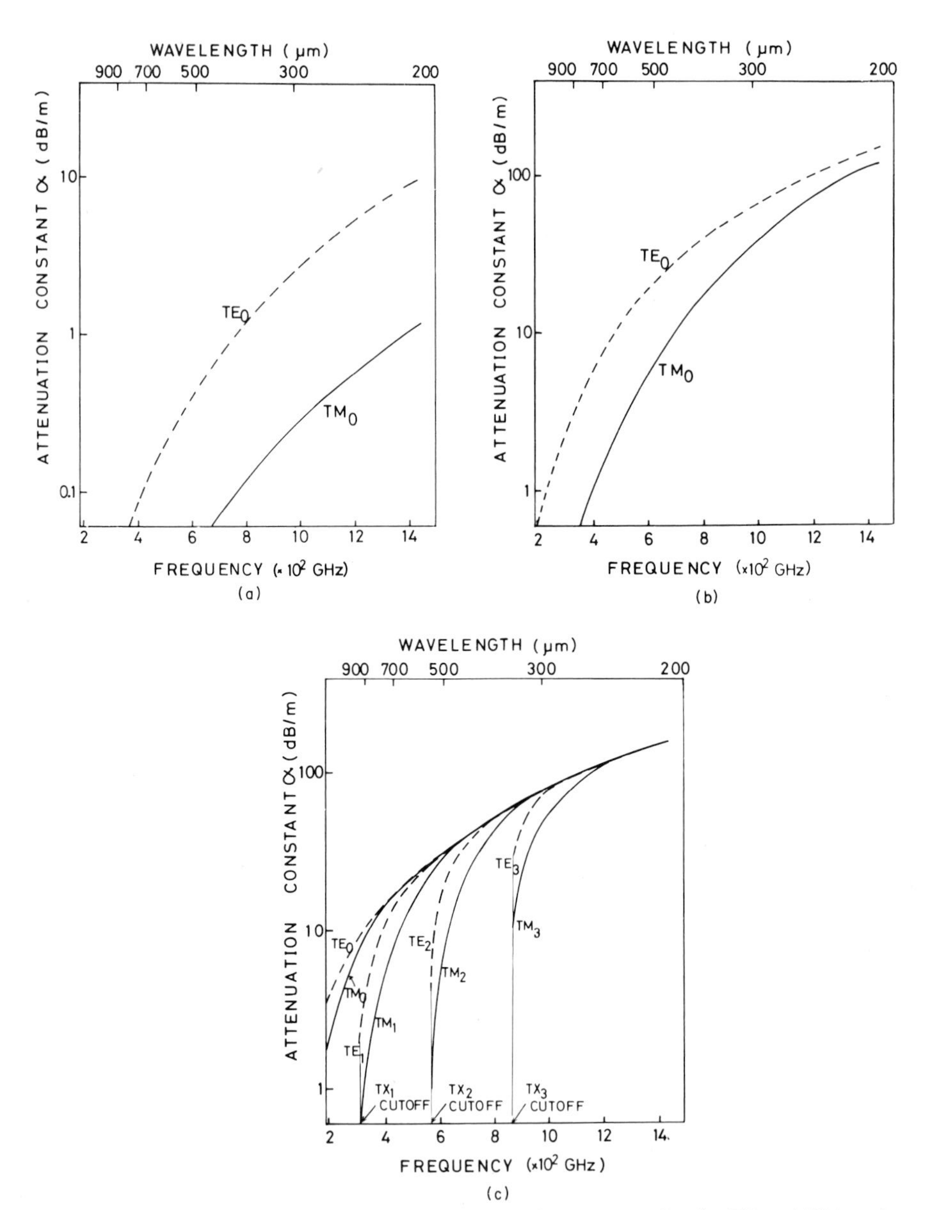

FIG. 4 Frequency dependence of the attenuation constant for the TE and TM modes. Slab thickness: (a) 10 μm, (b) 100 μm, and (c) 500 μm; $TX_n = TE_n$ or TM_n.

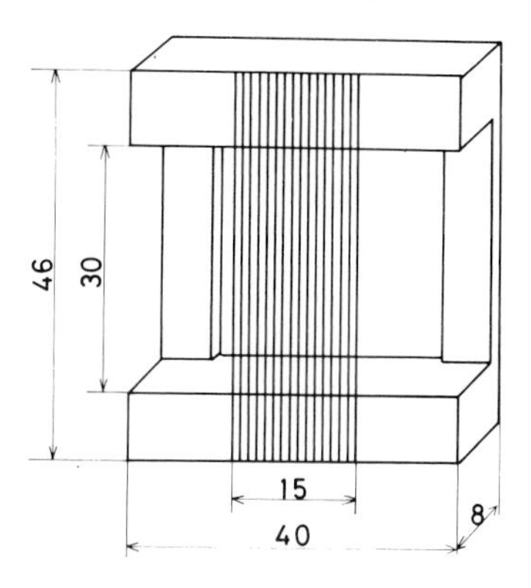

FIG. 5 Basic structure of the movable metal-grating coupler unit: millimeter. (From Tsuji *et al.*, 1979c. Reprinted from *IEEE Transactions on Microwave Theory and Techniques* **27**, 873–878. © 1979 IEEE.)

with $n = -1$ to excite selectively each of all guided modes, Eq. (14) yields the reasonable periods from $L = 168$ to 230 μm (note that the maximum of β_m/k_0 is 1.461 in a polyethylene slab waveguide). Table I shows the calculated results of the incident angle for several kinds of slab thickness and modes. It is clear from this table that the grating with $L =$ 200 μm will be able to couple with all propagating modes in slabs having the thickness between 10 and 500 μm. Figure 7a shows our hand-made coupler in which the molybdenum wires of 100-μm diameter are arranged with the period of 200 μm, as shown in Fig. 7b.

2. *Measurements of Phase Constants*

In order to investigate the characteristics of our grating coupler and to measure the phase constants of slab waveguides, the experimental set-up shown in Fig. 8a is employed, where the thin polyethylene film of 3×30 cm^2 in size is used as a slab and is fully stretched in practice over a metal frame, as shown in Fig. 8b. Two movable metal grating couplers are mechanically contacted on its surface so that the beam from an HCN laser

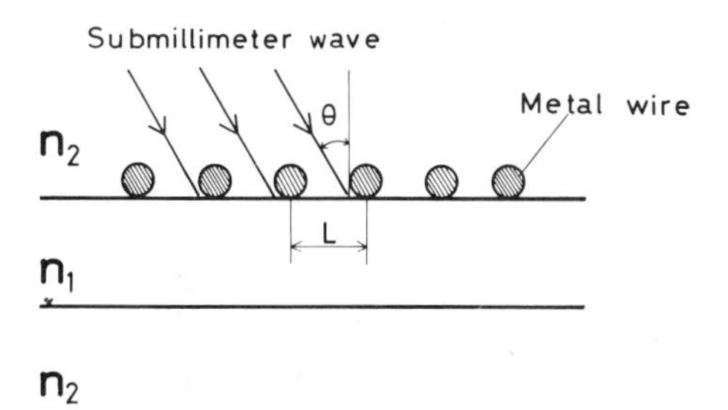

FIG. 6 Side view of a metal-grating coupler having period L. The submillimeter wave is launched with the angle θ. (From Tsuji *et al.*, 1979c. Reprinted from *IEEE Transactions on Microwave Theory and Techniques* **27**, 873–878. © 1979 IEEE.)

TABLE I

CALCULATED LAUNCHING ANGLES

Slab thickness (μm)		500			400	300	200	100	50	35	10
Grating period (μm)	220	200	180	165	200	200	200	200	200	200	200
TE_0 mode	5.0°	14.3°	25.0°	37.0°	15.3°	16.9°	19.3°	27.0°	35.7°	40.9°	42.8°
TM_0 mode	5.7°	14.7°	26.0°	37.6°	15.9°	18.0°	22.1°	34.3°	40.8°	42.0°	43.1°
TE_1 mode	10.0°	19.5°	31.0°	43.7°	22.4°	27.9°	38.7°				
TM_1 mode	11.9°	21.1°	33.2°	45.9°	24.6°	31.9°	41.8°				
TE_2 mode	20.0°	28.6°	42.0°	56.7°	37.4°						
TM_2 mode	21.6°	31.4°	45.0°	61.5°	39.1°						
TE_3 mode	30.0°	41.3°	58.0°								
TM_3 mode	31.2°	42.2°	60.0°								

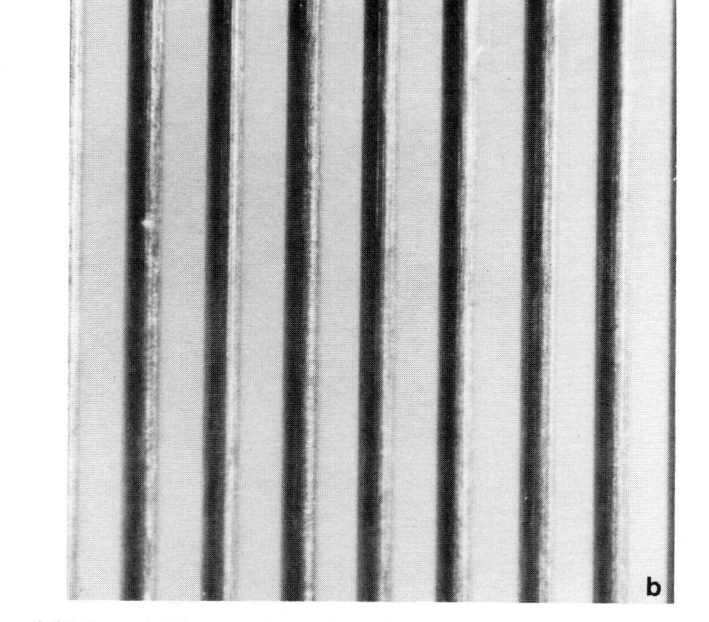

FIG. 7 Photographs of (a) the hand-made grating coupler and (b) the molybdenum wires (100 μm in diameter) arranged with the period of 200 μm. (From Tsuji *et al.*, 1979c. Reprinted from *IEEE Transactions on Microwave Theory and Techniques* **27**, 873–878. © 1979 IEEE.)

can be coupled to and out of the guide. The laser beam focused by a spherical mirror has an approximately Gaussian cross section with an e^{-1} power diameter of 1 cm. As shown in these figures, the slab waveguide with both couplers is mounted on a rotrary stage that allows to vary the angle of incidence θ_{in} of the laser beam onto the input grating coupler.

On the other hand, a Golay cell detector is mounted on the rotating arm in the output side so as to measure the direction θ_{out} of an output beam peculiar to each guided mode. The distance between the Golay cell detector and the grating coupler is 40 mm, and the aperture of the Golay cell detector is 1 mm in diameter.

The radiation angles from the coupler give us the information about the phase constant of guided mode through Eq. (14). The output power is then measured by varying the angle θ_{out}, while the angle θ_{in} is kept constant, to excite only one of the guided modes. The results for the 500-μm thick polyethylene slab are shown in Fig. 9. Although the eight modes can propagate along this guide, the maximum output is obtained at the output angle θ_{out} nearly equal to the angle of incidence θ_{in} of the launched mode, and hence we may conclude that little mode conversion occurs through propagation. The relations between these angles (θ_{in} and θ_{out}) are summarized in Table II to compare with the theoretical angles calculated from Eq. (14). It is clear from this table that the measured angles are in good agreement with the calculated ones, and each guided mode can be selectively launched and detected by our grating coupler.

As the launching angles for the TE_n mode and TM_n mode are close to each other, and there is little difference between them, it is hard to distinguish one from the other. However, it is reasonable to assume the predominant launching of the TM modes in our experiments, because there will be high reflection at the metal grating coupler for the TE incident wave with the electric field vector parallel to the metal wires, and also because the attenuation constant is higher for the TE mode than for the TM mode. Although more detailed study about our grating coupler should be performed, the coupling efficiency of about 40% may be attained at the maximum.

3. *Measurements of Attenuation Constants*

The movable metal grating coupler discussed previously is very useful for the measurement of the attenuation constant. The attenuation constant of a dielectric slab waveguide is measured by moving the output coupler along a slab and varying the transmission length between both couplers.

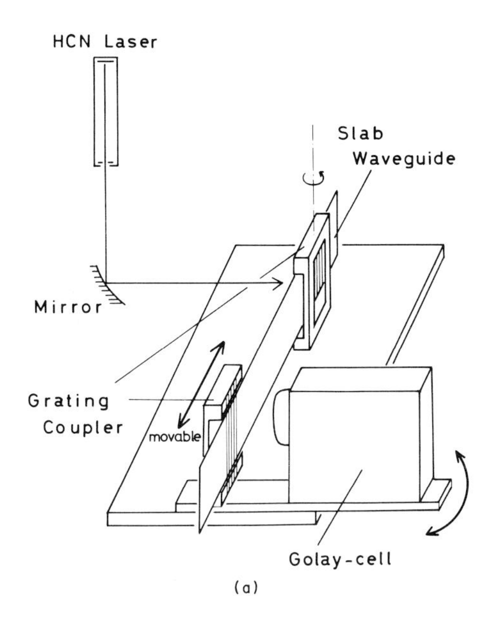

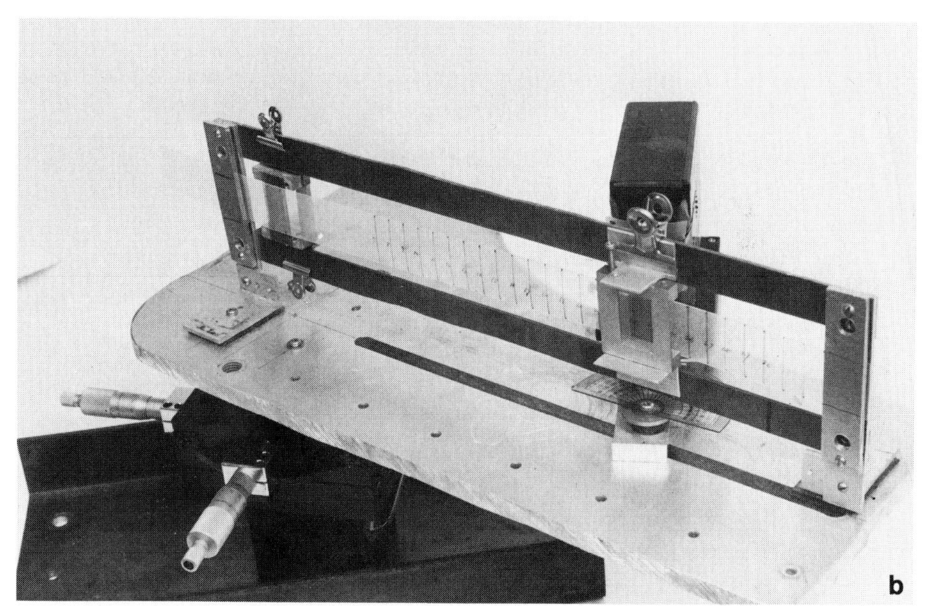

FIG. 8 (a) Illustration of experimental setup, (b) its external view. (From Tsuji *et al.*, 1979c. Reprinted from *IEEE Transactions on Microwave Theory and Techniques* **27,** 873–878. © 1979 IEEE.)

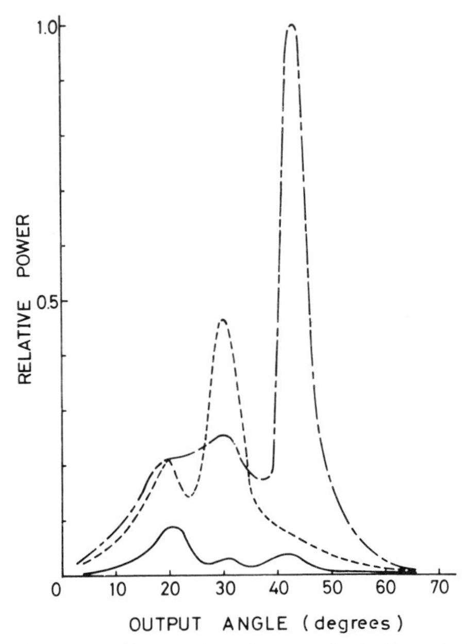

FIG. 9 Behavior of the movable metal-grating coupler in the selective excitation of each propagating mode. ——, first-order mode; incident angle, 21°; --, second-order mode, 28.5°; ————; third-order mode, 41.5°. (From Tsuji *et al.*, 1979c. Reprinted from *IEEE Transactions on Microwave Theory and Techniques* **27,** 873–878. © 1979 IEEE.)

The experimental setup is the same as that in Fig. 8, except that both the output grating coupler and the rotating arm on which the Golay cell detector is mounted are moving along a slab simultaneously.

The relative output power for various modes in the slab waveguide with 500-μm thickness is shown in Fig. 10 as a function of the length of a guide.

TABLE II

MEASURED[a] AND THEORETICAL LAUNCHING ANGLES[b]

Mode number	Theoretical values (TE)	Theoretical values (TM)	Measured values: Incident angle	Measured values: Output angle
0	14.3°	14.7°		
1	19.5°	21.1°	21.0°	20.5°
2	28.6°	31.4°	28.5°	30.0°
3	41.3°	42.2°	41.5°	42.5°

[a] In this measurement, the propagation of the zeroth-order mode is confirmed, but its measured angle is eliminated because its attenuation is too great to measure accurately the launching angle.

[b] (From Tsuji *et al.*, 1979c. Reprinted from *IEEE Transactions on Microwave Theory and Techniques* **27,** 873–878. © 1979 IEEE.)

The attenuation constant then is found as 130, 110, and 52 dB/m for the TM_1, TM_2, and TM_3 modes, respectively. These measured values are found to be about twice as high as the theoretical ones shown in Fig. 3b.

So far, the multimode slab waveguide has been discussed. Let us now examine a thin slab waveguide in which only the fundamental modes (the TE_0 and TM_0 modes) can propagate. As seen from Fig. 3a, such a waveguide can be realized for the slab thickness less than about 160 μm at $\lambda_0 = 337$ μm, and then the attenuation constants for the TM_0 mode are measured for several kinds of slab thicknesses (100, 50, 35, and 10 μm). All these results are shown in Figs. 11a–11b, and Fig. 12 summarizes these results in conjunction with the theoretical ones indicated by the solid curve.

It is noted from this figure that the measured results show indeed a relatively good agreement with the calculated results, but are slightly greater than the calculated ones. This difference may be caused by the radiation losses due to bends, irregularities, and inhomogeneities in a waveguide, and also by the underestimation of the loss tangent tan δ of a polyethylene in the calculation, since the commercial polyethylene often involves the various impurities having much effect on its absorption in the submillimeter-wave region (Chantry, 1971).

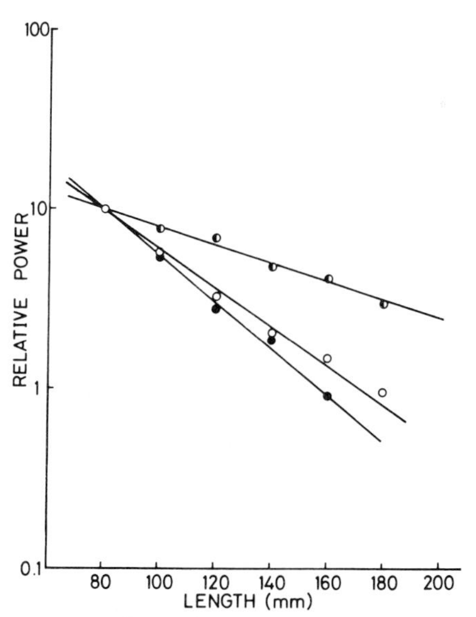

FIG. 10 Loss measurements for various modes in the slab waveguide with 500-μm thickness. (Relative output powers are measured as a function of its length.) ◐, TM_3 mode; ○, TM_2 mode; ●, TM mode. (From Tsuji *et al.*, 1979c. Reprinted from *IEEE Transactions on Microwave Theory and Techniques* **27**, 873–878. © 1979 IEEE.)

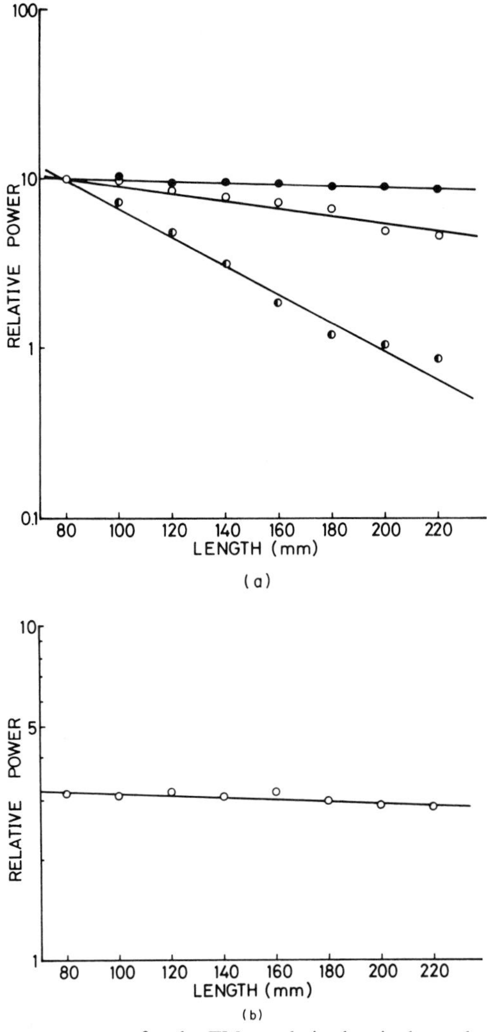

FIG. 11 Loss measurements for the TM_0 mode in the single-mode slab waveguides. (a) Slab thickness: ◐, 100; ○, 50; and ●, 35 μm. (b) Slab thickness 10 μm. (Relative output powers are measured as a function of its length.)

The TM_0 mode in a slab with 10-μm thickness shows the measured attenuation of only 1.3 dB/m. But, in such a thin slab waveguide, the rate of concentrated power in the slab becomes very low (only about 3%) as seen from calculated results of Fig. 13, and we must always discuss the attenuation characteristics by considering the excess losses due to bends in a waveguide. These problems will be treated in Section II.B.5.

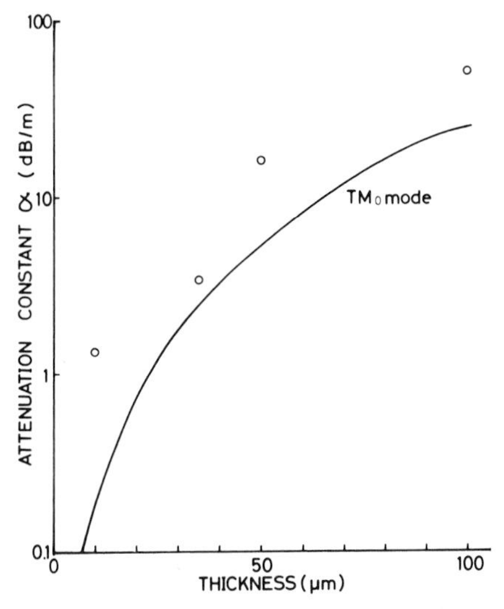

FIG. 12 Attenuation constant for the TM_0 mode. ——, Theory; ○, experiment. (From Tsuji *et al.*, 1979c. Reprinted from *IEEE Transactions on Microwave Theory and Techniques* **27**, 873–878. © 1979 IEEE.)

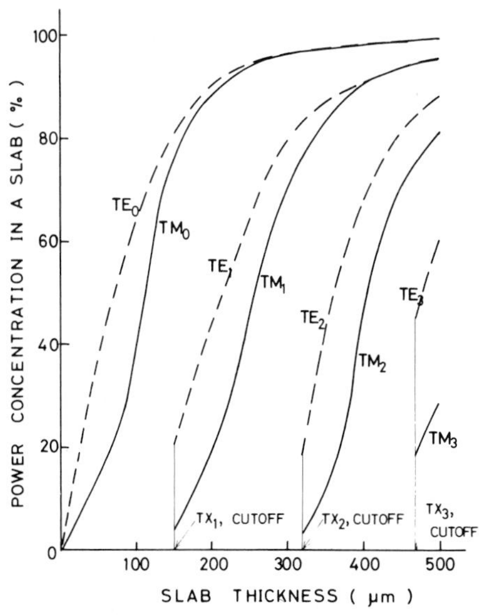

FIG. 13 Power concentration in a slab for various modes. $TX_n = TE_n$ or TM_n.

4. *Effects of the Transverse Broadening of Fields on Measured Data*

A dielectric slab waveguide provides confinement of the field in only one dimension, i.e., the x direction in Fig. 1. Thus to make the accuracy more evident for the loss measurements in the previous section, we should prove that the transverse broadening of the untrapped field in the y direction has no influence on the measured attenuation constants, even though the guide length, i.e., the distance between the input and output couplers, becomes over 300 mm. This transverse broadening of the field is then measured along the propagating direction of a slab waveguide by means of two simple methods.

In the first method, the output power is measured when a small submillimeter-wave absorber (3 mm² in area) is placed on the surface of a

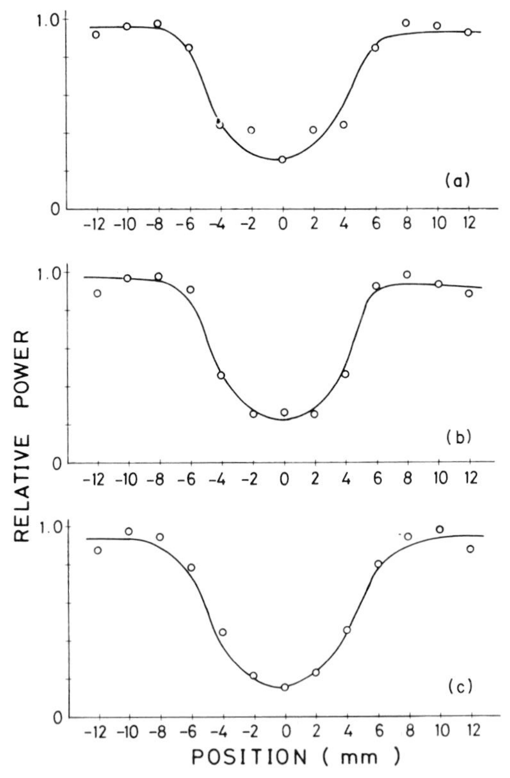

FIG. 14 Measurement of the transverse broadening of the field using the absorber. (a) l = 30 mm, (b) l = 100 mm, (c) l = 200 mm. (From Tsuji *et al.*, 1979c. Reprinted from *IEEE Transactions on Microwave Theory and Techniques* **27**, 873–878. © 1979 IEEE.)

slab and is moved along the transverse direction. Such a measurement is performed at the three positions l mm away from the input coupler (l = 30, 100, and 200 mm). The results are shown in Fig. 14 as a function of the displacement of the absorber from the center of a slab. It is clear from these results that the half-width of the power dip caused by the absorber coincides with that of the input beam (i.e., about 10 mm), and varies little with the increase of l.

In the second method, the near-field patterns of radiation from the end of a slab waveguide are measured by scanning the Golay cell detector along the transverse direction (y) of a slab. In this experiment, the distance between the input coupler and the end of a slab is adjusted to 100 and 190 mm, and the near-field pattern is measured at the distance of 50 mm from the end of a slab. The result shown in Fig. 15 indicates that the transverse broadening of the input beam varies little in its half-width over the transmission range of about 200 mm.

Consequently, it is proved by these results that the field of a guided wave is so well confined to a polyethylene slab within 30 mm in width that its side edges have no effect on the propagation of a guided mode. Thus, as for the length of the slab used in our measurement, it may be concluded that the transverse broadening of the field has negligible effects on the measured attenuation constants.

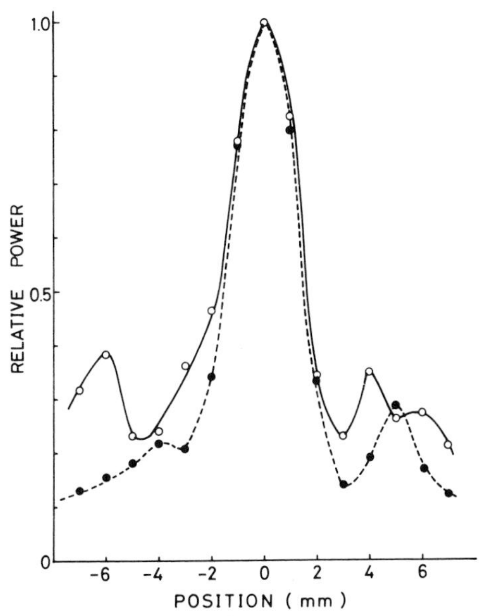

FIG. 15 Measurement of the transverse broadening of the field by the near-field pattern. ○, 190 mm; ●, 100 mm. (From Tsuji *et al.*, 1979c. Reprinted from *IEEE Transactions on Microwave Theory and Techniques* **27**, 873–878. © 1979 IEEE.)

5. *Measurements of Bending Losses*

Let us consider the radiation loss due to the bend in the x direction of a slab (see Fig. 1). When the field is loosely bound to a dielectric, e.g., the polyethylene slab guide 10 μm thick as shown in Fig. 13, the bending loss will have much effect on the attenuation of a slab guide. So the bending loss for the TM_0 mode of the slab guide with several kinds of thicknesses is measured. The experimental setup is the same as that in Fig. 8a except that the metal frame fixing a polyethylene film can be bent in an arbitrary curvature as shown in Fig. 16.

The bending losses are obtained from the ratio of the output in a curved slab guide to that in a straight one and are shown in Fig. 17, where the solid line indicates the theoretical values α_b that are calculated by suitably determining the constant C in the following equation (Neumann *et al.*, 1975):

$$\alpha_b = C \exp\{-2(q/a)^3 R/3k_0^2\}, \tag{15}$$

where R is the radius of curvature, k_0 the free-space wave number, jq/a the wave number outside a slab in the transverse direction, and C the constant independent of R. It is found that the measured attenuation is in good agreement with the theoretical one with regard to the dependence of the bending loss on R. Moreover, our measurements show that the guide with 10-μm slab thickness has the bending loss of about 10 dB/m in

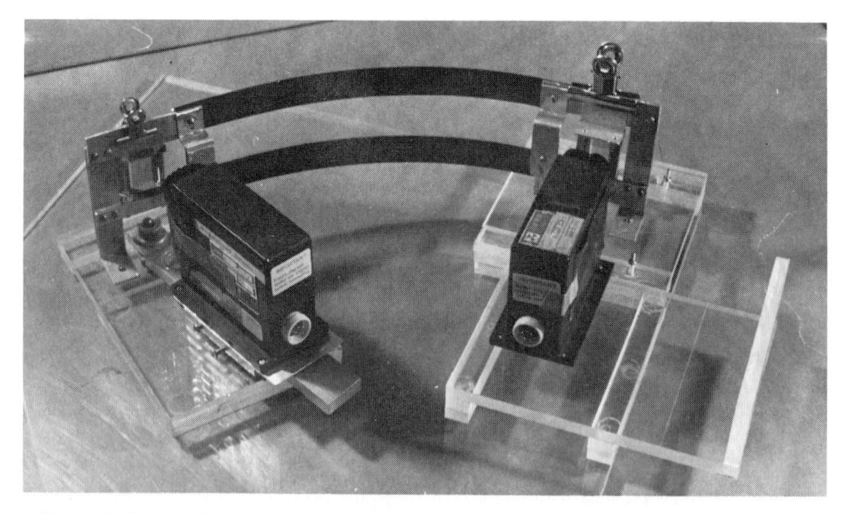

FIG. 16 External view of the experimental setup for measuring the bending loss.

$R = 1$ m. While the guide with 35-μm thickness has the same bending loss in $R = 0.5$ m and in a guide with slab thickness of more than 50 μm, its power attenuation is little affected by bends with the radius of curvature larger than 0.5 m.

In this section, the attenuation constant of symmetric slab waveguides is measured by using the movable metal grating coupler in the submillimeter-wave region. As a result, the attenuation constant for the TM_0 mode is found realiably to be as low as 1.3 dB/m for a slab of 10 μm in thickness. It is also confirmed that the transverse broadening of the field along the propagating direction of a slab has little influence on the measurement of the attenuation constant. However, the polyethylene slab waveguide the thickness of which is as thin as about 10 μm confines only the transmission energy of less than a few percent in the polyethylene, and so the radiation losses due to the bends are also measured. These results encourage us to discuss the transmission characteristics of an advanced waveguide that has a three-dimensional structure.

III. Dielectric Rib Waveguides

The dielectric slab waveguide discussed in Section II is a useful model for developing more advanced waveguides, and we have obtained valuable information about the properties of submillimeter-wave propagation in dielectric waveguide without the accompanying tedious mathematical

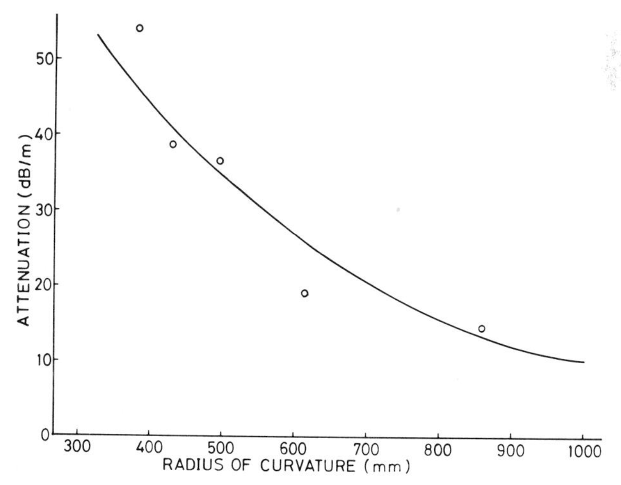

FIG. 17 Measurement of the bending loss for the TM_0 mode with 10-μm thickness. ——, theory; ○, experiment.

treatments. However, in most practical applications, more complicated structures that confine the fields horizontally as well as vertically are used.

Such waveguides used in integrated circuits are usually rectangular strips of dielectric material that are placed on or embedded in a dielectric substrate. A typical and basic structure of this type of waveguide is a dielectric rib waveguide (Goell, 1973) in which the material of strips is the same as that of the film on the substrate.

This waveguide has indeed practical applications in the spectral range from millimeter to optical frequencies because of its simplicity, mechanical stability, and precision of construction. However, its open structure, having irregular and unbounded boundaries, makes it difficult to perform its precise analysis. Hence several kinds of approximate analyses have been discussed.

The most simplified method will be the equivalent refractive index method discussed by Rasaswamy (1974) and McLevige *et al.* (1975). But this approximation is based only on the conservation relation of wave numbers and has no consideration for the continuity condition of fields through boundaries. It is then impossible to discuss the problems that are closely related to the eigenfunctions of a guide.

On the other hand, the most advanced method will be the mode-matching method (Yasuura, 1977) based on the Rayleigh principle in the Fourier transforms of wave fields. This method is indeed worthy to be used exactly for the treatment of the transmission characteristics of waveguides having unbounded boundaries, but its complicated field expressions in a homogeneous medium make it difficult to consider a waveguide consisting of several kinds of dielectrics.

We then propose here a new approximate method, which is indeed based on the mode-matching method, but is rather simplified in its field expressions. In our method, the cross section of a rib waveguide is subdivided into several subsections having a simple geometry of the boundaries, and the boundary conditions between these constituent sections are satisfied in the least-squares sense.

Many experiments have been performed in both millimeter-wave (Shigesawa *et al.*, 1979b) and submillimeter-wave regions by following the same techniques as those developed in Section II.B. These results, together with our analysis, prove the usefulness of this waveguide in both spectral regions.

A. Analysis

Figure 18 shows the geometry of a dielectric rib waveguide. As expected in this waveguide, all the modes are indeed hybrid in the rigorous sense, but it will be allowed to consider that there are two types of

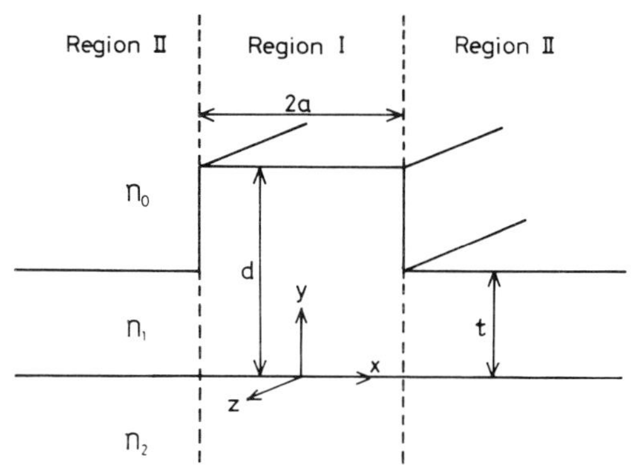

FIG. 18 Cross section of a rib waveguide.

approximate modes that the waveguide can support. One type, which we call TM_y mode, is polarized predominantly in the y direction. The other mode, TE_y, is polarized predominantly in the x direction and will be treated hereafter in this section.

Let us now subdivide this guide into two constituent regions, i.e., regions I and II, as shown in Fig. 18, and consider separately each of these regions as if region I consists of a rectangular dielectric waveguide and region II of semiinfinite slab waveguides. We then assume that all the field components can be expanded only into the propagating eigen modes in each region. Considering the energy concentration in the rib portion, the leading component of wave fields and the related components in each region can be expressed for the symmetric TE_y mode with respect to the y–z plane as follows:

$$E_x^{\mathrm{I}} = \sum_m A_m f_m^{\mathrm{I}}(y) \cos \kappa_m x \exp[j(\omega t - hz)] \qquad \text{(region I)}, \tag{16}$$

$$E_x^{\mathrm{II}} = \sum_m B_m f_m^{\mathrm{II}}(y) \exp(-\delta_m|x - a|) \exp[j(\omega t - hz)] \qquad \text{(region II)}, \tag{17}$$

$$\begin{aligned} E_z &= -\frac{j}{h} \cdot \frac{\partial E_x}{\partial x}, & H_x &= \frac{1}{\omega\mu_o h} \cdot \frac{\partial^2 E_x}{\partial x\, \partial y}, \\ H_y &= -\frac{1}{\omega\mu_o h}\left(\frac{\partial^2}{\partial x^2} - h^2\right) Ex, & H_z &= -\frac{j}{\omega\mu_0} \cdot \frac{\partial E_x}{\partial y}, \end{aligned} \tag{18}$$

where A_m and B_m mean the modal expansion coefficients to be defined later and h denotes the phase constant to be found. The functions f_m^{I} and f_m^{II} mean the modal functions in the y direction described as

$$f_m^{\mathrm{I}}(y) = \begin{cases} \sin(\beta_m^{\mathrm{I}} d + \Phi_m^{\mathrm{I}}) \exp\{-\alpha_m^{\mathrm{I}}(y - d)\}, & y \geq d, \\ \sin(\beta_m^{\mathrm{I}} y + \Phi_m^{\mathrm{I}}), & 0 \leq y \leq d, \\ \sin \Phi_m^{\mathrm{I}} \exp(\gamma_m^{\mathrm{I}} y), & y \leq 0; \end{cases} \tag{19}$$

$$f_m^{\mathrm{II}}(y) = \begin{cases} \sin(\beta_m^{\mathrm{II}} t + \Phi_m^{\mathrm{II}}) \exp\{-\alpha_m^{\mathrm{II}}(y - t)\}, & y \geq t, \\ \sin(\beta_m^{\mathrm{II}} y + \Phi_m^{\mathrm{II}}), & 0 \leq y \leq t, \\ \sin \Phi_m^{\mathrm{II}} \exp(\gamma_m^{\mathrm{II}} y), & y \leq 0, \end{cases} \tag{20}$$

with the conservation relations of wave numbers:

$$(n_1^2 - n_0^2)k_0^2 = \alpha_m^2 + \beta_m^2, \tag{21}$$

$$(n_1^2 - n_2^2)k_0^2 = \gamma_m^2 + \beta_m^2, \tag{22}$$

where the superscripts I and II are both available in Eqs. (21)–(22) and the phase constant h in the z direction is given by

$$h^2 = (n_1 k_0)^2 - \beta_m^{\mathrm{I}2} - \kappa_m^2 = (n_1 k_0)^2 - \beta_m^{\mathrm{II}2} + \delta_m^2. \tag{23}$$

First, let us determine these fields in each region so as to fulfill the boundary conditions on the core-substrate boundary at $y = 0$ and on the plane at $y = d$ in region I or at $y = t$ in region II.

As a result, it is easily found that the wave numbers α_m, β_m, and γ_m in the y direction can be found independently of those in the x direction from the following well-known eigenvalue equations of asymmetric slab waveguide:

$$\tan \beta_m^{\mathrm{I}} d = \beta_m^{\mathrm{I}}(\alpha_m^{\mathrm{I}} + \gamma_m^{\mathrm{I}})/(\beta_m^{\mathrm{I}2} - \alpha_m^{\mathrm{I}}\gamma_m^{\mathrm{I}}), \tag{24}$$

$$\tan \beta_m^{\mathrm{II}} t = \beta_m^{\mathrm{II}}(\alpha_m^{\mathrm{II}} + \gamma_m^{\mathrm{II}})/(\beta_m^{\mathrm{II}2} - \alpha_m^{\mathrm{II}}\gamma_m^{\mathrm{II}}). \tag{25}$$

Next, we must impose the continuity condition of the tangential fields through the infinite plane at $x = a$. However, in our approximation, the fields in each region have different functional forms in the whole region of $|y| < \infty$, so that it is impossible to match analytically the fields in both regions on the y–z plane at $x = a$.

Hence let us fit the fields to this boundary condition in the least-squares sense. For this purpose, we introduce the quantity F defined by the following equation:

$$\begin{aligned} F &= \int_{-\infty}^{\infty} |\mathbf{n} \times [\mathbf{E}^{\mathrm{I}}(x = a, y) - \mathbf{E}^{\mathrm{II}}(x = a, y)]|^2 \, dy \\ &\quad + (\omega\mu_0/n_i k_0)^2 \int_{-\infty}^{\infty} |\mathbf{n} \times [\mathbf{H}^{\mathrm{I}}(x = a, y) - \mathbf{H}^{\mathrm{II}}(x = a, y)]|^2 \, dy \\ &= \int_{-\infty}^{\infty} |E_z^{\mathrm{I}}(x = a, y) - E_z^{\mathrm{II}}(x = a, y)|^2 \, dy \\ &\quad + (\omega\mu_0/n_i k_0)^2 \int_{-\infty}^{\infty} [|H_y^{\mathrm{I}}(x = a, y) - H_y^{\mathrm{II}}(x = a, y)|^2 \\ &\quad + |H_z^{\mathrm{I}}(x = a, y) - H_z^{\mathrm{II}}(x = a, y)|^2] \, dy, \end{aligned} \tag{26}$$

where $\mathbf{n}$ is the unit vector in the x direction and $(\omega\mu_0/n_i k_0)$ is the intrinsic impedance of medium n_i. This quantity F means, of course, the mean-square error in the continuity condition of approximated fields. Thus the eigenvalue equation of a rib waveguide will be obtained by minimizing F with respect to both expansion coefficients A_m, B_m, and the phase constant h.

It is now easy to solve the eigenvalue equation, considering the problem in the most rough approximation that the only dominant mode [i.e., only $m = 0$ in Eqs. (16)–(26)] is employed. In this case, the unknown variables to be defined in relation to the boundary condition become A_0, B_0, and h. But we have the freedom to define arbitrarily one of the two expansion coefficients, so that by setting B_0 as unity, the quantity F can be expressed as follows:

$$F = A_0 P(h) + 2A_0 Q(h) + R(h), \tag{27}$$

where P, Q, and R are the functions of only the phase constant h, and the explicit but complicated forms may be found in Appendix I.

The dispersion relation in this approximation can be obtained by solving numerically the following simultaneous equations:

$$\partial F/\partial A_0 = 0, \qquad \partial F/\partial h = 0. \tag{28}$$

Once the eigenvalues are found, eigenfunctions can be easily defined by using A_0 in our method, and so the perturbation technique may be utilized to calculate the attenuation characteristics of this waveguide.

A typical example of calculated dispersion relation is shown in Fig. 19 for dominant TE_y and TM_y modes. The waveguide is assumed to have such dimensions and refractive indices as shown in the inset. The solid curves show our results, while the dotted ones show the results calculated by the equivalent refractive index method.

Another comparison between our method and the mode-matching method is also shown in Fig. 20. Again the guide dimensions and refractive indices are shown in the inset of the figure, where the solid curve shows our result while the crosses indicate the result of mode-matching method.

More careful comparisons have been made between our method and others in many kinds of guide dimensions and refractive indices. These results conclude that we cannot judge the quality of each method as far as the dispersion characteristics are concerned.

On the other hand, Fig. 21 shows the calculated results of the attenuation constants of the dominant TE_y and TM_y modes. The loss tangent tan δ of the dielectric material is assumed to be 2×10^{-4}, and this value is chosen from the fact that the loss tangent of a polyethylene in 50-GHz

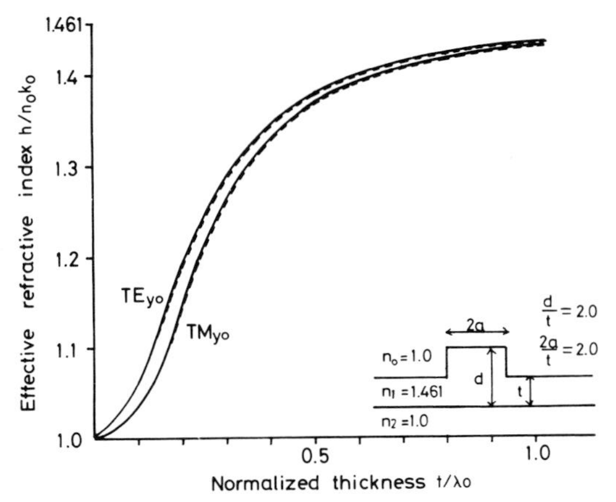

FIG. 19 Phase constant for the TE_{y0} and TM_{y0} modes as a function of the normalized thickness t/λ_0. ——, present analysis; --, effective refractive index method.

region is $1 \times 10^{-4} \sim 2 \times 10^{-4}$ reported by Chantry (1971) and Cook *et al.* (1974). These calculated results will be discussed again in relation to the experimental results.

B. EXPERIMENTS

1. *Millimeter-Wave Experiments*

First, the model experiments in a 50-GHz region are performed to confirm the accuracy of our analytical results. The experimental setup is shown in Fig. 22, where the rib waveguide consists of the polyethylene ($n_1 = 1.461$) and the air ($n_0 = n_2 = 1.0$), and the movable metal-grating

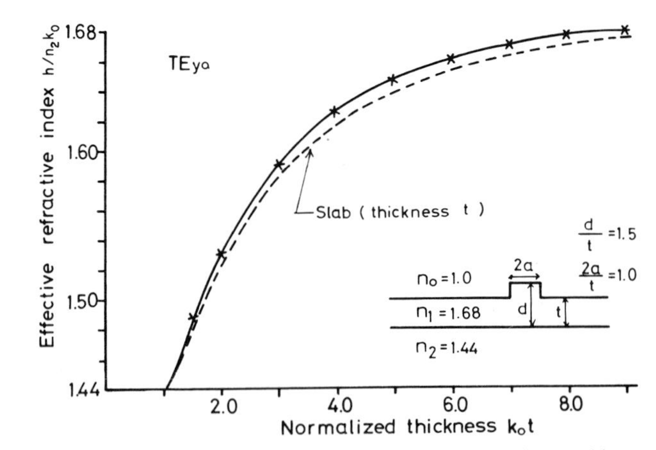

FIG. 20 Comparison between this method (——) and the mode-matching method (x).

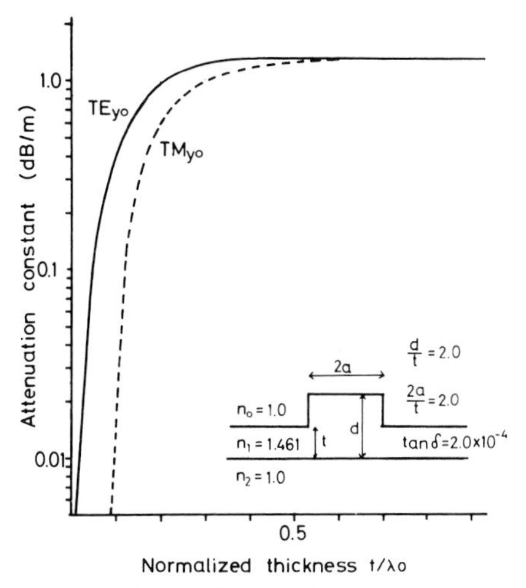

FIG. 21 Attenuation constant for the TE_{y0} and TM_{y0} modes as a function of the normalized thickness t/λ_0.

coupler proposed for the experiments of a slab waveguide in Section II.B.1 is used for the same purpose. This coupler is constructed of steel wires 1.4 mm in diameter arranged with the period of 3.2 mm on the rigid frame and is movable along the propagating direction on a rib waveguide. The reception of the wave as well as its launching are performed by a pyramidal horn, and one of the horns is mounted on the rotatable arm so as to measure the direction θ_{out} of a radiated wave from the metal-grating coupler peculiar to each guided mode. The distance between the horn and the coupler is 40 cm. Figure 23 shows the photograph of our experimental setup.

Now, the phase constants for the TM_{y0} mode of polyethylene rib waveguides with various values of $2a/t$ are measured from the radiation angle θ_{out} through Eq. (14). The results for $d/t = 1.5$ and 2.0 are plotted in

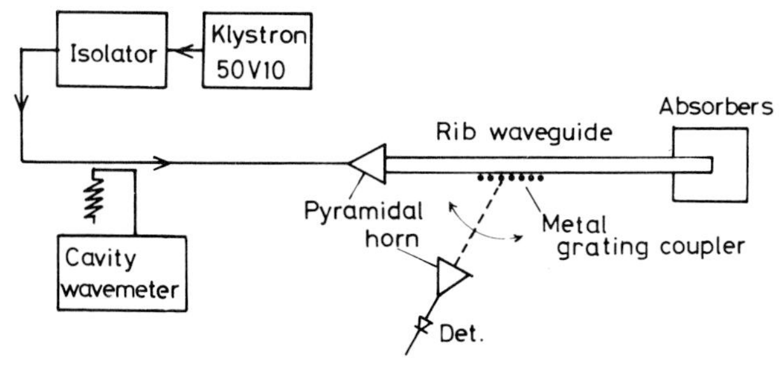

FIG. 22 Schematic diagram of the experimental setup in 50-GHz region.

FIG. 23 External view of the experimental setup in 50-GHz region.

Figs. 24a–24b, respectively, as a function of the normalized thickness t/λ_0, where the solid and dashed curves indicate the theoretical values calculated by our analytical method. It is noted from these figures that the measured phase constants show good agreement with the theoretical ones.

On the other hand, the attenuation constant is measured by moving the grating coupler along a rib waveguide and varying the transmission length between the launching horn and the grating coupler. In Fig. 25, the attenuation constants for the TM_{y0} mode of rib waveguides with $d/t = 2.0$ and $2a/t = 3.0$ are shown as a function of the normalized thickness t/λ_0. This figure also indicates an example of the relative output power that is shown as a function of the length of guide, and so it is evident that the accurate and reliable measurement of the attenuation constant is performed. The solid curves denote the theoretical values in our method when assuming the loss tangent tan δ of a polyethylene to be 1×10^{-4} or 2×10^{-4}. Con-

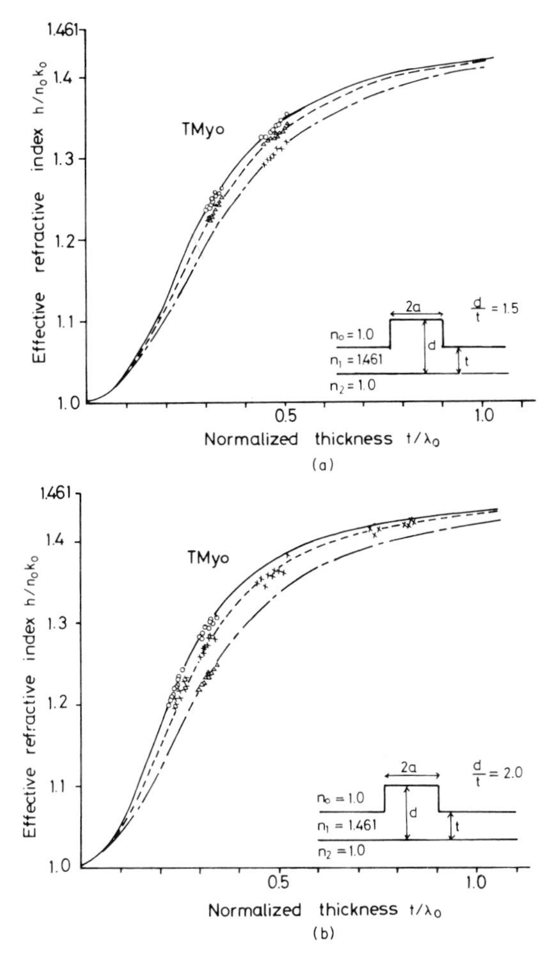

FIG. 24 Measurements of the phase constant for the TM_{y0} mode in the rib waveguide with various values of $2a/t$. Thickness ratio (a) $d/t = 1.5$ and (b) $d/t = 2.0$. $2a/t = 3.0$; 2.0; 1.0. ○, △, ×, experiment; ——, – –, ———–——, theory.

sidering the fact that the loss tangent of 2×10^{-4} is overestimated, the measured attenuation constants may be slightly greater than the theoretical ones. This slight difference between the measured and the theoretical values may be mainly caused by lossy adhesives, which are used to fix the rib portion to the flat dielectric sheet.

Consequently, these experiments prove the validity of our analytical method, and in the following section this method will be successfully applied to compare with the experimental values in the submillimeter-wave region.

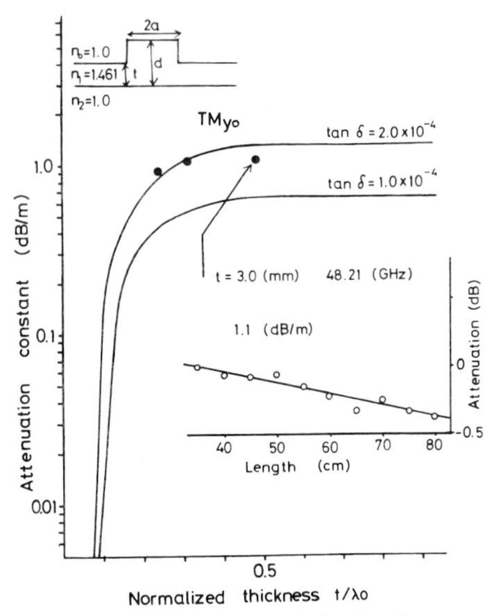

FIG. 25 Loss measurements for the $TM_{\nu 0}$ mode in the rib waveguides with $2a/t = 3.0$ and $d/t = 2.0$. (An example of relative output powers measured as a function of the length of the guide is also shown.) ●, Experiment.

2. *Submillimeter-Wave Experiments*

The experimental setup is shown in Fig. 26, and the measurements of both phase and attenuation constants of rib waveguides in the submillimeter-wave region are performed in the same manner as those in Section III.B.1. The rib waveguide consists of only a polyethylene (n_1 = 1.461) whose cross section is shown in Fig. 27, and the movable metal-

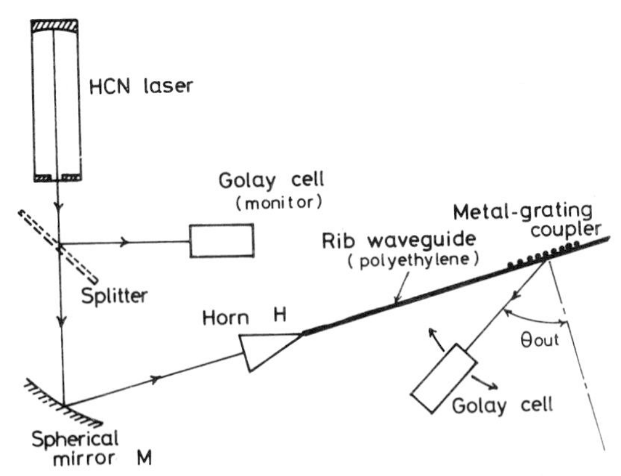

FIG. 26 Schematic diagram of the experimental setup.

FIG. 27 Photograph of the cross section of a rib waveguide with t = 170 μm, d = 350 μm, and $2a$ = 400 μm.

grating coupler described in Section II.B.1 is again used. On the incident side, the output beam from an HCN laser (λ_0 = 337 μm) is focused onto the end of a rib waveguide by a spherical mirror M and a horn H as shown in Fig. 26, while the construction in the output side is identical with that in Fig. 8a. Figure 28 shows our experimental setup.

First, the radiation angle θ_{out} from the metal-grating coupler, which is

FIG. 28 External view of the experimental setup.

related to the phase constant through Eq. (14), is measured for the TM_{y0} mode of several kinds of rib waveguides. The results, together with the dimensions of each guide, are summarized in Table III. In this table the theoretical angles that are calculated by our analytical method are also presented, and the agreement between the measured and the theoretical angles is very good.

On the other hand, the attenuation constant of a rib waveguide is measured by moving the metal-grating coupler along a guide. As an example, the relative output power for the TM_{y0} mode of the multimode rib waveguide with $t = 250$ μm, $d = 550$ μm, and $2a = 500$ μm are shown in Fig. 29 as a function of the length of a guide. Then the attenuation constant is found as 116 dB/m, which is about twice as much as the theoretical values (62.0 dB/m) calculated by our analytical method. However, in this multimode guide, the attenuation for the TM_{y0} mode becomes nearly equal to the attenuation of a bulk polyethylene because the field of the TM_{y0} mode is well confined into a polyethylene guide. Let us examine the attenuation constants for the higher-order mode the field of which is loosely bound to a polyethylene. The results for two higher-order modes of the same guide are shown in Fig. 30, and the attenuation constants are found to be 108 and 88 dB/m. It is clear from these results that the higher-order mode has the lower attenuation constant, but these measured values of the attenuation constant and of the phase constant have not been compared with the theoretical ones because there are as yet no theoretical results for the higher-order modes.

The calculation of the transmission properties for the higher-order modes of a rib waveguide and the fabrication of a low-loss rib waveguide, i.e., a single mode, are our current task, and their calculated and experimental results will be reported in another paper.

TABLE III

MEASURED AND THEORETICAL RADIATION ANGLES OF THE TM_{y0} MODE

Dimension (μm)			Values	
t	d	$2a$	Measured	Theoretical
250	500	1000	16°	15.1°
200	450	1000	15°	15.5°
250	550	500	15°	15.5°
150	350	400	18°	18.0°
120	330	500	18°	17.9°

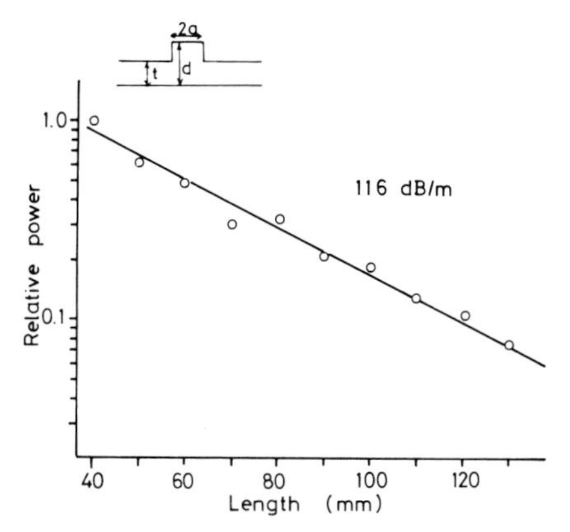

FIG. 29 Loss measurements for the $TM_{\nu 0}$ mode in the rib waveguide at $\lambda_0 = 337$ μm. (Relative output powers are measured as a function of its length.) $t = 250$, $d = 550$, $2a = 500$ μm.

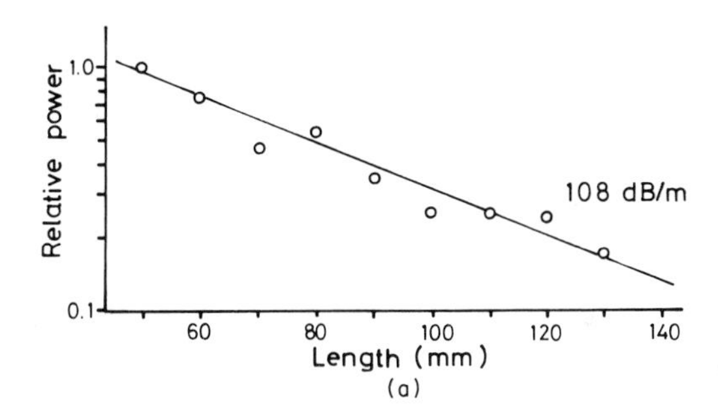

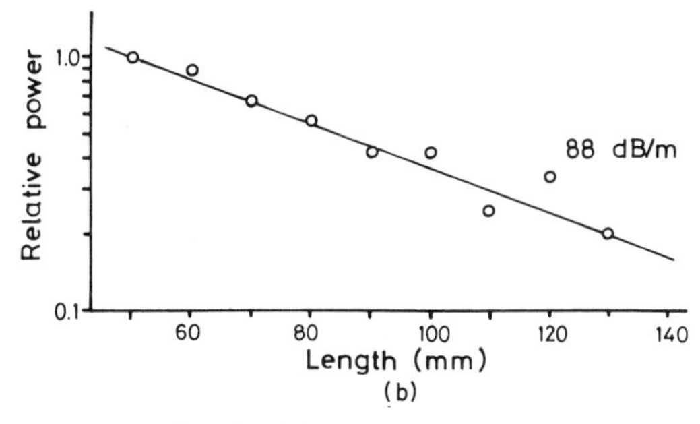

FIG. 30 Loss measurements for the higher-order modes in the rib waveguide at $\lambda_0 = 337$ μm. (a) Output angle, $\theta_{out} = 21°$; (b) output angle $\theta_{out} = 29°$. (Relative output powers are measured as a function of its length.)

IV. Dielectric Cylindrical Waveguides

The slab and rib waveguides will be useful for integrated circuits in the millimeter- and submillimeter-wave regions. We now turn our attention to submillimeter waveguides usable for rather long distances that are applied to the connection between integrated devices.

For this purpose, it is better to use a dielectric cylindrical waveguide having the same structure as an optical fiber that consists of a high-index core surrounded by a cladding of low-index material. In such a structure, most of transmitting power is concentrated in and near its core region, so that the exceptionally low-loss dielectric materials are essential to a core material. At present it is not clear what materials have the low-loss properties in the submillimeter-wave region.

Thus, at this stage, we shall try to realize a low-loss waveguide by transmitting a fractional power through a dielectric material. A typical model of such a waveguide will be an O-type guide (Unger, 1954; Sugi and Nakahara, 1959), which consists of a thin-film dielectric tube. Then, not only multimode guides but also single-mode O-type guides are fabricated by using a polyethylene, and their attenuation constants are measured at $\lambda_0 = 337\ \mu$m.

On the other hand, this waveguide is practically difficult to wire and support without excess losses, so that an improved O-type guide is proposed and discussed numerically. This guide has a metal wire coated with a low-index material in its center, and an outer cylindrical layer with a higher index is isolated from the metal wire to keep the low-loss properties of an O-type guide. However, our present estimation tells us the inevitable increase of the attenuation constant, which is about a few times as much as that of an O-type guide.

A. Analysis

1. *O-Type Waveguides*

To make clear our discussions, let us consider an O-type guide having its cross section shown in Fig. 31 (Shigesawa *et al.*, 1978). The analysis of such a waveguide is well understood and was described by Sugi and Nakahara (1959). Its essentials are discussed briefly in Appendix I.

We now assume that medium 1 has a refractive index n_1 slightly larger than unity as lossless gaseous materials, and medium 2 consists of a polyethylene with $n_2 = 1.461$ and $\tan \delta = 5.3 \times 10^{-4}$ [refer to Eq. (13)] at $\lambda_0 = 337\ \mu$m.

Applying the perturbation techniques, the attenuation constants for the dominant HE_{11} mode are calculated as a function of a/λ_0 at 337 μm. Typical results are shown in Fig. 32, where both $\Delta = b/a$ and refractive index

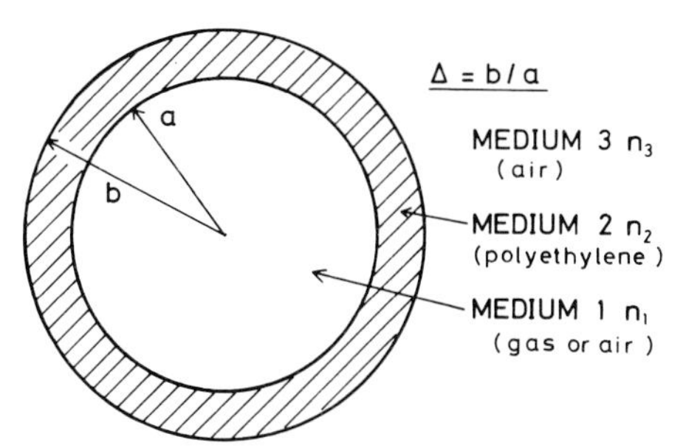

FIG. 31. Cross section of an O-type guide.

n_1 of gaseous medium 1 are varied to control a fractional power in medium 2. It is noted from Fig. 32 that the transmission loss for $n_1 = 1$ increases monotonically with a/λ_0 independent of Δ, while increasing n_1 results in the remarkable reduction of loss for the large values of a/λ_0. This interesting effect, which depends on Δ remarkably and disappears in the small values of a/λ_0, is closely related to the variation of a fractional power in medium 2 as shown in Fig. 33.

Our purpose is to operate such a guide with the HE_{11} mode, and discussions are limited, hereafter, to such a structure. The single-mode transmission in an O-type guide will be determined by the cutoff point of the next higher-order (TE_{01}) mode and depends strongly on n_1. It is found that the cutoff values for a/λ_0 decrease rapidly with increasing n_1 as shown in Fig. 34. As a result, the attenuation characteristics for single-mode transmission are given by the solid lines in Fig. 35 in case of $\Delta = 1.01$. It

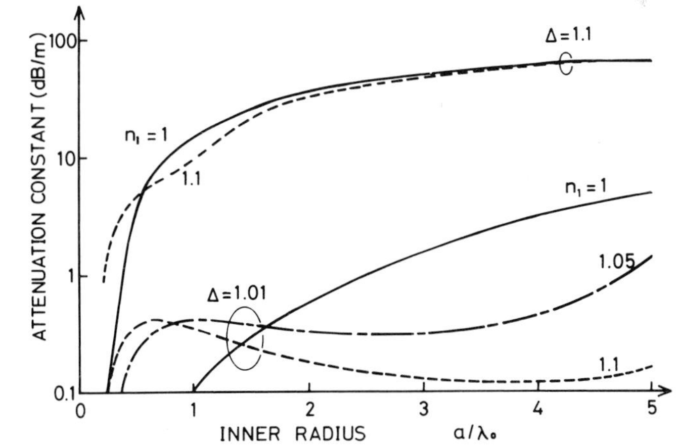

FIG. 32 Attenuation constant for the HE_{11} mode as a function of the normalized inner radius a/λ_0; $\lambda_0 = 337\ \mu m$.

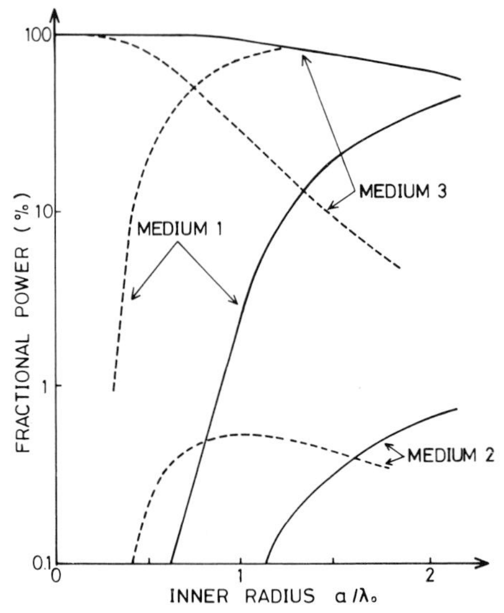

FIG. 33 Fractional power as a function of the normalized inner radius a/λ_0 at $\Delta = 1.01$. ——, $n_1 = 1$; – –, $n_1 = 1.05$.

proves that increasing n_1 in a single-mode O-type guide no longer has an effect on reducing losses, and its losses increase monotonically with a/λ_0. Thus a low-loss single-mode O-type waveguide may be realized in case of a relatively small a/λ_0 when $n_1 = 1$.

On the other hand, such a single-mode surface waveguide suffers a

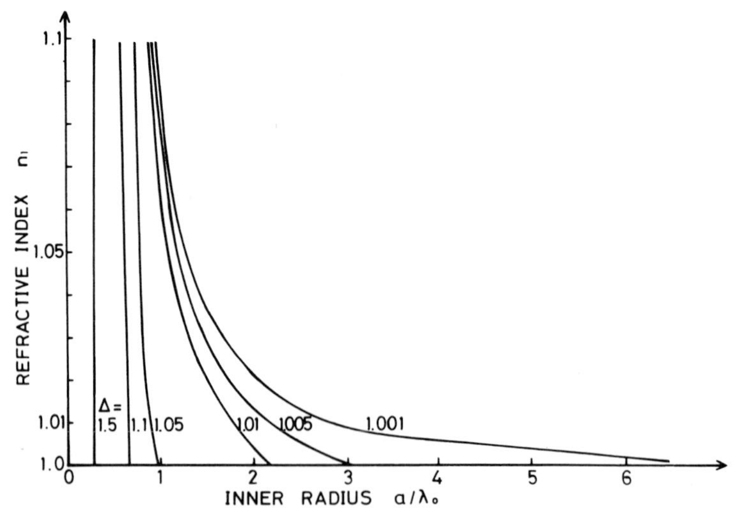

FIG. 34 Cutoff value of the TE_{01} mode in an O-type guide; $\lambda_0 = 337\ \mu m$.

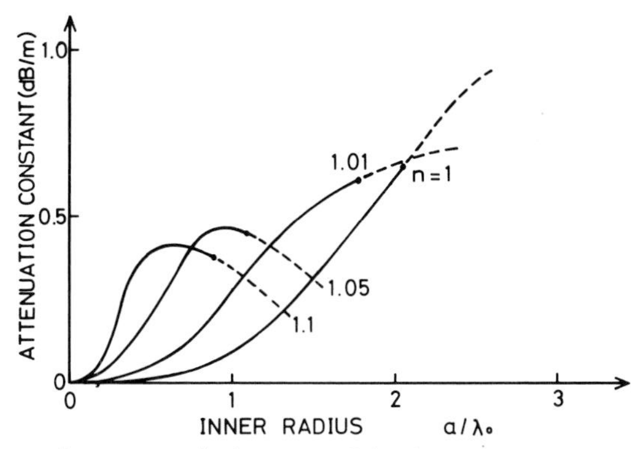

FIG. 35 Attenuation constant in the range of the single-mode transmission at $\Delta = 1.01$. $\lambda_0 = 337$ mm.

remarkable loss at its bend. This loss is closely related to a fractional power in medium 3, which decreases monotonically with the increase of a/λ_0 as shown in Fig. 33. Thus, in order to reduce a bend loss, it is effective to increase both a/λ_0 and n_1. But increasing n_1 is in contradiction with the above-mentioned result, and thus we must find the optimum value of a/λ_0 for each n_1. There is much difficulty in calculating the bending loss of a round waveguide, and this is precisely why such an optimum value has not yet been found.

2. *Improved O-Type Waveguides*

An O-type guide has many difficulties in supporting or wiring without serious excess losses. These difficulties are resolved by introducing a metal wire in its center. The cross section of such a guide is shown in Fig. 36. To support an outer cylindrical dielectric tube, medium 1 must be filled up with a solid, lossless, and lower-index material. The formed polyethylene that is now examined for submillimeter uses (Yamamoto, 1978) is an example of what can be used for this purpose. But in this sec-

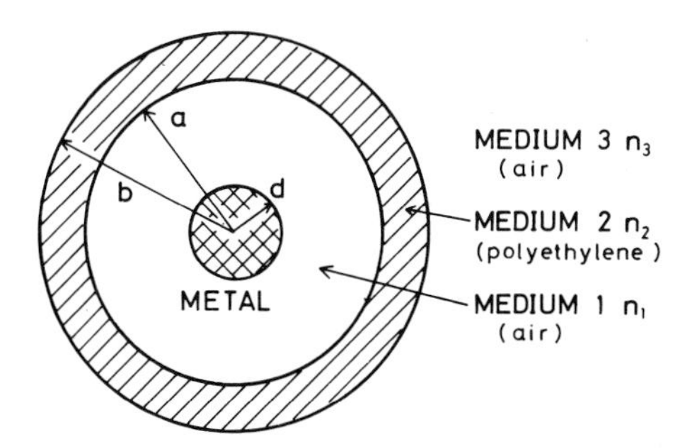

FIG. 36 Cross section of an improved O-type guide.

tion the lossless air of $n_1 = 1$ is considered as medium 1, and we discuss the advantages or disadvantages of introducing a metal wire.

Analysis of this waveguide is again well understood, and the brief summary is presented in Appendix III.

In our calculations, we assume that losses from both medium 2 of a polyethylene and a copper wire with conductivity $\sigma = 0.5 \times 10^8$ [S/m].

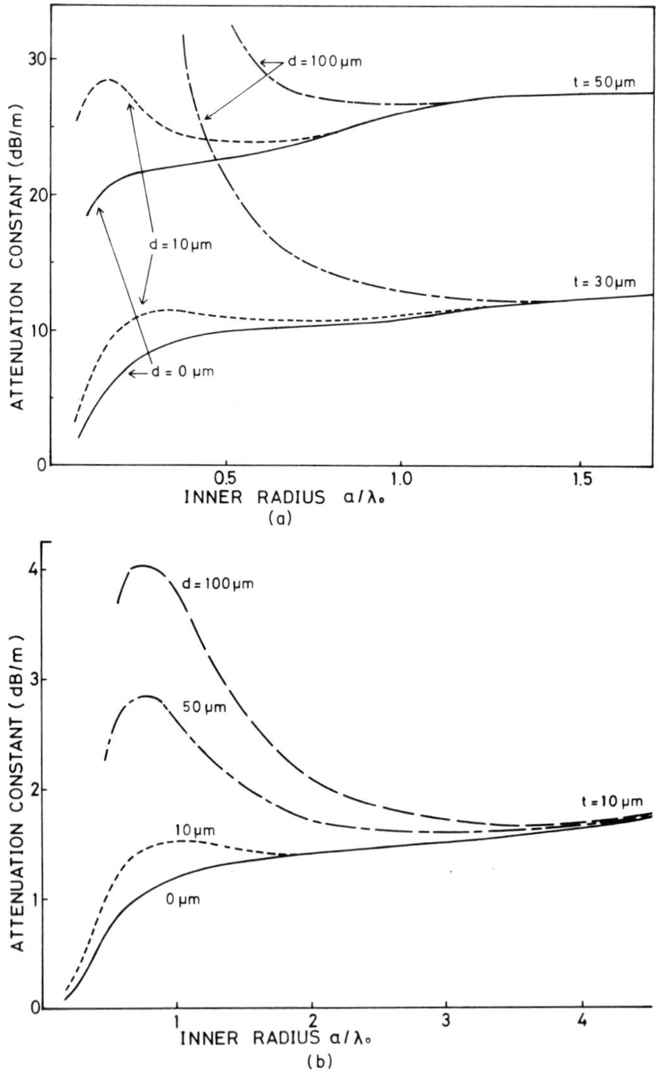

FIG. 37 Attenuation constant for the HE_{11} mode as a function of the normalized inner radius a/λ_0. Thickness of a polyethylene (a) t = 30 and 50 μm, λ_0 = 337 μm; and (b) t = 10 μm, λ_0 = 337 μm.

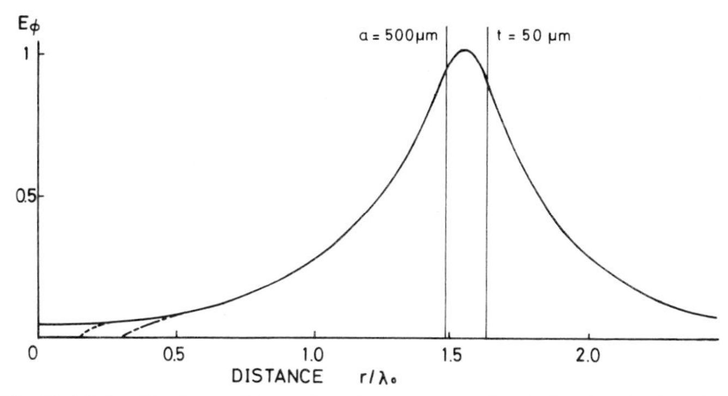

FIG. 38 Field distributions of the azimuth component E_ϕ of the electric field for the HE_{11} mode. $\lambda_0 = 337\ \mu m$; ——, $d = 0\ \mu m$; --, $d = 50\ \mu m$; —— – ——, $d = 100\ \mu m$.

Figures 37a–37b show the calculated results of the attenuation constants for the HE_{11} mode, where both the thickness t of a polyethylene and the radius d of a copper wire are varied. Comparing these results with those for $d = 0$, i.e., a conventional O-type guide, it is found that the loss properties are affected by the presence of a copper wire; there are peak values of attenuation constant in the range of small a/λ_0 and their behavior approaches asymptotically that of an O-type guide with increasing a/λ_0. This asymptotic characteristic is explained by the large gap between a copper wire and a polyethylene, and the field distributions given in Fig. 38 make it clearer. Figure 38 shows the distributions of the azimuth component E_ϕ of the electric field for $a = 500\ \mu m$ and $t = 50\ \mu m$ as a function of

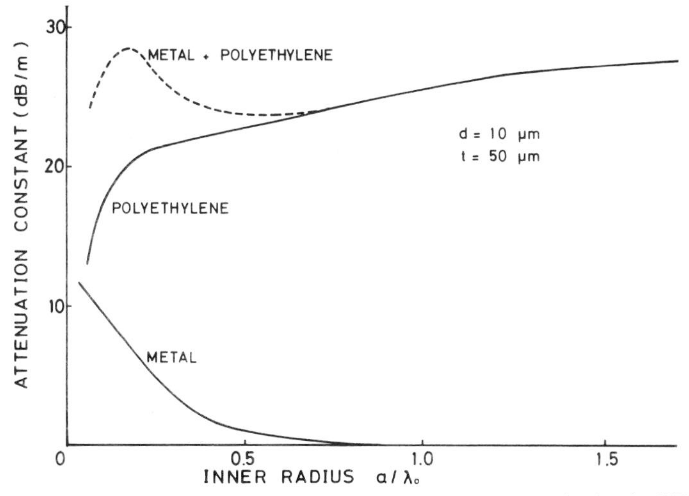

FIG. 39 Behavior of losses due to a polyethylene and a copper wire for the HE_{11} mode. $\lambda_0 = 337\ \mu m$.

the distance r from the center of a guide normalized by the wavelength λ_0. It is evident from this figure that the copper wire has little influence on the field distribution.

On the other hand, each behavior of losses caused by a polyethylene and a copper wire for a guide with $t = 50\ \mu$m and $d = 10\ \mu$m is depicted in Fig. 39 and explains the loss characteristics having the peak value.

We shall now discuss the single-mode guide in the same way as we discussed an O-type guide in Section IV.A.1. The single-mode transmission in an O-type guide is related to the cutoff of the next higher-order (TE_{01}) mode, and its cutoff values for a/λ_0 are found to depend on both t and d as shown in Fig. 40. As a result, loss curves for single-mode transmission are given by solid lines in Fig. 41, where the only case of $t = 10\ \mu$m is illustrated. We note from this figure that the single-mode operation of an improved O-type guide suffers always from excessive losses caused by the metal wire and is favorable near the cutoff TE_{01} mode.

B. Experiments

This section is concerned only with the experiments on O-type guides; and single-mode O-type guides, as well as multimode ones, have been fabricated by using a polyethylene. The features of guide cross section are shown in Fig. 42. The experimental setup is shown in Fig. 43. A spherical mirror M focuses the output beam from an HCN laser ($\lambda_0 = 337\ \mu$m) onto

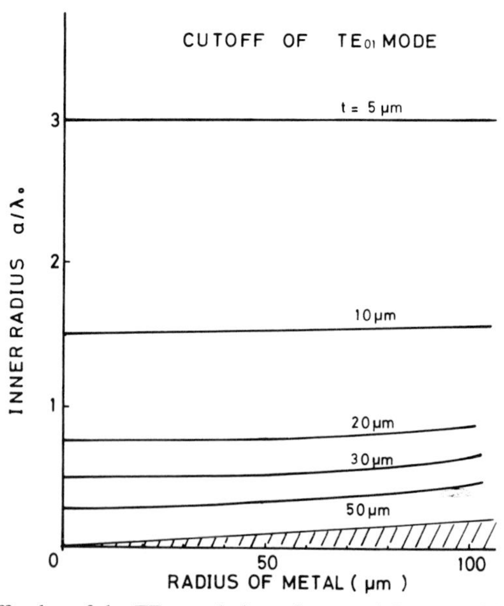

FIG. 40 Cutoff value of the TE_{01} mode in an improved O-type guide. $\lambda_0 = 337\ \mu$m.

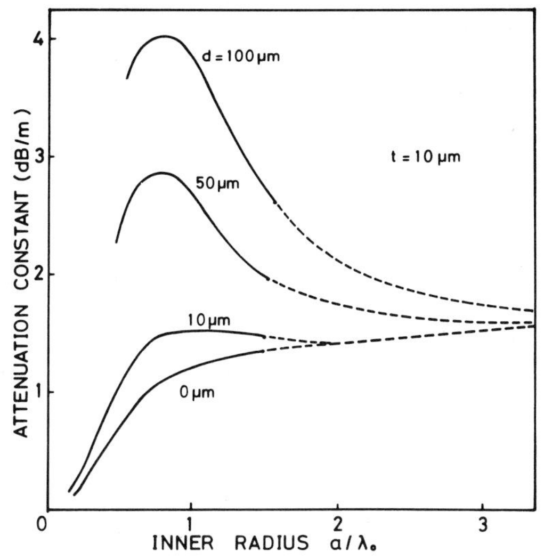

FIG. 41 Attenuation constant in the range of the single-mode transmission at $t = 10\ \mu m$; $\lambda_0 = 337\ \mu m$.

the entrance end of a launching horn H. A polyethylene sheet ($3 \times 3\ cm^2$ in area) with projection in its center is attached to the entrance end of the horn, and the projection is thrust into the center region of an O-type guide for mounting and fixing it to the horn (Fig. 44). The appearances of both the launching horn and the polyethylene sheet with a projection are shown in Figs. 45 and 46, respectively. The laser output is always monitored, and the output from the end of the guide is detected by a Golay cell detector. In our experiments, the O-type guide under test is slightly bent in the range of a few centimeters on the incident side in order to avoid the

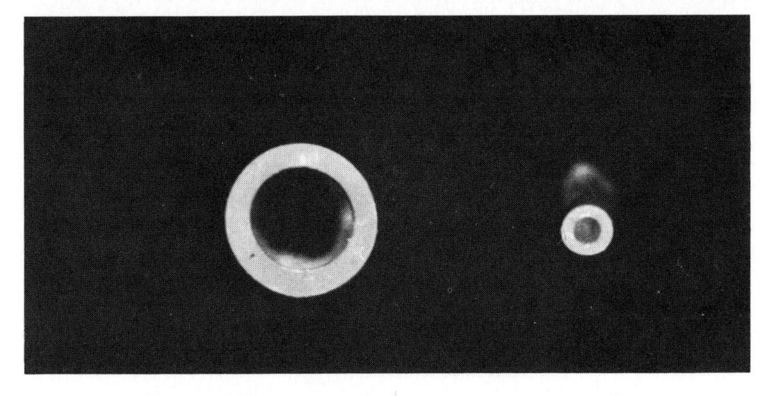

FIG. 42 Photograph of the cross section of an O-type guide.

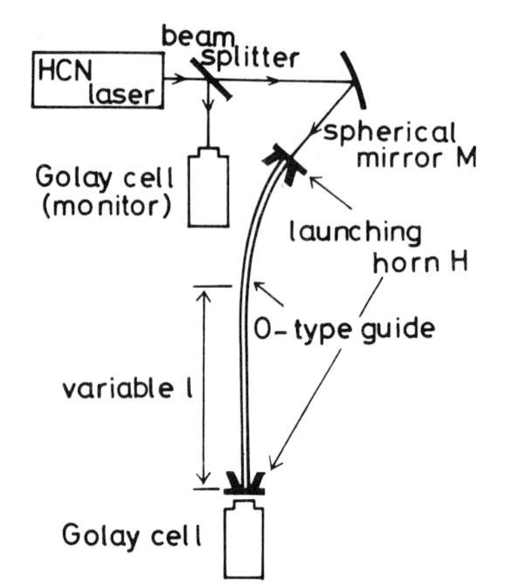

FIG. 43 Schematic diagram of the experimental setup.

direct reception of radiated waves from the launching horn (shown in Fig. 43), and the attenuation constants are measured by varying the guide length in its straight portion.

First, the multimode O-type guides were tested. The results are shown in Fig. 47, in which the relative output powers are plotted as a function of the length of the guide. The dimensions and the measured attenuation constants of each mode are summarized in Table IV. Note that the attenuation constant decreases as inner radius a of the guide becomes small, but these results are not compared with the theoretical values because of the simultaneous propagation of many modes.

As mentioned before, we are interested in operating a guide with the fundamental (HE_{11}) mode. Figure 48 shows the results for this case. The measured attenuation constants, together with the cut-off inner radii of the next higher-order (TE_{01}) mode are also summarized in Table V. It is found from this table that the measured attenuation constants are about twice as much as the theoretical values of the HE_{11} mode. This difference may be caused by the understimation of the loss tangent of a polyethylene

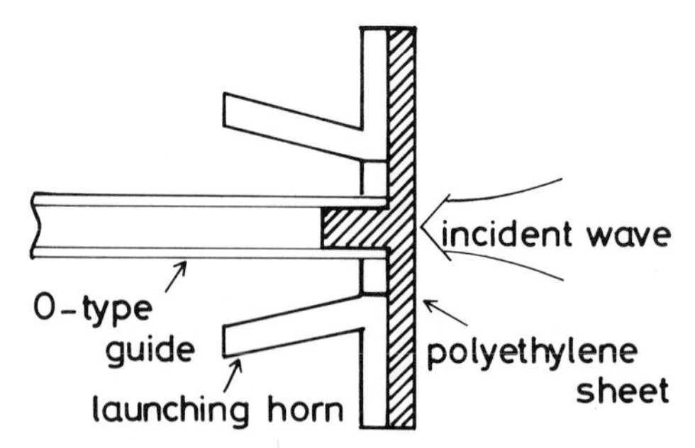

FIG. 44 Launching horn supporting an O-type guide.

FIG. 45 Photograph of the launching horn.

($\tan \delta = 5.3 \times 10^{-4}$) in theoretical calculation and by the radiation losses due to irregularities and inhomogeneities in a waveguide. Such a tendency of the measured attenuation constants has also been obtained for the polyethylene slab waveguide, which has been discussed theoretically and experimentally in Section II.B.3.

V. Conclusion

In this chapter, the experimental investigations of the transmission characteristics of three kinds of dielectric waveguides, i.e., slab waveguides, rib waveguides, and cylindrical waveguides, are performed in the submillimeter-wave region.

First, a slab waveguide is treated as a basic structure common to all di-

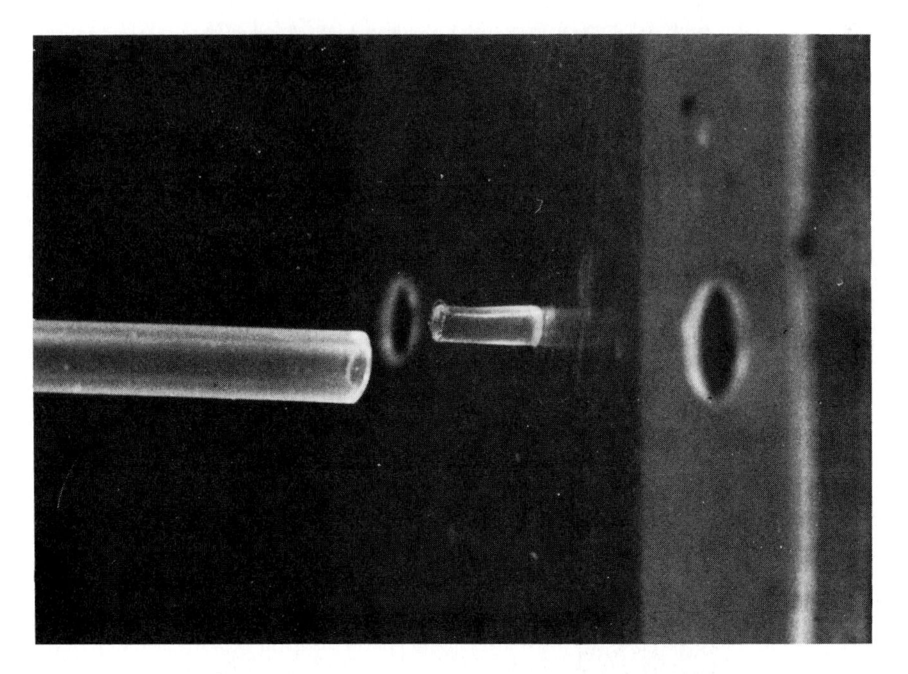

FIG. 46 Photograph of the polyethylene with a projection.

electric waveguides, and its phase and attenuation constants are successfully measured by using the movable metal-grating coupler. As a result, the typical attenuation constants for the TM_0 mode is found to be as low as 1.3 dB/m for a polyethylene slab 10 μm thick. Also, both the transverse broadening of a guided wave and the radiation loss due to the bend having influences on the accuracy of our loss measurements are experimentally discussed.

Next a rib waveguide, which is a more practical waveguide, i.e., confines the fields horizontally and vertically, are investigated. The theoretical considerations with respect to this guide encounter many difficulties

TABLE IV

MEASURED ATTENUATION CONSTANT OF THE MULTIMODE O-TYPE GUIDE

Radius (mm)			
Inner a	Outer b	b/a	Attenuation constant (dB/m)
1.00	1.40	1.4	87.7
0.41	0.82	2.0	62.8
0.28	0.50	1.8	51.0

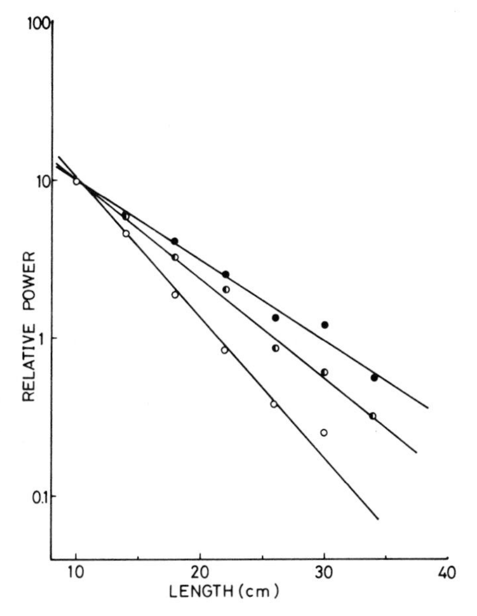

FIG. 47 Loss measurements of the multimode O-type guide. (Relative output powers are measured as a function of its length.) Inner radius (in millimeter): ○, 1.00; ◐, 0.41; ●, 0.28; with b/a = 1.4, 2.0, and 1.8, respectively.

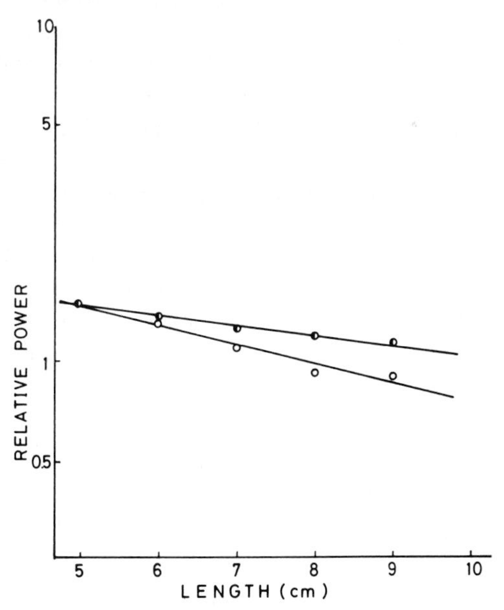

FIG. 48 Loss measurements of the single-mode O-type guide. (Relative output powers are measured as a function of its length.) Inner radius (in micrometer): ○, 70; ◐, 57; with b/a = 1.8 and 1.7, respectively.

TABLE V

MEASURED ATTENUATION CONSTANT OF THE SINGLE-MODE O-TYPE GUIDE

Radius (μm)				Attenuation constant (dB/m)	
Inner a	Outer b	b/a	Cutoff of TE_{01} mode (μm)	Measured	Theoretical
70	126	1.8	73	54.4	25.9
57	97	1.7	79	26.5	13.0

caused by the irregular and unbounded structure of its boundaries. Its transmission characteristics are analyzed by a new approximate method that satisfies the boundary conditions in the least-squares sense, and the validity of our analytical method is proved by the model experiments in the 50-GHz region. On the other hand, the measurement of the phase and attenuation constants of a rib waveguide is again successfully performed by using the same coupler as that of a slab waveguide. However, at present their measured values are obtained only for multimode rib waveguides, and hence the fabrication of a single-mode rib waveguide as well as the analysis of the higher-order modes of a rib waveguide are our current task.

Finally, dielectric cylindrical waveguides are treated as a transmission line to connect a number of isolated integrated circuits that are constructed by a rib waveguide, and so on. For the purpose of realizing a low-loss transmission line, O-type and improved O-type guides, which are transmitting only a fraction of power through the dielectric, are theoretically and experimentally investigated. In the measurement of the attenuation constant of a polyethylene O-type guide, a typical value is found to be 26 dB/m, which is about twice as much as the theoretical one.

For practiced uses of dielectric waveguides in the submillimeter-wave region, the development of a low-loss material is indispensable, but the experimental techniques and results described in this chapter will be useful for developing a submillimeter-integrated circuits.

Appendix I

The functions $P(h)$, $Q(h)$, and $R(h)$ are expressed in the following equations:

$$P(h) = (n_0^2 I_1 + n_1^2 I_2 + n_2^2 I_4) P_1^2 + (I_1 + I_2 + I_4) P_2^2 + (\alpha_0^{\prime 2} I_1 + \beta_0^{\prime 2} I_3 + \gamma_0^{\prime 2} I_4) h^2, \tag{A1}$$

where

$$I_1 = \frac{\sin^2(\beta_0^{\rm I} d + \Phi_0^{\rm I}) \cos^2 \kappa_0 a}{2n_0^2 k_0^2 h^2 \alpha_0^{\rm I}}, \tag{A2}$$

$$I_2 = \frac{\cos^2 \kappa_0 a}{n_1^2 k_0^2 h^2}\left\{\frac{d}{2} - \frac{\cos(\beta_0^{\rm I} d + 2\Phi_0^{\rm I}) \sin \beta_0^{\rm I} d}{2\beta_0^{\rm I}}\right\}, \tag{A3}$$

$$I_3 = \frac{\cos^2 \kappa_0 a}{n_1^2 k_0^2 h}\left\{\frac{d}{2} + \frac{\cos(\beta_0^{\rm I} d + 2\Phi_0^{\rm I})\sin \beta_0^{\rm I} d}{2\beta_0^{\rm I}}\right\}, \tag{A4}$$

$$I_4 = \frac{\sin^2 \Phi_0^{\rm I} \cos^2 \kappa_0 a}{2n_2^2 k_0^2 h^2 \gamma_0^{\rm I}}, \tag{A5}$$

$$P_1 = k_0 \kappa_0 \tan \kappa_0 a, \tag{A6}$$

$$P_2 = \kappa_0^2 + h^2, \tag{A7}$$

$$\begin{aligned} R(h) = {} & (n_0^2 J_1 + n_1^2 J_2 + n_1^2 J_3 + n_2^2 J_5)\delta_0^2 k_0^2 \\ & + (J_1 + J_2 + J_3 + J_5) R_1^2 \\ & + (\alpha_0^{\rm II2} J_1 + \alpha_0^{\rm II2} J_2 + \beta_0^{\rm II2} J_4 + \gamma_0^{\rm II2} J_5) h^2, \end{aligned} \tag{A8}$$

where

$$J_1 = \frac{\sin^2(\beta_0^{\rm II} t + \Phi_0^{\rm II}) \exp\{-2\alpha_0^{\rm II}(d - t)\}}{2n_0^2 k_0^2 h^2 \alpha_0^{\rm II}}, \tag{A9}$$

$$J_2 = \frac{\sin^2(\beta_0^{\rm II} t + \Phi_0^{\rm II})}{2n_1^2 k_0^2 h^2 \alpha_0^{\rm II}}[1 - \exp\{-2\alpha_0^{\rm II}(d - t)\}], \tag{A10}$$

$$J_3 = \frac{1}{n_1^2 k_0^2 h^2}\left\{\frac{t}{2} - \frac{\cos(\beta_0^{\rm II} t + 2\Phi_0^{\rm II}) \sin \beta_0^{\rm II} t}{2\beta_0^{\rm II}}\right\}, \tag{A11}$$

$$J_4 = \frac{1}{n_1^2 k_0^2 h^2}\left\{\frac{t}{2} + \frac{\cos(\beta_0^{\rm II} t + 2\Phi_0^{\rm II}) \sin \beta_0^{\rm II} t}{2\beta_0^{\rm II}}\right\}, \tag{A12}$$

$$J_5 = \frac{\sin^2 \Phi_0^{\rm II}}{2n_2^2 k_0^2 h^2 \gamma_0^{\rm II}}, \tag{A13}$$

$$R_1 = \delta_0^2 - h^2, \tag{A14}$$

$$\begin{aligned} Q(h) = {} & (n_0^2 K_1 + n_1^2 K_2 + n_1^2 K_3 + n_2^2 K_4) P_1 \delta_0 k_0 \\ & + (K_1 + K_2 + K_3 + K_4) P_1 R_1 \\ & - (\alpha_0^{\rm I} \alpha_0^{\rm II} K_1 - \beta_0^{\rm I} \alpha_0^{\rm II} K_5 - \beta_0^{\rm I} \beta_0^{\rm II} K_6 + \gamma_0^{\rm I} \gamma_0^{\rm II} K_4) h^2, \end{aligned} \tag{A15}$$

where

$$K_1 = \frac{\sin(\beta_0^{\rm I} d + \Phi_0^{\rm I}) \cos \kappa_0 a \sin(\beta_0^{\rm II} t + \Phi_0^{\rm II}) \exp\{-\alpha_0^{\rm II}(d - t)\}}{n_0^2 k_0^2 h^2 (\alpha_0^{\rm I} + \alpha_0^{\rm II})}, \tag{A16}$$

$$K_2 = \frac{\sin(\beta_0^{\mathrm{II}}t + \Phi_0^{\mathrm{II}}) \cos \kappa_0 a}{n_1^2 k_0^2 h^2(\alpha_0^{\mathrm{II}2} + \beta_0^{\mathrm{I}2})},$$

$$\times [\alpha_0^{\mathrm{II}} \sin(\beta_0^{\mathrm{I}}t + \Phi_0^{\mathrm{I}}) + \beta_0^{\mathrm{I}} \cos(\beta_0^{\mathrm{I}}t + \Phi_0^{\mathrm{I}})$$
$$-\exp\{-\alpha_0^{\mathrm{II}}(d - t)\}\{\alpha_0^{\mathrm{II}} \sin(\beta_0^{\mathrm{I}}d + \Phi_0^{\mathrm{I}})$$
$$+ \beta_0^{\mathrm{I}} \cos(\beta_0^{\mathrm{I}}d + \Phi_0^{\mathrm{I}})\}], \tag{A17}$$

$$K_3 = -\frac{\cos \kappa_0 a}{2n_1^2 k_0^2 h^2}\left[\frac{\sin\{(\beta_0^{\mathrm{I}} + \beta_0^{\mathrm{II}})t + \Phi_0^{\mathrm{I}} + \Phi_0^{\mathrm{II}}\} - \sin(\Phi_0^{\mathrm{I}} + \Phi_0^{\mathrm{II}}}{\beta_0^{\mathrm{I}} + \beta_0^{\mathrm{II}}}\right.$$
$$\left. - \frac{\sin\{(\beta_0^{\mathrm{I}} - \beta_0^{\mathrm{II}})t + \Phi_0^{\mathrm{I}} - \Phi_0^{\mathrm{II}}\} - \sin(\Phi_0^{\mathrm{I}} - \Phi_0^{\mathrm{II}})}{\beta_0^{\mathrm{I}} - \beta_0^{\mathrm{II}}}\right], \tag{A18}$$

$$K_4 = \frac{\sin \Phi_0^{\mathrm{I}} \sin \Phi_0^{\mathrm{II}}}{n_2^2 k_0^2 h^2(\gamma_0^{\mathrm{I}} + \gamma_0^{\mathrm{II}})}, \tag{A19}$$

$$K_5 = \frac{\sin(\beta_0^{\mathrm{II}}t + \Phi_0^{\mathrm{II}}) \cos \kappa_0 a}{n_1^2 k_0^2 h^2(\beta_0^{\mathrm{I}2} + \alpha_0^{\mathrm{II}2})},$$

$$\times [\alpha_0^{\mathrm{II}} \cos(\beta_0^{\mathrm{I}}t + \Phi_0^{\mathrm{I}}) - \beta_0^{\mathrm{I}} \sin(\beta_0^{\mathrm{I}}t + \Phi_0^{\mathrm{I}})$$
$$- exp\{- \alpha_0^{\mathrm{II}}(d - t)\}\{\alpha_0^{\mathrm{II}} \cos(\beta_0^{\mathrm{I}}d + \Phi_0^{\mathrm{I}})$$
$$-\beta_0^{\mathrm{I}} \sin(\beta_0^{\mathrm{I}}d + \Phi_0^{\mathrm{I}})], \tag{A20}$$

$$K_6 = \frac{\cos \kappa_0 a}{2n_1^2 k_0^2 h^2}\left[\frac{\sin\{(\beta_0^{\mathrm{I}} + \beta_0^{\mathrm{II}})t + \Phi_0^{\mathrm{I}} + \Phi_0^{\mathrm{II}}\} - \sin(\Phi_0^{\mathrm{I}} + \Phi_0^{\mathrm{II}})}{\beta_0^{\mathrm{I}} + \beta_0^{\mathrm{II}}}\right.$$
$$\left. + \frac{\sin\{(\beta_0^{\mathrm{I}} - \beta_0^{\mathrm{II}})t + \Phi_0^{\mathrm{I}} - \Phi_0^{\mathrm{II}}\} - \sin(\Phi_0^{\mathrm{I}} - \Phi_0^{\mathrm{II}})}{\beta_0^{\mathrm{I}} - \beta_0^{\mathrm{II}}}\right]. \tag{A21}$$

Appendix II

The longitudinal field components for the HE_{11} mode of an O-type guide in Fig. 31 follow from Maxwell's equations as

$$E_{z1} = A_e I_1(\gamma r) \cos \phi, \qquad \text{or} \qquad E_{z1} = A_e J_1(\delta r) \cos \phi, \tag{A22}$$

$$H_{z1} = A_h I_1(\gamma r) \sin \phi, \qquad \text{or} \qquad H_{z1} = A_h J_1(\delta r) \sin \phi, \tag{A23}$$

$$E_{z2} = B_e J_1(\beta r) + C_e N_1(\beta r) \cos \phi, \tag{A24}$$

$$H_{z2} = B_h J_1(\beta r) + C_h N_1(\beta r) \sin \phi, \tag{A25}$$

$$E_{z3} = D_e K_1(\rho r) \cos \phi, \tag{A26}$$

$$H_{z3} = D_h K_1(\rho r) \sin \phi, \tag{A27}$$

where A_e, A_h, B_e, B_h, C_e, C_h; D_e, and D_h are the amplitude constants related by the boundary conditions; I_1 and K_1 are the first and second kinds of modified Bessel functions of the first order, respectively; J_1 a Bessel function of the first order; and N_1 a Neumann function of the first order; the subscripts denote the mediums in Fig. 31; $j\gamma$ or δ, β, and $j\rho$ are the wave numbers in the radial direction. For brevity, we omit the common factor $\exp j(\omega t - hz)$ from Eqs. (A22)–(A27).

The following equations show the relation among γ, δ, β, and ρ:

$$(n_1 k_0)^2 = h^2 - \gamma^2, \quad \text{or} \quad (n_1 k_0)^2 = h^2 + \delta^2, \tag{A28}$$

$$(n_2 k_0)^2 = h^2 + \beta^2, \tag{A29}$$

$$(n_3 k_0)^2 = h^2 - \rho^2. \tag{A30}$$

Therefore, the characteristic equation of an O-type guide is derived by considering the boundary conditions at $r = a$ and $r = b$.

Appendix III

The longitudinal field components for the fundamental HE_{11} mode of an improved O-type guide in Fig. 36 are the same as those of an O-type guide except that the logitudinal components in medium 1 are shown as follows, instead of Eqs. (A22)–(A23),

$$E_{z1} = A_e I_1(\gamma r) + G_e K_1(\gamma r) \cos \phi, \tag{A31}$$

$$H_{z1} = A_h I_1(\gamma r) + G_h K_1(\gamma r) \sin \phi. \tag{A23}$$

Therefore, in the derivation of the characteristics equation, the boundary conditions must be considered not only at $r = a$ and $r = b$, but also at $r = d$ on the metal surface.

Acknowledgments

The authors wish to thank the publishers who kindly gave permission to reproduce figures and data. The authors also wish to thank K. Kawai and S. Suhara for their assistance and discussion.

References

Burke, J. J. (1970). *Appl. Opt.* **9,** 2444–2452.

Chantry, G. W. (1971). "Submillimeter Spectroscopy." Academic Press, New York.

Chantry, G. W., Fleming, J. W., Smith, P. M., Cudby, M., and Willis, H. A. (1971). *Chem. Phys. Lett.* **10,** 473–477.

Cook, R. J., Jones, R. G., and Rosenberg, C. B. (1974). *IEEE Trans. Instrum. Meas.* **23,** 438–442.

Danielwitz, E. J., and Coleman, P. D. (1977). *IEEE J. Quantum Electron.* **13,** 310–317.

Goell, J. E. (1973). *Appl. Opt.* **12,** 2797–2798.

Kao, K. C., and Hockham, G. A. (1966). *Proc. IEE* **113,** 1151–1158.

Kapany, N. S., and Burke, J. J. (1972). "Optical Waveguide." Academic Press, New York.

McLevige, W. V., Itoh, T., and Mittra, R. (1975). *IEEE Trans. Microwave Theory Tech.* **23,** 788–794.

Neumann, E. G., and Rudoluph, H. D. (1975). *IEEE Trans. Microwave Theory Tech.* **23,** 142–149.
Ogawa, K., Chang, W. S. C., Sopori, B. L., and Rosenbaum, F. J. (1973). *IEEE J. Quantum Electron.* **9,** 29–42.
Rasaswamy, V. (1974). *Bell Syst. Tech. J.* **53,** 697–704.
Shigesawa, H., Tsuji, M., and Takiyama, K. (1978). *Int. Conf. Submillimeter Waves Their Appl., 3rd, Dig., Guilford,* SB1-3.
Shigesawa, H., Tsuji, M., Suhara, S., and Takiyama, K. (1979a). *URSI Natl. Radio Sci. Meet., Boulder, Colo.* B-5.
Shigesawa, H., Tsuji, M., Suhara, S., and Takiyama, K. (1979b). *Int. Conf. Infrared Millimeter Waves Their Appl., 4th Dig., Miami Beach, Fla.* F-1-2.
Sugi, T., and Nakahara, R. (1959). *IRE Trans. Microwave Theory Tech.* **7,** 366–369.
Tacke, M., and Ulrich, R. (1973). *Opt. Commun.* **8,** 234–238.
Tien, P. K., Ulrich, R., and Martin, R. J. (1969). *Appl. Phys. Lett.* **14,** 291–292.
Tsuji, M., Shigesawa, H., and Takiyama, K. (1979a). *Infrared Phys.* **19,** 669–671.
Tsuji, M., Shigesawa, H., and Takiyama, K. (1979b). *Int. Conf. Infrared Millimeter Waves Their Appl., 4th, Dig., Miami Beach, Fla.* F-1-4.
Tsuji, M., Kawai, K., Shigesawa, H., and Takiyama, K. (1979c). *IEEE Trans. Microwave Theory Tech.* **27,** 873–878.
Unger, H. G. (1954). *Arch. Elektron. Ubertragungstech.* **8,** S241.
Yamamoto, K. (1978). *Pap., Tech. Group Microwaves, IECE Jpn.* MW78-8.
Yasuura, K. (1977). *Int. Conf. Integrated Opt. Opt. Fiber Commun., Dig., Tokyo* A2-3.

CHAPTER 7

Imaging-Mode Operation of Active NMMW Systems

Gary A. Gordon

R & D Associates
Marina del Rey, California

Richard L. Hartman

U.S. Army Missile Command, Research Directorate
Redstone Arsenal, Alabama

Paul W. Kruse

Honeywell Corporate Technology Center
Bloomington, Minnesota

I. INTRODUCTION 327
II. PROPAGATION OF NMMW RADIATION 330
III. SYSTEM DESIGN CONSIDERATIONS 333
A. *Surveillance Volume and Rate* 333
B. *Speckle Averaging* 337
C. *System Sensitivity* 339
IV. SYSTEM LIMITS AND WAVELENGTH TRADEOFFS 340
V. IMAGE QUALITY CONSIDERATIONS 344
A. *Resolution Requirements* 344
B. *Resolution Enhancement* 345
C. *Image Speckle* 346
D. *Image Glint* 350
REFERENCES 351

I. Introduction

Visible light and forward-looking infrared (FLIR) imaging systems provide detailed imagery suitable for acquisition of tactical targets at ranges up to several kilometers, as long as suitable propagation conditions exist. However, when visibility is limited by fogs or battlefield smokes, such visible and IR imaging systems are blinded and other means must be provided for tactical target acquisition. Here we consider the performance of an active near-millimeter-wave (NMMW) system, operating in an imaging

ISBN 0-12-147704-5

mode, to satisfy this requirement. Such a system would scan a transmitted beam in a rasterlike fashion over a given total angular field of view (FOV). The focal plane contains one or more receiver elements that detect radiation backscattered from the scene. The received signals are displayed to provide a representation of the angular variation in reflectivity of the target and surroundings over which the beam is scanned. The concept, depicted in Fig. 1, has been previously discussed by Hartman and Kruse (1976, 1979) and Kruse and Garber (1978).

It is likely that imaging operation of an active NMMW target acquisition system would be provided along with more conventional radar operating modes of the system. An imaging mode may be characterized by the essential feature wherein an image display is examined by an operator who attempts to perform one or more of the desirable acquisition functions (e.g., target detection, recognition, identification, and damage assessment) on the basis of the displayed information. The quality of this information, which we shall represent in terms of the signal-to-noise ratio (S/N) and resolution of the image, depends on the various parameters of the system, the target, and the propagation path.

We imagine that target acquisition may be performed on the basis of information provided by the NMMW system operating in its various modes under operator control, and that other sensor inputs may be present. Different targets, surveillance conditions, and acquisition goals may be expected to call for different sensors and/or modes. For example, we expect that for rapidly closing targets, detection in an imaging mode may be inferior to that provided by a Doppler radar mode. Even stationary targets

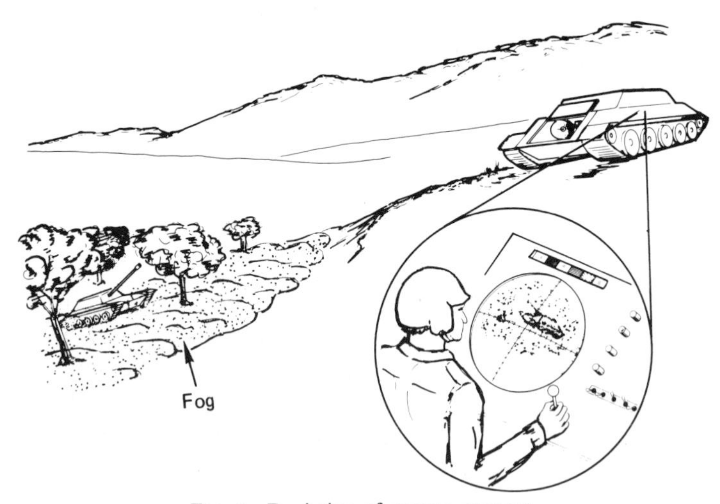

FIG. 1 Depiction of system concept.

may be detected with radar operation, provided that clutter conditions are favorable. Furthermore, target identification may be assisted by separate, cooperative identification-friend-or-foe equipment, acoustic sensors, or what have you. While such alternate operation will be referred to below, we are here chiefly concerned with those contributions, in many situations unique, that can be provided by imaging operation.

Imaging-mode operation is well suited to the detection of stationary or slowly closing targets in the presence of clutter that would render conventional radar modes ineffective. Also, independent of the means of target detection, an image of sufficient quality may permit target recognition. The promise of obtaining adequate image resolution for the detection and recognition of tactical targets is the principal motivation for considering NMMW operation. Propagation of near-millimeter waves through the clear air (where water vapor is the principal attenuation factor), as well as through fogs and smokes, improves with increasing wavelength. On the other hand, target resolution (for fixed target range and imaging optics diameter) improves with decreasing wavelength. This tradeoff between propagation and target resolution dominates other system design considerations and underlies much of the discussion below.

Since detailed imagery out to the longest possible range is desirable, there is a premium on the use of image-enhancement processing techniques that improve the image resolution obtained. Although there are difficulties in implementing such techniques on the complicated and noisy images that are expected in practice, recent work in image-resolution enhancement gives us some basis for guarded optimism regarding this high payoff approach. A more straightforward approach to achieving improved resolution is to increase the effective dimensions of the optics. Since the system platform is conceived to be a highly mobile tactical vehicle, such as a battlefield tank, practical considerations will limit antenna sizes. The use of conformal arrays or interferometric approaches may be worthy of consideration.

Here we consider a system with a 1-m-diameter aperture that is rapidly scanned (e.g., using rotating mirrors) in raster fashion over the FOV. The transmitted signal radiated as the antenna scans is reflected off the target and background scene elements and is detected in an array of superheterodyne receivers. An exemplar system is assumed to have a 1-W average power transmitter, a 30-Hz frame rate (with 10 subframes averaged per frame for reasons discussed below), and an array of as many as 92 high-quality receivers. Multiple receivers are necessary in order to simultaneously achieve a high frame rate, a wide FOV, and a reasonable range window depth. The target is modeled as a diffuse reflector with 10% reflectivity.

Results are obtained for operation in the three atmospheric transmission windows at nominal wavelengths of 0.73, 0.88, and 1.3 mm and for various propagation conditions including clear air at 0°C and 100% relative humidity, a 100-m visibility fog under the same conditions, and a 1-cm/hr rain at 20°C and 100% relative humidity. (The latter case is the most stressing for the system.) It is evident from these results that with reasonable component technology, the system will achieve adequate sensitivity at tactically useful ranges under adverse propagation conditions. The range achieved is dominated by the propagation and is rather insensitive (compared to systems enjoying good propagation) to system parameters such as transmitter power or receiver sensitivity. In addition, when empirically derived criteria are considered for the detection and recognition of typical battlefield targets based on observed imagery, it appears that for target recognition (and in some cases, detection) the system range will be limited by resolution rather than sensitivity.

Along with resolution limitations, images produced by the system will have a somewhat spotty look, referred to as a "speckle" effect. The speckles are caused, in a system that employs coherent illumination, by the interference with random phase of the many scattering centers generally contained within a resolution element. In addition, some resolution elements may contain scattering centers (due to specular glint) considerably larger than the average diffuse return, leading to bright spots on the display. Both the overall speckle effect and the specular glints in particular give the images produced a somewhat unconventional appearance. If recognition is to be on the basis of a conventional diffuse image, the speckles must be smoothed by averaging techniques, while the specular glints may be suppressed by using a nonlinear (i.e., compressive) gray scale. These approaches are illustrated below.

The following discussion begins with a brief review of NMMW propagation considerations. Then system design is discussed, leading to expressions describing the sensitivity limits on system performance. These are compared to resolution limits, and the wavelength tradeoffs are considered. Finally, image-quality considerations are discussed, including resolution, speckle, and glint.

II. Propagation of NMMW Radiation

The choice of operating wavelength in the spectral region around 1 mm is greatly restricted by atmospheric properties. This is in contrast with the situation in the visible and at radio frequencies, where the atmosphere is quite transparent. Even in clear weather, the atmospheric constituents, principally water vapor, severely attenuate electromagnetic radiation over most of the NMMW region. Inclement weather (i.e., fog, rain, or snow) and smoke produce increased attenuation through absorption and scattering mechanisms.

It will be clear below that NMMW system performance is dominated by propagation effects, and thus subject to two related sources of uncertainty. The first corresponds to the naturally occurring variability in the climatic conditions, even at a particular location and season. This variability implies that we can at most specify the system performance statistically, with statistics derived from those describing the climatology. An alternative (and far simpler) approach is to specify the performance attained under particular (e.g., "worst-case") climatic conditions of interest. The second source of uncertainty follows from shortcomings in our present ability to predict the attenuation accurately for specified atmospheric conditions (including micrometeorology). Presumably this latter problem will be remedied through ongoing laboratory and field experiments.

Figure 2 shows the attenuation of electromagnetic radiation produced by the clear atmosphere under standard conditions over a broad range of wavelengths. It is evident that one would not choose to operate in the NMMW region were it not for the compelling reasons outlined in the In-

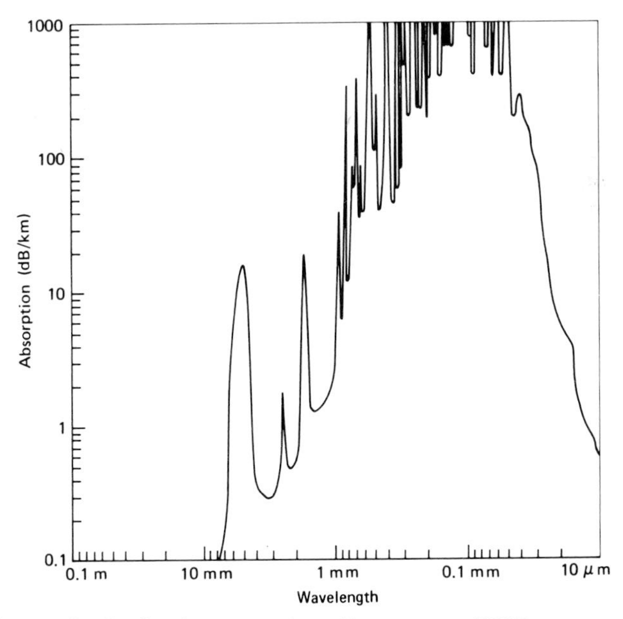

FIG. 2 Attenuation by the clear atmosphere. Temperature, 293°K; pressure, 1 atm; absolute humidity, 7.5 g/m³.

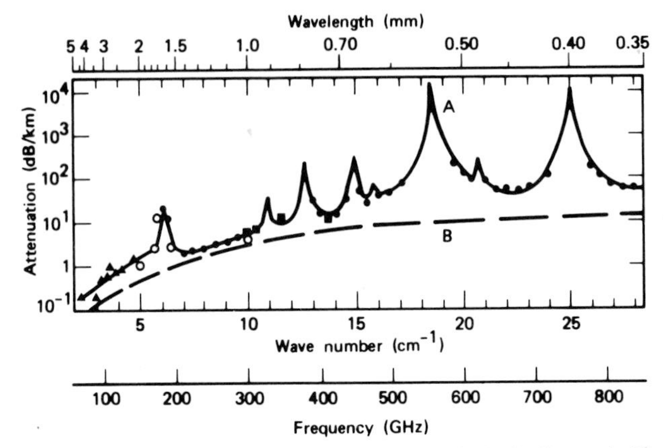

FIG. 3 Attenuation of NMMW radiation by atmospheric H_2O. Curve A: Theoretical and experimental results; curve B: empirical continuum included in theoretical results; temperature, 293°K; pressure, 1 atm; absolute humidity, 5.91 g/m^3. (After Burch, 1968.)

troduction. When the NMMW region is examined in more detail, as in Fig. 3, we see that there are a few windows of potential interest. A considerable body of theoretical and experimental work in the NMMW region is discussed in the literature, much of which is referenced in Guenther (1976b), and Kulpa and Brown (1979).

Quantum mechanical calculations yield attenuation predictions that are surprisingly inaccurate in the windows, where slight variations in the representation of the line shapes lead to a big change in the attenuation. In order to reconcile experimental and theoretical values, it has thus far been necessary to introduce empirical corrections for "continuum absorption" (Burch, 1968). It has also been suggested, amid much controversy, that

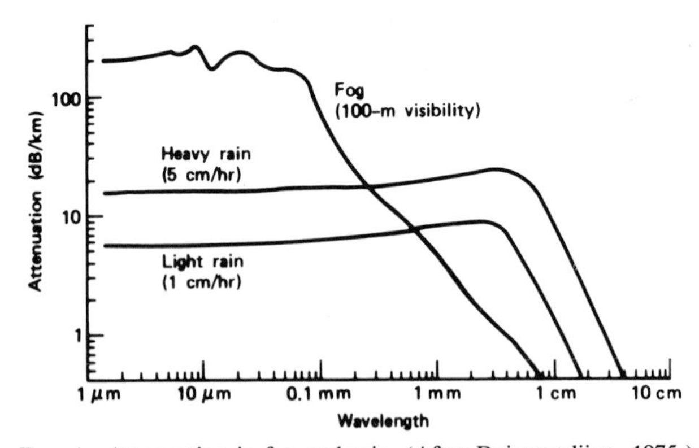

FIG. 4 Attenuation in fog and rain. (After Deirmendjian, 1975.)

TABLE I

ATTENUATION RATES (DB/KM) FOR THREE CLIMATIC CONDITIONS

	Wavelength (mm)		
	0.73	0.88	1.30
Clear (0°C, 100% humidity)	10	7	2
100-m fog (0°C, 100% humidity)	12.6	9	3.4
Light rain (20°C, 100% humidity)	40	27	9

dimers (i.e., formed from two water molecules) contribute significantly to the observed absorption, an effect that has been ignored in theoretical calculations and in scaling experimental measurements (Kulpa and Brown, 1979). If dimers contribute, we would expect a variation in attenuation with humidity of the form

$$\alpha = \alpha_0 H + \alpha_1 H^2, \tag{1}$$

rather then a purely linear dependence. Also, we would expect attenuation to be affected by sunlight (which can dissociate the dimer molecules), a dependence for which some unconfirmed evidence exists (Tanton *et al.*, 1979).

We have defined a number of climatic conditions for the prediction of system performance. These are clear and 100-m visibility fog at 0°C, and light rain at 20°C. The relative humidity is taken as 100% in all three cases. We use the results of theoretical calculations of the attenuation produced by fog and rain such as are illustrated in Fig. 4. Attenuation is assumed to scale linearly with water vapor density in deriving the total attenuation rates for the climatic conditions of interest. These are summarized in Table I and used in the system calculations below.

III. System Design Considerations

A. SURVEILLANCE VOLUME AND RATE

When an active NMMW imager is used for the detection of targets, it is desirable, within limits, for the system to examine a surveillance volume that is as large as possible, and to do so as frequently as possible. Indeed, we may expect that there is a minimum surveillance volume and rate for which a system of this type will be considered useful. The required surveillance volume and rate for target detection have important design implications, in particular the number of receivers required and the system sensitivity attained with a given transmitter average power and receiver quality. Recognition of a detected target requires only that the immediate neighborhood of the target be examined, thus generally leading to relaxed design parameters, with of course the exception of image resolution.

We begin by considering the raster scan performed by the transmitted beam within the sensor FOV. Although the transmitted beam is undoubtedly scanned continuously along each bar of the raster, it is convenient to imagine that the raster consists of a number of discrete beam positions, where a new beam position is generated corresponding to some defined separation along a bar. If we define this separation so that the beam positions form a hexagonal array, then the number of beam positions is given by

$$N_B = \Omega/4\sqrt{3}f^2(1 - \cos\theta/2), \tag{2}$$

where Ω is the solid angle subtended by the FOV, θ the width of the beam formed by the system optics, and f a beam-spacing factor equal to the ratio of the center-to-center separation to the beamwidth.

In considering wavelength tradeoffs, it is logical to keep the FOV and optics diameter constant. However, then θ and N_B will vary. In the numerical examples below, we shall use the expression

$$\theta = 1.2\lambda/D, \tag{3}$$

where θ is the angular separation between half-power points in the beam pattern and D is the diameter of the optics. The factor 1.2 in this expression is intended to be representative of a nonuniform aperture illumination that might be realized in practice. (It is worth noting in passing that the near field of an antenna extends out to approximately a range of $2D^2/\lambda$, or about 2.7 km for $D = 1$ m and $\lambda = 0.73$ mm. Thus, in general, the beam will need to be focused on some nominal range, and θ is the width of this focused beam.)

Each new complete image generated by the system constitutes a frame. Due to speckle-averaging requirements discussed below, the frame rate may be less rapid than the rate at which complete raster scans are performed. The required frame rate is established in part based on the desire to minimize or eliminate flicker in the image display. This may be particularly important if the FOV is to be mechanically scanned at times or if the system platform may be in motion. Also, very slow frame rates lead to a concern that the system will not permit a response to transient target appearances or that target motion will seriously degrade its image appearance. We shall denote the system frame rate by the symbol n_f.

It is convenient to ignore, for the moment, any effects that speckle-averaging techniques might have on the required raster-scan rate and thus

TABLE II

SOME EXEMPLAR SCAN PARAMETERS[a]

	Nominal wavelength (mm)		
	1.30	0.88	0.73
Beam width (mrad)	1.560	1.056	0.876
Number of beam positions	7.20×10^3	1.57×10^4	2.29×10^4
Beam-scan rate ($\sec^{-1}$)	2.16×10^5	4.71×10^5	6.86×10^5
Beam dwell time (μsec)	4.63	2.12	1.46
Range window/dwell (m)	695	318	219
Number of receivers	3	7	10

[a] Field of view: $10° \times 5° = 0.0152$ sr; range window depth: 2 km; frame rate: 30/sec; optics diameter: 1 m.

to assume that the raster-scan rate and frame rate are equal. Then the beam-scan rate n_s at which the transmitted beam sweeps through the N_B positions in the FOV is given by

$$n_s = n_f N_B. \tag{4}$$

This rate may be quite rapid, and this is an opportune point at which to introduce some numerical examples.

Table II lists some exemplar values for parameters just discussed, as well as others to be described below. We assume an FOV that has horizontal and vertical dimensions of 10° and 5°, respectively, giving a solid angle of 0.0152 sr. An optics diameter of 1 m is considered, and beamwidths are evaluated from Eq. (3) for operation in the three atmospheric transmission windows at nominal wavelengths of 1.30, 0.88, and 0.73 mm. The number of beam positions is then calculated with Eq. (2), where the beam-spacing factor is taken to be unity. Finally, the beam-scan rate is evaluated, assuming a 30/sec frame-rate characteristic of some visible and FLIR imaging systems.

The resulting rapid beam scan rates may give rise to the requirement for multiple receivers. To see how this can occur, consider that the transmitter beam dwells on a given beam position for a time t_d given simply by

$$t_d = 1/n_s = 1/n_f N_B. \tag{5}$$

If a receiver beam scans as fast as the transmitter beam, then during the dwell time t_d the receiver can respond to reflections from ranges within a window of depth R_d given by

$$R_d = ct_d/2, \tag{6}$$

where c is the speed of light. Of course, this range window can be positioned at a desired nominal range by suitably lagging the receiver beam scan behind the transmitter beam scan. Continuing our example, the transmitter beam dwell times and range windows generated by these dwell times are shown in the table.

If the desired system range-window depth (i.e., the third dimension of the surveillance volume) is greater than that permitted by the transmitter beam dwell time, then multiple receivers are needed so that each can "listen" for reflections in a given beam position for the required time. If the system range-window depth is R_w, then the number of required receivers is

$$N_R = R_w/R_d = (2R_w/c)n_f N_B. \tag{7}$$

In our example we assume that a 2-km range window is required and find that 3, 7, or 10 receivers are needed for operation at the indicated wavelengths. Speckle-averaging requirements can increase this number of receivers, as discussed below.

If a linear array of receivers is envisioned, then the scan might be realized as suggested schematically in Fig. 5. The receiver beams and transmitter beam move across the FOV to the right at the same speed, but the transmitter beam traverses all r receiver beam positions while the scan moves one bar to the right.

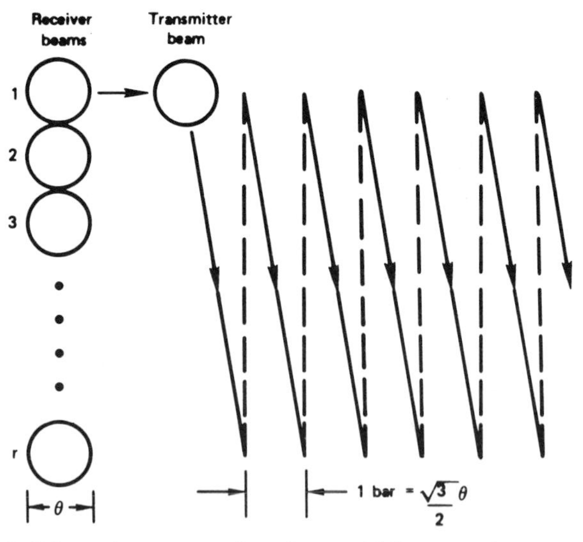

FIG. 5 Schematic representation of scan with linear receiver array.

B. Speckle Averaging

Image speckle arises due to the interference with arbitrary phase of the reflections from the multiple scattering centers within each beam position. As illustrated below, it is desirable to reduce the speckle effect in order to produce an image with a more conventional appearance, thus enhancing detection and/or recognition of the target. Image speckle may be reduced through use of averaging approaches in which each beam position is illuminated multiple times. It is necessary that the relative phases of the scattering center reflections are effectively randomized from one time to another so that the averaging tends to smooth out the effects of arbitrary phase.

Although relative motion (in particular, rotations) of scene elements with respect to the sensor will produce phase randomization, the most practical approaches to speckle averaging make use, in one form or another, of frequency change effects. A frequency change Δf results in a change

$$\Delta\phi \cong (4\pi/c)\,\Delta R\,\Delta f \tag{8}$$

in the relative phase between reflections from two scatterers separated in range by ΔR. Thus, in order to produce a phase change of 2π, we need to change the frequency by

$$\Delta f = c/2\,\Delta R. \tag{9}$$

If scatterers in a given beam are distributed over a range interval ΔR, then the frequency change Δf given by Eq. (9) results in a "shuffling" of the various relative phases in the sense that each relative phase is changed by as much as 2π. In this sense, we may take Eq. (9) to define the nominal frequency shift between "independent" looks at the beam position for the purposes of speckle averaging. When the principal scatterers in a beam arise from parts of the surface of a target or scene background object, the corresponding value of ΔR may be relatively small, requiring substantial frequency change Δf for phase randomization. For example, if ΔR is 0.1 m, then Δf must be ~ 1.5 GHz.

We shall discuss three different methods of performing speckle averaging that are equivalent in effect but quite different in implementation, with diverse implications for component requirements. The first method is the easiest of the three to visualize. For each frame, the system performs some number N_a of complete raster scans. (These may be termed subframes.) The transmitter frequency is changed between subframes, and the subframe images are averaged to produce the displayed image. One disadvantage of the approach is that the optics must scan N_a

times as fast. As discussed below, a reasonable value for N_a would be 10, which would correspond to a subframe rate of 300/sec in our previous example. Since the raster scan is performed mechanically and the (at least primary) optics are relatively large, the mechanical problems may be nontrivial.

A second drawback of this approach is that the more rapid subframe rate leads to a proportionately larger number of required receivers, since the transmitter beam-scan rate will be N_a times greater, and thus the beam dwell time and range window per dwell will be N_a times smaller. In our previous example, the number of receivers for operation in the three spectral windows considered would be increased to 29, 63, or 92.

The second speckle-averaging implementation method considered avoids the more rapid scan requirement by sequentially transmitting the N_a different frequencies during the original transmitter beam dwell time [i.e., given by Eq. (5)]. In this case, the number of receiver elements (i.e., feeds) in the focal plane is not increased either. However, since all N_a frequencies must be received simultaneously in each beam position, there must be N_a receiver channels for each receiver element in the focal plane. (These channel outputs are averaged to produce the image pixel.) Furthermore, the transmitter must change frequencies very rapidly, specifically every t_d/N_a sec. In our example, this switching time would be 0.46, 0.21, or 0.15 μsec for the three windows, which clearly excludes any mechanical tuning of the transmitter source. However, this method might be interesting if rapidly tunable sources can be obtained.

The third implementation approach is very similar to the second, and in fact may be considered to be essentially a generalization. However, the differences in point of view and implementation possibilities are nontrivial. Again, all N_a frequencies are transmitted during the original beam dwell time. However, here we conceive that they are all transmitted simultaneously rather than sequentially as in method 2. Furthermore, we do not restrict the transmitted waveform to consist of a series of narrow lines (in the spectrum) but rather permit essentially any waveform that extends over the same broad bandwidth. Thus, for example, the frequency-stepped waveform of method 2 is perfectly acceptable here. Other possibly interesting waveforms include frequency-swept (i.e., chirp) waveforms and phase-modulated waveforms.

The receiver must be matched to whatever waveform is used. (Note that in method 2 the bank of N_a receiver channels constitutes a composite receiver that is matched to the transmitted frequency-stepped waveform, provided that suitable time delays are introduced to the channels to compensate for the delays between the frequency steps and that the channel outputs are coherently added.) Since the waveform bandwidth is $N_a\,\Delta f$, the range resolution of the output of the matched receiver is

$$\Delta R_{\mathrm{res}} = c/2N_{\mathrm{a}}\,\Delta f = \Delta R/N_{\mathrm{a}}, \tag{10}$$

where we have used Eq. (9).

Recall that ΔR is defined as the range extent over which significant scatterers in the beam are distributed. Thus the system produces outputs for N_{a} independent range resolution elements within this scatterer interval. These outputs are (noncoherently) added (i.e., averaged) to produce the image pixel. Thus with this third method only a single receiver channel is necessary for each receiver element in the focal plane, the only requirement being that this receiver be sufficiently wide band. If N_{a} is 10 and Δf is 1.5 GHz, then a 15-GHz bandwidth would be required.

Finally, we note that with all three speckle-averaging methods the requirements (e.g., scan rate, number of receivers and/or channels, or bandwidth) can be reduced at the expense of compromising the surveillance volume or frame rate of the system.

C. System Sensitivity

The range equation for this system is quite similar to that for ordinary radar. However, there are substantial differences. In this system, the beam is normally smaller than the target. If the target is viewed as a diffuse scatterer, or as a distribution of point scatterers, there is no R^2 loss in target illumination, and it is convenient to represent the target signature by a reflectivity rather than a cross section. Furthermore, atmospheric attenuation dominates the usual sources of radar loss, and results in a range dependence that ultimately becomes more an exponential than a power law.

Specifically, if R is the range of a target of ideal diffuse reflectivity ρ, the received S/N is given by

$$\mathrm{S/N} = \frac{E\rho}{\pi R^2 N_{\mathrm{a}}} A_{\mathrm{R}} \frac{10^{-0.2\alpha R}}{kT_0FL} N_{\mathrm{a}}^{1/2}, \tag{11}$$

where E is the energy transmitted in each beam position (corresponding to a single frame), A_{R} the receiver aperture area, F the receiver noise figure, k Boltzmann's constant, T_0 a 290°K reference temperature, α the propagation attenuation coefficient (in units of decibels/distance), and L represents all nonpropagation system losses. The equation includes the effects of speckle averaging and is applicable to all three implementation methods discussed above. The factor N_{a} in the denominator corresponds to the division, in method 1, of the energy among the N_{a} subframes; in method 2, of the energy among the N_{a} frequency steps; and in method 3, of the target scattering centers among the N_{a} range resolution cells. The factor $N_a^{1/2}$ in the numerator represents the improvement resulting from the (noncoherent) integration of the N_a independent outputs for each frame.

Note that we have assumed that the receiver is operated using heterodyne methods and is matched to the transmitted waveform. A direct (video) detection approach could be used but would probably not provide comparable performance unless heroic cooling methods were employed. The receiver noise representation in Eq. (11) ignores quantum effects, as is justified at the wavelengths and temperatures of interest.

It is convenient to rewrite Eq. (11) in terms of the transmitter average power P_{av}. Since energy E is transmitted at the rate n_s, we have $P_{av} = En_s$, so that

$$\mathrm{S/N} = \frac{P_{av}\rho}{\pi R^2 n_s N_a^{1/2}} A_R \frac{10^{-0.2\alpha R}}{kT_0FL}. \tag{12}$$

Finally, some mention should be made regarding the effects of processing that takes place between the receiver output and the display. Nonlinear transformations, such as the use of a compressive gray scale, will have some effect on the noisiness of the resulting image. Potentially much more serious, however, is the method used to collapse the data in range so as to form the two-dimensional display. If the method used is simply to integrate the signal over range, then the image will be degraded by a "collapsing loss" caused by the integration over ranges that contribute noise but not signal. This loss may be substantial (e.g., as much as 10 dB in the example of Table II). The loss may be essentially eliminated by suitable processing, such as threshold or "greatest of" detection, to identify the range interval occupied by the target or background object, which is then selectively displayed. Another advantage of these processing approaches is that they permit display compensation for sensitivity variations on the image caused by reflections from objects within the FOV at different ranges.

IV. System Limits and Wavelength Tradeoffs

Here we illustrate, by example, the interplay of limits to system performance due to sensitivity and resolution effects, and how these change with operating wavelength. We define an exemplar system with parameters summarized in Table III. The assumed average power and receiver noise figure are characteristic of currently available laboratory devices. Some projection is required to justify these values for production components, particularly in the case of linear receiver arrays.

The results of a sensitivity calculation using Eq. (12) are shown in Fig. 6. The propagation conditions considered and attenuation coefficients

TABLE III

EXEMPLAR SYSTEM PARAMETERS

Field of view	$10^\circ \times 5^\circ$
Range window depth	2 km
Frame rate	30/sec
Speckle-averaging number	10
Antenna diameter	1 m
Average power	1 W
Noise figure	8 dB
Nonpropagation losses	3 dB
Signal-to-noise (power)	14 dB
Target reflectivity	0.1

used are those in Table I corresponding to clear and fog conditions at 0°C and rain at 20°C. Shown on the same figure are the Johnson criteria (see discussion further below) for target detection and recognition, as applied to an M-48 tank. The minimum dimension (height) of the tank is 3.28 m. Since recognition requires about seven resolution elements (3.5 line pairs) across this dimension, the required resolution is about 0.5 m. The resolution required for detection is about a factor of five less (about 2.5 m). Finally, we show a relaxed recognition limit that might be applicable if a factor of two enhancement of resolution could be attained through techniques discussed in Section V.

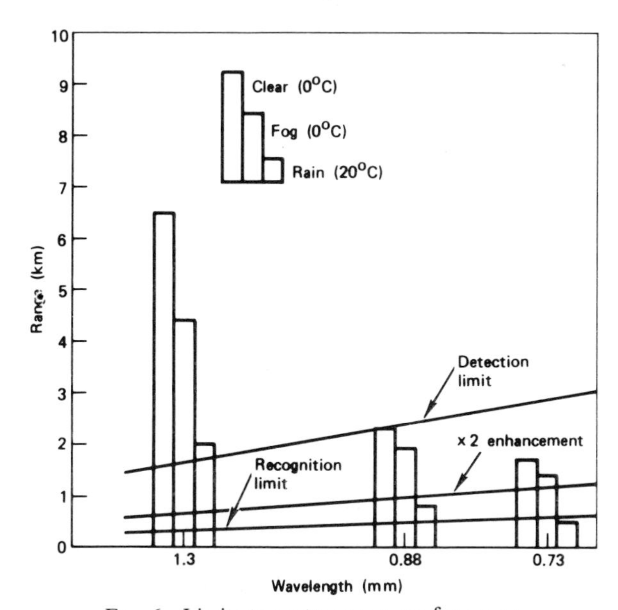

FIG. 6 Limits to system range performance.

For target detection, the system is resolution-limited at 1.3 mm and sensitivity-limited at the shorter wavelengths. If the system were to be optimized for target detection through fog, then 0.88 mm would be the best window, with a range of just under 2 km. In rain, however, 1.3 mm is best. For target recognition, we see that the system is resolution-limited in all three windows (assuming no resolution enhancement is available) and the shortest wavelength is best.

If resolution enhancement is achieved, then recognition optimization becomes more complex. If the enhancement is a factor of about two, then 0.73 mm is still the optimum window for operation through fog, while 0.88 mm is optimum in rain. If greater enhancement (but less than about a factor of five) were possible, then 0.88 mm would be optimum in fog, while 1.3 mm would be optimum in rain. Clearly, if one wavelength must be selected (multiple-wavelength designs would be nice of course), the choice must consider resolution enhancement availability.

It is evident from Fig. 6 that changes in the propagation conditions and wavelength result in substantial variations in the sensitivity-limited range. This variation is illustrated in an alternate format in Fig. 7, which shows range as a function of attenuation coefficient for our exemplar system designs at the three wavelength windows. The parameter K in the figure incorporates all of the relevant parameters with the exception of R and α. Specifically, from Eq. (12) we define

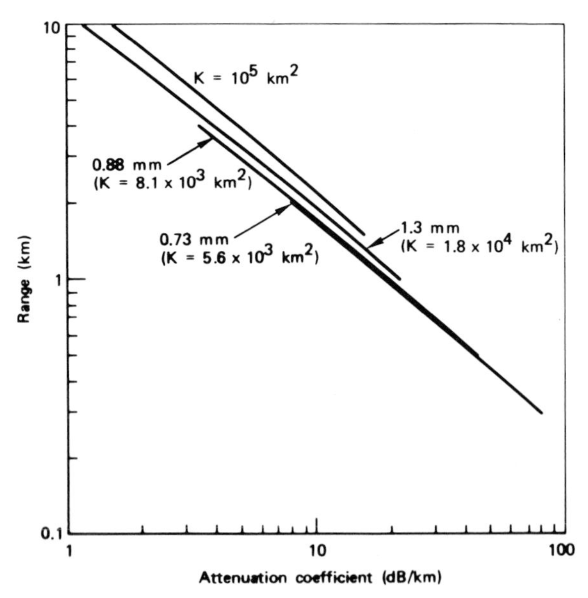

FIG. 7 Variations in range with attenuation coefficient.

$$K \equiv R^2 10^{0.2\alpha R} \tag{13}$$

$$= \frac{P_{\mathrm{av}} \rho A_{\mathrm{R}}}{\mathrm{S/N} \pi n_{\mathrm{s}} \sqrt{N_{\mathrm{a}}} k T_0 F L}. \tag{14}$$

The figure makes it clear that the differences in sensitivity-limited range in the three windows are due almost completely to the differences in attenuation coefficient, and not to the assumed system differences. Furthermore, even when K is increased by about an order of magnitude as shown, the effect on range is relatively minor compared to variations due to attenuation.

Sensitivities of system range to various parameters are illustrated explicitly in Fig. 8 for the 0.88-mm system. The center line represents the baseline system. The range of the actual system can be determined by moving along the axis of the parameter of interest. Since the axis is logarithmic, the effect of changing two parameters can be superimposed by adding the displacements along the x axis. It is again obvious that system

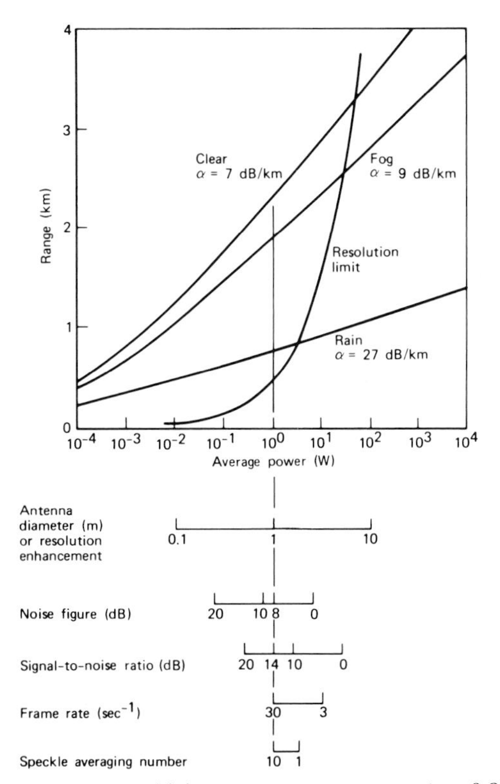

FIG. 8 Range sensitivity to system parameters. λ = 0.88 mm.

range is not very sensitive to the system parameters, but that attenuation changes can have pronounced effects. The resolution limit, also shown, may be used with the axis labeled "antenna diameter or resolution enhancement."

V. Image Quality Considerations

A. Resolution Requirements

Through experiments with observers, Johnson (1958) determined the number of line pairs across the minimum target dimension required to detect, determine the orientation, recognize, and identify tactical targets. These results are illustrated in Fig. 9. It can be seen that the requirements in each category are relatively independent of the target. Recognition, which is the most frequently used figure of merit for system performance

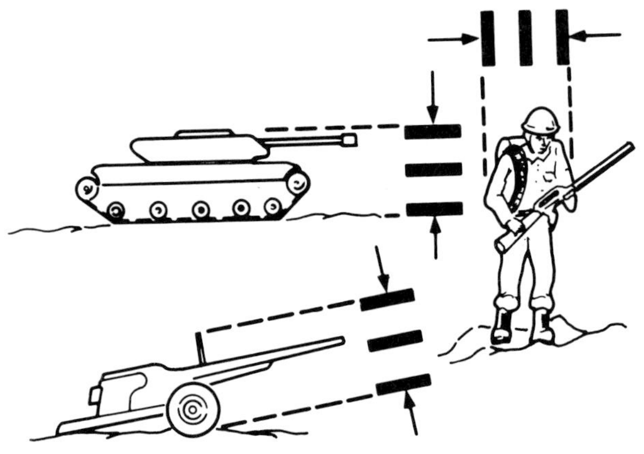

Target	Resolution per minimum dimension in line pairs			
Broadside view	Detection	Orientation	Recognition	Identification
Truck	0.90	1.25	4.5	8.0
M-48 tank	0.75	1.20	3.5	7.0
Stalin tank	0.75	1.20	3.3	6.0
Centurion tank	0.75	1.20	3.5	6.0
Half-track	1.00	1.50	4.0	5.0
Jeep	1.20	1.50	4.5	5.5
Command car	1.20	1.50	4.3	5.5
Soldier (standing)	1.50	1.80	3.8	8.0
105 Howitzer	1.00	1.50	4.3	6.0
Average	1.0 ± 0.25	1.4 ± 0.35	4.0 ± 0.8	6.4 ± 1.5

FIG. 9 Required resolution for detection, orientation, recognition, and identification. (After Johnson, 1958.)

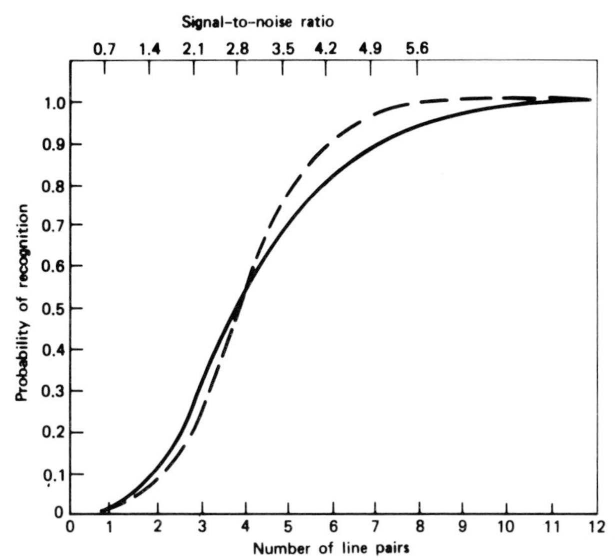

FIG. 10 Probability of recognition as a function of resolution and signal-to-noise ratio. --, S/N with four line pairs; ——, line pairs with S/N = 2.8. (After Rosell and Wilson, 1973.)

calculations, requires four line pairs across the minimum target dimension. The results of a more complete analysis (Rosell and Willson, 1973), which takes into account not only resolution but also S/N, are illustrated in Fig. 10. This shows that the Johnson recognition criterion of four line pairs per minimum target dimension is that value for which the probability of recognition is 0.5 when the S/N is 2.8. An increase in either the resolution or the S/N increases the probability of recognition.

B. RESOLUTION ENHANCEMENT

The classical (Rayleigh) diffraction limit for the angular resolution provided by an aperture of diameter D is given by $1.22\lambda/D$. Since it is often the resolution that limits the system range, it is useful to consider whether the Rayleigh limit can be circumvented by signal processing. Of interest is the concept of "superresolution," or resolution enhancement through deconvolution processing. This approach has been studied extensively, initially for noise-free imagery (Harris, 1964; Barnes, 1966; Frieden, 1967) and later for noisy images (Buck and Gustincic, 1967; Rushforth and Harris, 1968; Toraldo di Francia, 1969; Goodman, 1970; Frieden, 1975). Because the image formed by an optical system is the convolution of the object with the point spread function of the optics, in principle the object can be restored from the image by mathematically inverting the convolution process. The Rayleigh limit is surpassed by extending the deconvolution process to spatial frequencies not present in the image through analytic continuation methods.

In theory, an arbitrarily high resolution can be restored in the noise-free case. In practice, it is noise in the image and uncertainties in the point spread function that limit the deconvolution process. When resolution beyond the Rayleigh limit is attempted, the effect is to increase the image noise level. The degree of increase is related to the complexity of the original image. Images containing only a few independent resolution elements can be significantly resolution-enhanced with reasonable noise degradation. Complex images, however, such as those likely to be obtained from the NMMW imaging system may permit only small improvements in resolution. (When a detected target is to be processed for recognition, we suppose that the system operator might specify a limited image region for resolution enhancement.) Frieden (1975), who has reviewed the various mathematical approaches to resolution enhancement, estimates that a factor of two improvement in resolution beyond the Rayleigh limit should be possible for complex imagery in practice. Within a small region of the display, containing only a few resolution elements, it may be possible to obtain more than a factor of two.

It may also be possible to obtain improved target information without true resolution enhancement. Because a tank target includes many unresolved scattering centers, it is possible to measure the size, shape, and orientation of the ensemble of scatterers in terms of the spatial second-central-moments of the target's radar cross-section distribution. Estimation algorithms for these moments based on raster-scan data have been derived (Gordon and Casowitz, 1972). The estimated second-central-moments may be visualized as representing the target by an ellipsoid of similar size, shape, and orientation. Thus this approach provides a form of cueing for target recognition. Diversity is required to reduce speckle effects.

C. Image Speckle

Although no images have been as yet produced by active systems at near-millimeter wavelengths, some have been obtained (Guenther, 1976a; Baird, 1976) with active imaging systems at both 10.6-μm and 3.2-mm wavelengths. Photographs of targets used in the experiments are shown in Fig. 11a–11d. Shown are the actual target range, the resolution at that range of the active 3.2-mm system, and the simulated range of an imaging system operating in the 0.73-mm wavelength window with a 1-m aperture diameter.

Typical results obtained at the nearest range at both wavelengths are shown in Fig. 12. Both images are characterized by speckle arising from scattering center interference. Speckle is a potential problem in any

FIG. 11 Targets and operating ranges for imaging experiment. (a) Actual range, 50 ft; resolution, 10 cm; simulated range, 143 m. (b) Actual range, 100 ft; resolution, 48 cm; simulated range, 0.7 km. (c) Actual range 200 ft; resolution, 146 cm; simulated range, 2.1 km. (d) Actual range, 400 ft; resolution, 341 cm; simulated range, 4.9 km. (After Guenther, 1976a.)

system with coherent illumination, be it laser or radar, since it can degrade the ability of an observer to recognize a target. (Note that the speckles in Fig. 12 are smaller in the 10.6-μm image than in the one at 3.2 mm because of the higher resolution of the former system.) In the images of Fig. 12 the targets are readily recognizable despite the unconventional image appearance. However, as the resolution is reduced by increasing the target range, the target recognition suffers. Figure 13 illustrates imagery obtained at simulated ranges (for a 0.73-mm wavelength system having a 1-m aperture diameter) of 0.7 km and 2.1 km. These ranges are approximately at the recognition and detection limits, respectively, for the tank. The objects in the 3.2-mm image on the left are, from left to right, a truck, a jeep, and a tank. A corner reflector is located between the jeep and tank. The objects in the 3.2-mm image on the right are, from left to right, a truck, a tank, and a gasoline fuel storage tank. It is evident from results such as these that the Johnson criteria give unrealistically optimistic predictions for the recognition of speckled images.

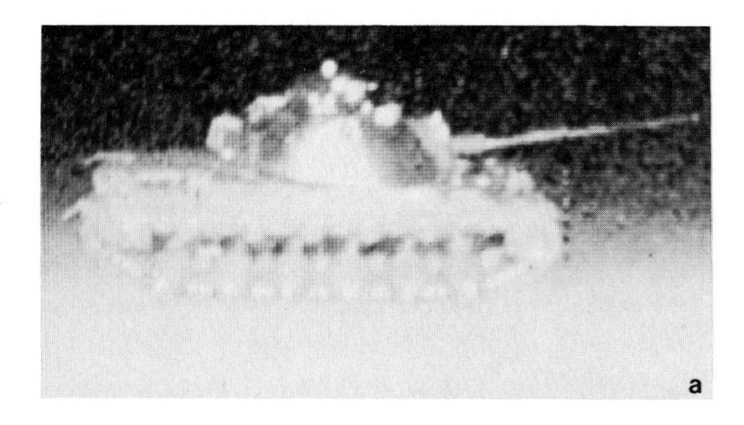

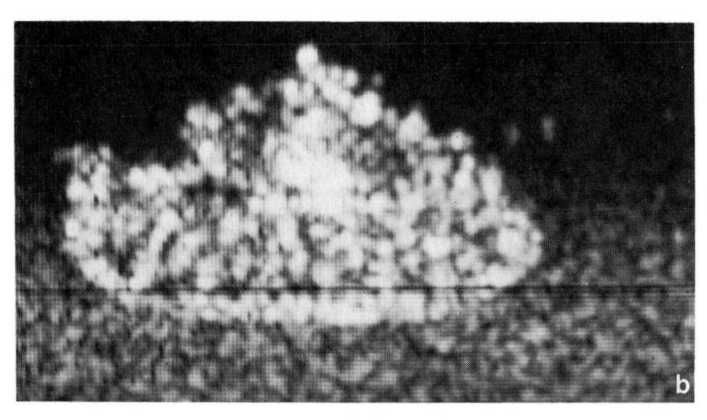

FIG. 12 10.6-μm (a) and 3.2-mm (b) images of a M48A5 tank, side view. (After Guenther, 1976a.)

The effects of image speckle may be reduced through the use of averaging techniques such as were discussed above as well as others discussed in the literature (McKechnie, 1975). The resulting benefits are illustrated in Fig. 14. An image taken in incoherent light at fairly low resolution is shown in Fig. 14a. The same image taken with coherent light is shown in Fig. 14b. This second image contains very drastic and high contrast speckle so that the target is virtually undetectable as well as unrecognizable. Images (c) and (d) result from averaging 18 or 40 images, with much of the information content restored.

If we view image speckle as a form of noise, we can use results such as those in Fig. 10 to determine the amount of averaging required to meet the system goals. When image noise is dominated by speckle, the S/N may be approximated as (George *et al.*, 1976)

$$S/N = (R_I/R_0)N_a^{1/2}, \qquad (15)$$

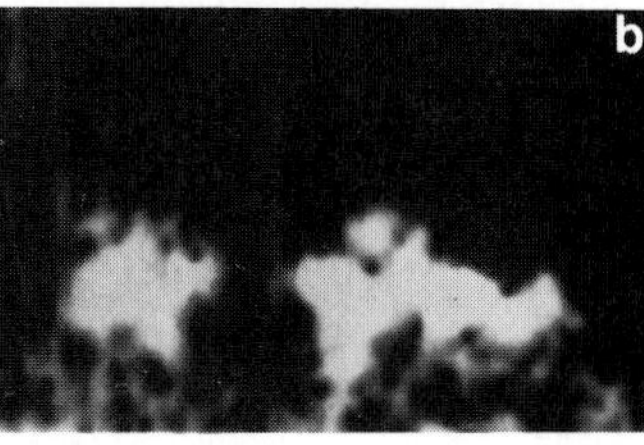

FIG. 13 3.2-mm simulation of 700-μm images at simulated (a) 0.7-km and (b) 2.1-km ranges. (After Guenther, 1976a.)

where R_I is the desired image resolution, R_0 the resolution provided by the optics ($R_0 > R_I$), and N_a the number of images averaged. This equation reflects the fact that speckle averaging may be performed spatially (at the expense of resolution) as well as temporally. Since in our application we do not have resolution to spare, we should have $N_a \geq 9$ to satisfy the usual Johnson criterion that, as we have seen, corresponds to $S/N \geq 2.8$. It is noted, however, that other workers (Christensen and Kozma, 1976) have maintained that a larger number (~ 100) of averages would be required. Possibly this disagreement reflects merely a difference in the image quality criteria used in the two cases.

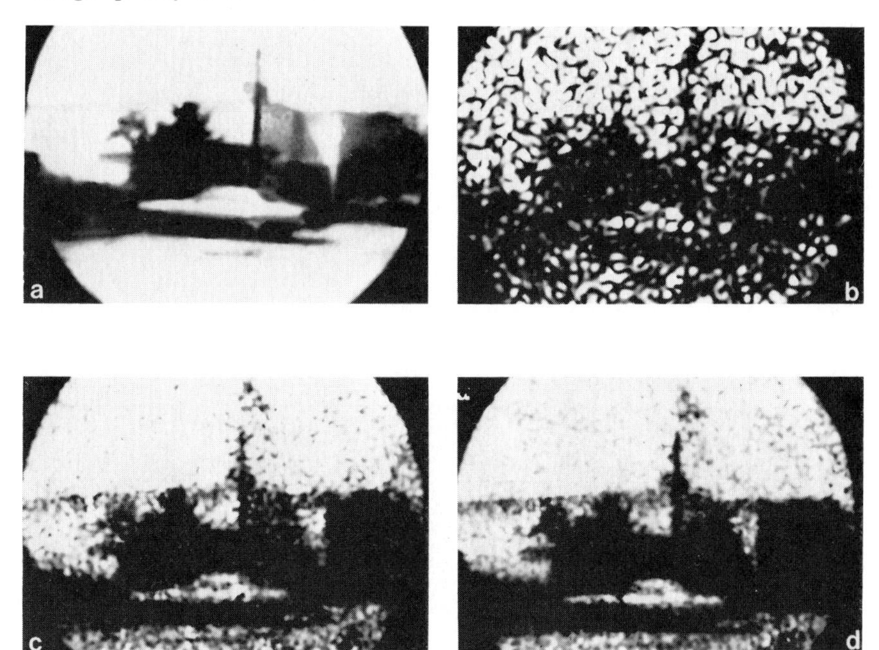

FIG. 14 Effect of speckle averaging. Scene with: (a) incoherent light; (b) coherent light; (c) 18 images averaged; (d) 40 images averaged. (After Christensen and Kozma, 1976.)

D. Image Glint

In addition to "ordinary" speckle, coherent images may contain "glint," bright spots on the display corresponding to specular returns from scattering centers (generally corners or normal surfaces) of amplitude much greater than the average diffuse return. The presence of glint is illustrated in Fig. 15, which shows the intensity histogram of a tank image produced at a 50-ft range by the 3.2-mm imaging system. The sharp peak in the histogram corresponds to background while the broader peak corresponds to diffuse returns from the tank. The specular glint returns appear in the high cross-section tail of the histogram, and are as much as 30 dB above the typical diffuse reflections.

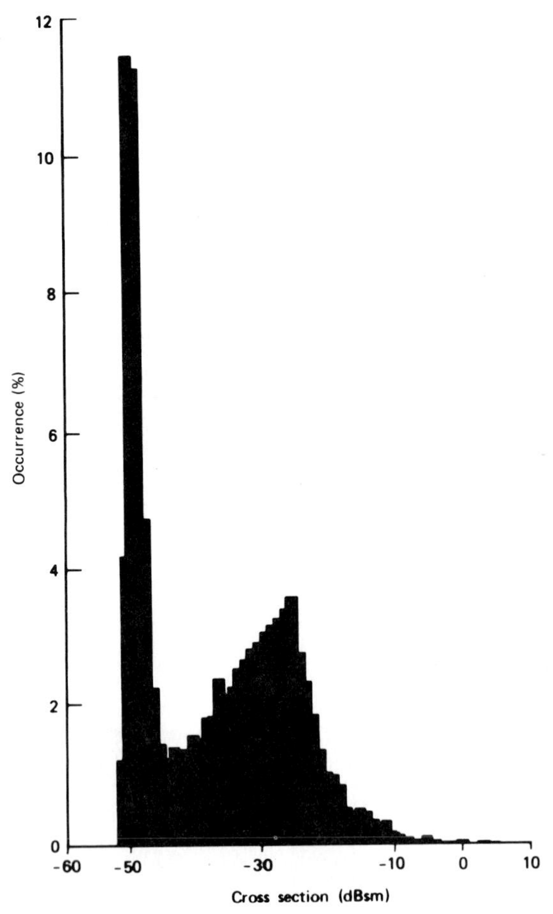

Fig. 15 Intensity histogram of tank image. (After Guenther, 1976a.)

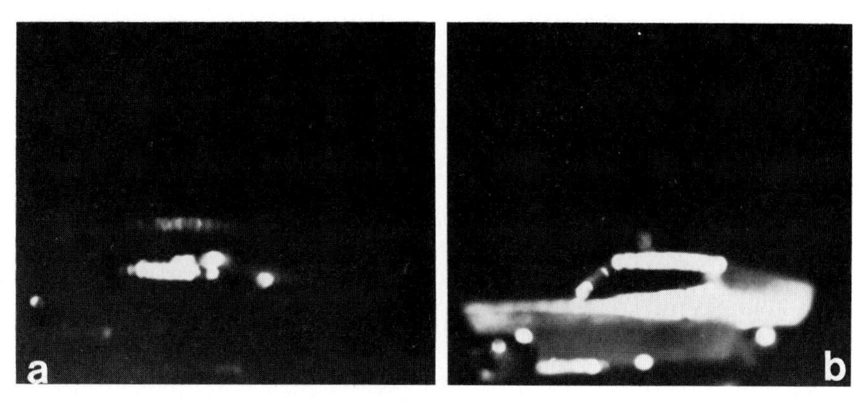

FIG. 16 Glint suppression by nonlinear recording. (a) Linear recording, (b) nonlinear recording.

Target detection may in some cases actually be improved by the presence of target glint, which greatly enhances the system sensitivity. Also, some very high resolution target images produced using only the glint returns (Guenther, 1976a) seem to permit target recognition. However, with the lower resolutions achievable in practice, we expect that the glint returns will have to be suppressed through use of a nonlinear gray scale in order to permit target recognition based on a more conventional diffuse image. Such suppression is illustrated in Fig. 16, which shows images produced with collimated laser illumination. Image (a), which uses a linear gray scale, contains a few bright speculars that dominate the image. Image (b), produced with logarithmic processing, shows significant improvement. (Note that glints have been similarly suppressed in Figs. 12–14.)

REFERENCES

Baird, J. M. (1976). Final Task Rep., Contract F30602-73-C-0191. U.S. Army MICOM, Redstone Arsenal, Alabama.

Barnes, C. W. (1966). *J. Opt. Soc. Am.* **56,** 575.

Buck, G. J., and Gustincic, J. J. (1967). *IEEE Trans. Antennas Propag.* **15,** 376.

Burch, D. E. (1968). *J. Opt. Soc. Am.* **58,** 1383.

Christensen, C. R., and Kozma, A. (1976). *J. Opt. Soc. Am.* **66,** 1257.

Deirmendjian, D. (1975). Rep. AD AOU-947. Rand Corp., Santa Monica, California.

Frieden, B. R. (1967). *J. Opt. Soc. Am.* **57,** 1013.

Frieden, B. R. (1975). *In* "Picture Processing and Digital Filtering" (T. S. Huang, ed.), Chapter 5, page 179. Springer-Verlag, Berlin and New York.

George, N., Christensen, C. R., Bennett, J. S., and Guenther, B. D. (1976). *J. Opt. Soc. Am.* **66,** 1282.

Goodman, J. W. (1970). *Prog. Opt.* **8,** 3.

Gordon, G. A., and Casowitz, P. (1972). *IEEE Trans. Aerosp. Electron. Syst.* **8,** 840.

Guenther, B. D. (1976a). Tech. Rep. RR-77-4. U.S. Army MICOM, Redstone Arsenal, Alabama.

Guenther, B. D. (1976b). Tech. Rep. RR-77-3. U.S. Army MICOM, Redstone Arsenal, Alabama.

Harris, J. L. (1964). *J. Opt. Soc. Am.* **54,** 931.

Hartman, R. L., and Kruse, P. W. (1976). *Int. Conf. Submillimeter Waves, 2nd,* San Juan, Puerto Rico, Paper F-3-1, p. 229.

Hartman, R. L., and Kruse, P. W. (1979). *Int. Conf. Infrared Millimeter Waves Their Appl., 4th, Dig., Miami Beach, Fla.* Paper number M-5-6, p. 57.

Johnson, J. (1958). *Image Intensifier Symp.,* Paper AD-220160, *Fort Belvoir, Va.*

Kruse, P. W., and Garber, V. (1978). *Proc. IRIS* **23,** p. 175.

Kulpa, S. M., and Brown, E. A., eds. (1979). Rep. HDL-SR-79-8. U.S. Army ERADCOM, Harry Diamond Labs, Maryland.

McKechnie, T. S. (1975). *In* "Laser Speckle and Related Phenomena" (J. C. Dainty, ed.), Topics in Applied Physics, Vol. 9, p. 123. Springer-Verlag, Berlin and New York.

Rosell, F. A., and Willson, R. H. (1973). *In* "Perception of Displayed Information" (L. M. Biberman, ed.), p. 167. Plenum, New York.

Rushforth, C. K., and Harris, R. W. (1968). *J. Opt. Soc. Am.* **58,** 539.

Tanton, G. A., Metra, S. S., Stettler, J. D., Morgan, R. L., Osmundsen, J. F., Castle, J. G. Jr. (1979). *Int. Conf. Infrared Millimeter Waves Their Appl., 4th, Dig.,* Paper W-2-7. *Miami Beach, Fla.*

Toraldo di Francia, G. (1969). *J. Opt. Soc. Am.* **59,** 799.

INDEX

A

Absorption bands, millimeter-wave radar, 27
Absorption of millimeter waves, atmospheric, 2, 37, 98–99
Active millimeter-wave seeker, 112–113, 123, 129–130
Active millimeter-wave sensor, 87
Active–passive missile-guidance system, 11, 112–113
Aerosols, atmospheric, effect on millimeter waves, 99, 100, 125
Airborne radiometry, 13
Air-to-ground missile guidance, 87
Air-to-surface missile guidance, 11, 95–96
All-solid-state technology, 7, 9, 11, 87
Amplifier, 191–194, *see also* specific amplifiers
  active conical scan radiometer, 119
  injection-locked IMPATT, 6
  gyrotron, 68
  millimeter-wave seeker, 145
  millimeter-wave traveling-wave tube, 154
  passive conical scan radiometer, 117
  pulse sensor, 136
Amplitude fluctuations, atmospheric, 5
Angle noise, 104
Angle of arrival, 5, 42
Antenna, 10–11, 16–17
  active conical scan radiometer, 118–120
  beamwidth, 98
    and radar resolution, 28–29
  dielectric millimeter-wave integrated circuits, 254, 259–260, 262–269
  millimeter-wave missile guidance system, 11, 123–126, 129–132, 141
  millimeter-wave radar, 25, 68–73, 87
  near-millimeter wave imaging system, 329, 334, 341
  passive conical scan radiometer, 113–117
  radio astronomy, 13
  size, and beamwidth and gain, 2
Antiballistic missile terminal homing seeker, 35
Applications Technology Satellites, 15
Arc plasma sprayed ferrite toroid, for millimeter-wave phase shifter, 75–76
Atmospheric propagation effects
  in millimeter-wave applications, 2, 4–5, 11, 14–17, 36–42, 98, 104
    missile guidance systems, 99–103, 125, 132
    radar, 25–26
    sky-warming effect, 111–112
  in near-millimeter wave imaging system, 329–333, 341
Atmospheric research, radiometry, 13
Atmospheric window, radar, 31
ATS, *see* Applications Technology Satellites
Attenuation, atmospheric
  active radiometer, 122
  millimeter-wave propagation, 2, 4, 14–15, 25, 31, 36–40, 42, 98–103
  near-millimeter wave imaging system, 329–333
  versus frequency, 27
Attenuation, in dielectric waveguides, 227–236
Attenuation constant
  O-type waveguide, 310–311, 313, 315, 318, 320
  rib waveguide, 301, 303–305, 308, 322
  slab waveguide, 277, 283–284, 288–293, 297, 320
Auto collision avoidance, radar application, 12
Automatic gain control, 5

Avalanche behavior, in IMPATT operation, 169–174, 177–179
Avalanche resonance frequency, 172
Azimuth scan, seeker antenna, 126–127, 130

B

Background reflectivity coefficient, 106
Backscattering
atmospheric, 2, 37, 40–42, 102
clutter, 42–56
metal targets, 123
Backward-wave oscillator, 6, 8, 58, 65
Balanced mixer, 243, 260–262, 270–271
Bandpass filter, 72, 74, 245–247
Band-reject filter, 245–248, 258
Bandwidth, millimeter-wave, 31, 97
Battery-powered solid-state terrestrial communications system, 14–15
Beam dilution factor, radiometer, 116
Beam-lead diode, 260
Beamrider millimeter-wave missile guidance system, 11, 87, 125–126
Beam-splitter type coupler, 245–246
Beam-terrain geometry, active conical scan radiometer, 120
Beamwidth, millimeter-wave systems, 2, 25, 28–29, 98
Bending loss, submillimeter guided waves, 278, 296–297, 313
Binocular radio, 14–15
Biological applications, of millimeter-wave technology, 13, 17
Boron nitride, dielectric-waveguide systems, 245, 260–262, 270–271
Bragg reflector oscillator, 259
Brillouin flow, 160
B-scope display, millimeter-wave radar, 81–82
BWO, *see* Backward wave oscillator

C

Cancer detection, 17
Carcinotron, *see* Backward-wave oscillator
Cassegrain antenna, 68–70, 87
Cavity resonator, 225
Ceramic material, 203, 263
Chirp, IMPATT devices, 118, 187–188
Circular dielectric waveguide, 219
Circular-electric mode, 15
Circulator, radar, 31, 35, 74
Clear air, effect of
on millimeter waves, 5, 37–38, 40, 99, 103
on near-millimeter wave imaging system, 329–330, 333, 341
and radiometric sky temperature, 112
Closed dielectric waveguide, 218–219
Clouds, effect on millimeter waves, 2, 37, 39–40, 100–101, 103
Clutter, 106, 109, 111
active conical scan radiometer, 118, 121
FMCW sensor, 135, 138–140
millimeter-wave seeker, 123
pulse sensor, 136, 140
of radar system, 25, 44–56, 78–80
Clutter radar cross section, 45, 78
Coast Guard obstacle avoidance radar application, 12
Coaxial probe, waveguide analysis, 226–227, 232–233
Conductor loss, image guide, 230–231
Conical scan antenna, 70, 87, 124, 126, 130–132, 141–142
Conical scan radiometer
active, 118–122
passive, 113–117
Continuous wave diode, 179–180, 183, 185–187, 195–196
Continuous-wave oscillator, 8, 62, 65
Continuum radiation detection, 12
Copper
heat sink, 179–180, 183, 185–186, 195
coupled-cavity tube, 165
wire, submillimeter waveguide, 314–316
Coupled-cavity tube, 152–157, 164–166
Coupled waveguide, 215–217, 243
Coupler
dielectric waveguide, 243–246
movable metal grating, 283–290, 303–304, 307–308
Cross range millimeter-wave seeker, 128
Current density, IMPATT diode, 169, 172
Curved slab waveguide, 234–236
Cylindrical resonator, 192–193
Cylindrical waveguide, 276, 310–322

D

Diamond heat sink, diode, 183, 185–186, 195–196
Dicke-switched radiometer, 115, 117
Dielectric image line, 10
Dielectric lens antenna, 69–70, 73
Dielectric loss, in image guide, 229–231
Dielectric waveguide, 199–201
attentuation characteristics, 227–236
millimeter-wave integrated circuits, 15, 201–207
active components, 253–262
antennas, 262–269
passive components, 243–253, 271
subsystems, 269–271
phase constant and field distributions, 207–227
submilimeter waves, 275–326
theoretical analysis, 236–243
Diffraction limited antenna, 28
Dimer, effect on near-millimeter wave imaging system, 333
Diode, *see also* specific diodes
dielectric waveguide oscillator, 257–258
mixers, 8–9
noise, 189–191
packaging, 180–183
performance, 183–191
Directional coupler, 243–246
DIRT-1 test, 5
Discontinuity problems, dielectric millimeter-wave integrated circuits, 239
Doppler frequency shift, 25, 29–30
Double-drift diode, 168–175, 181, 185–188, 195–196
Double sideband noise, Schottky barrier diode, 57, 60
Drift velocities, in IMPATT operation, 170
Dual-mode millimeter wave sensor and guidance techniques, 87, 98, 103–106
Dual-mode radar, tank fire control, 12
Dust, effect on millimeter-wave propagation, 5, 37, 42, 98, 125

E

ECCM, *see* Electronic Counter Counter Measures
ECM, *see* Electronic Counter Measures
Effective dielectric constant (EDC) 211–216, 237–239
EIA, *see* Extended interaction amplifier
Eigenvalue equations
waveguide analysis, 210–211, 214–215, 229–230, 278–279, 282, 300–301
EIO, *see* Extended interaction oscillator
Electric field effects, in IMPATT operation, 169–172
Electric probe, waveguide analysis, 222, 224
Electromagnetic radiation
atmospheric attenuation, 27, 330–331
frequency spectrum with band designation, 24
Electromechanical scanning antenna, 69
Electron beam, traveling-wave tube, 152, 158, 160–161
Electron current, in IMPATT operation, 169–171
Electron cyclotron maser, *see* Gyrotron
Electron gun, traveling-wave tube, 152, 161–163
Electron-hole pair, in IMPATT operation, 169
Electronic Counter Counter Measures (ECCM), 16, 25
Electronic Counter Measures (ECM), 16, 25
Electronic warfare, millimeter-wave applications, 16–17
Emissivity, of background materials, 109–111
Equivalent circuit
dielectric waveguide oscillator, 257–258
diode, 181, 184–185
Equivalent refractive index method, submillimeter waveguide analysis, 298
Error signals, target tracking, 131
Extended interaction amplifier (EIA), 8
Extended interaction oscillator (EIO), 6–7, 31, 34, 58, 61, 64–65
Extremely far infrared frequency range, 2

F

False alarm rate (FAR), millimeter-wave seeker, 143, 145
Faraday rotation isolator, millimeter-wave radar, 74

Fence waveguide, 204–205
Ferrite, 6, 10, 74–76, 250–252
Ferrule, of coupled-cavity structure, 159, 163, 165
Ferruless coupled-cavity circuit, 165–166
Field distributions, dielectric waveguide, 207–227
Field of view (FOV), near-millimeter wave imaging system, 328–329, 334–335, 341
Fill factor, *see* Beam dilution factor
Filter, 245–251, *see also* specific filters
  millimeter-wave radar, 72, 74
Fine-edge blanking, 165
Fin line, 10
Fire control system, 35, 88–91
FLIR imaging system, *see* Forward-looking infrared imaging system
Fluctuation power, active conical scan radiometer, 118–120
FMCW sensor, 133–136, 138–140
Fog, effect of
  on millimeter waves, 2, 4, 37–40, 98–99, 100–101, 103
  on near-millimeter wave imaging system, 329–333, 341
  and radiometric sky temperature, 112
Forward-looking (FLIR) imaging system, 327, 335
Forward scattering
  atmospheric, 102
  multipath effects, 42–44
Four-horn monopulse antenna feed horn, 70, 73
Four-pole filter, 247
FOV, *see* Field of view
Frame rate, near-millimeter wave imaging system, 334–335, 339, 341
Free-fall missile guidance system, 95
Frequency, 1–2, 24
  atmospheric attenuation as a function of, 27, 40, 100–101, 111
  millimeter-wave radar, 27, 104
Frequency-band designations, 3, 24
Frequency-change effects, near-millimeter wave imaging system, 337–338
Frequency-doubled Gunn oscillator, 6
Frequency-scanned antenna, 263, 267
Frequency-stabilized millimeter-wave sources, 6
Frequency-tunable diode oscillator, 188–189
Fresnel zone plate, 10
Front-fed reflector antenna, 68

G

Galerkin's method, in waveguide analysis, 222
Gallium arsenide
  dielectric waveguide, 206, 257
  Gunn devices, 7, 151, 185–186, 191, 196
  IMPATT diode, 175–178
  Schottky beam-lead diode, 260
Generalized telegrapher's equation method, waveguide analysis, 220–222
Geodesic lens, 10, 69, 80–81, 85
Gimbal design, millimeter-wave seeker, 126, 130, 145–146
Glint, target, 104, 123, 131–132, 330, 350–351
Goell's method, waveguide analysis, 210–212
Golay cell detector, submillimeter waveguides, 288–290, 295, 317–318
Grating structure
  dielectric waveguide oscillator, 258–259
  filter made from, 247–251
  leaky-wave antenna, 265–268, 270
  movable metal coupler, 283–290, 302–304, 306–308
Ground-to-ground antiarmor beamrider, 125
Gunn amplifier, 191
Gunn diode, 123, 151, 184–186, 196
  packaging, 182
Gunn effect, in gallium arsenide diode, 175–176
Gunn oscillator, 5–7, 61–62, 224–225, 257, 260
  noise, 190–191
Gunn radar transmitter, 34, 58
Gunn Varactor-controlled oscillator, 188–189
Gyrotron, 5–6, 8
  jamming, 17

radar-transmitter power source, 34, 58, 65, 68

H

Harmonic mixing, 9, 57, 60–61
HCN laser, submillimeter guided wave experiments, 285, 288–289, 307, 316–317
Heat sink, diode, 169–170, 179–180, 183, 185–186, 195–196
Helix tube, 152–153
HE_{11} mode, submillimeter guided wave experiments, 310–311, 315, 318, 324–325
Heterodyne receiver, near-millimeter wave imaging system, 340
H-guide, 10, 204–205
High-K 707L, 263
High-pass filter, millimeter-wave radar, 74
Hole current, in IMPATT operation, 169–171
Hollow metal waveguide, 15
Hybrid technology, 74, 194, 207, 271
Hydrometers, *see* Fog; Rain

I

Ice cloud, effect on millimeter-wave propagation, 2, 39–40
IF signal frequency, FMCW sensor, 133, 136
Illuminator, millimeter-wave missile guidance system, 123–125
ILO, *see* Injection-locked oscillator
Image waveguide, 201–202, 204–205, 207, 210, 213, 218–219, 222, 225, 229–233
  balanced mixer, 260–261, 270–271
  directional coupler, 245
  Gunn oscillator, 262
  isolator, 251–252
  two-pole filter, 248
Impact ionization, in IMPATT operation, 169–171, 174
IMPATT amplifier, 6–7, 16, 191
IMPATT diode, 58, 62, 151, 168–196
  active conical scan radiometer, 118
  design considerations, 173–180
  millimeter-wave missile guidance system, 123
  operation, 168–173
  packaging, 182–183
  performance, 184–189
  pulse sensor, 136
  reliability, 194–197
IMPATT oscillator, 5–7, 31, 58, 61–64, 257
  noise, 190–191
IMPATT radar transmitter, 34, 58, 87, 90
IMPATT Varactor-controlled oscillator, 188
Impedance
  amplifier, 191–192
  diode, 183–184
    noise, 189
Indium phosphide diode, 7, 185, 191
Infrared radiation, effect of atmosphere on, 27, 327
Inhomogeneously filled waveguide, 218, 220
Injection-locked IMPATTs, 6–7
Injection-locked oscillator (ILO), 191–194
Instrumentation radar system, 36, 82–85
Insulated image (insular) waveguide, 10, 202–203, 207, 213, 245, 263–264, 270
Integrated circuit, *see* Microwave integrated circuit; Millimeter-wave integrated circuit; Submillimeter-wave integrated circuit
Integration time, millimeter-wave seeker, 129–130
Interference
  and beamwidth, 98
  radar-radiometer input signal, 104–105
Inverted strip (IS) waveguide, 203–204, 207, 213, 215–216, 236–239, 243–245, 247–250, 265–266
Ionization, in IMPATT operation, 169–171, 174–177
IRIG-B compatible time code generator, 103
Isolator, 74, 250–253
IS waveguide, *see* Inverted strip waveguide

J

Jamming, 14, 16–17, 97–98, 104
Johnson criteria, near-millimeter wave imaging system, 341, 344–347, 349

Junction circulator, millimeter-wave radar, 74
Junction problems, dielectric millimeter-wave integrated circuits, 239–243

K

Klystron, 6
  radar transmitter, 58
  tuning range, 31

L

Land clutter, 45–53, 78–80
Laser, *see* HCN laser; Optically pumped laser
Launch-and-leave targeting, missile-guidance systems for, 95
Leaky-wave antenna, 254, 259–260, 265–270
Lincoln Experimental Satellites, 15–16
Line-of-sight guidance command, millimeter-wave seeker, 146
LOAL, *see* Lock-on-after-launch
LOBL, *see* Lock-on-before-launch
Local oscillator, 57, 262, 271
  FMCW sensor, 133
  noise, 190–191
  pulse sensor, 136
  tuning ranges, 31
Lock-on-after-launch (LOAL) operation, 112, 124
Lock-on-before-launch (LOBL) operation, 124
Lumped-element solid-state devices, 271
Luneberg lens antenna, 10, 69–71

M

Magnetic field, traveling-wave tube, 160–161, 163
Magnetron, 5–7, 34, 58, 61–62, 64–65
Marcatili's method, waveguide analysis, 208–213, 218, 220
Massachusetts, University of, radio astronomy, 13
Master oscillator, power amplifier (MOPA), 34
Matched filter radar, range delay resolution, 29
Measurements radar system, 36, 82–86
Metallic target, 123
MIC, *see* Microwave integrated circuit
Microstrip techniques, 10, 74
Microwave dielectric waveguide, 206–207
Microwave integrated circuit, 200
Microwave radar, 25, 98–99
Microwave radar range equation, 31
Millimeter beamrider guidance concept, 87–88
Millimeter-wave integrated circuit (MMIC), 10, 15, 146
  dielectric-waveguide type, 199–273
Millimeter-wave missile guidance, 2, 11, 35, 85–88, 95–150
  advantages, 97–99, 122
  atmospheric effects, 99–103
  countermeasures, 148–150
  options, 123–126
  range equations, 112–122
  seeker design, 122–148
  seeker search and track, 126–133
  target and background, 102–112
  terminal guidance, 2, 11–12, 85, 87, 95–97
  waveforms and processing, 132–145
Millimeter-wave radar, 12, 23–94, 123
  advantages and limitations, 26, 97–99
  applications, 35–36, 80–91
  bandwidth, 31
  characteristics, 25–31
  clutter characteristics, 44–56
  components and technology, 33–35
    antennas, 68–73
    passive components, 72–76
    receivers, 56–57
    transmitters, 34, 58, 61–68
  example, 76–80
  frequencies of operation, 27
  millimeter-wave propagation effects, 36–44
  multipath effects, 42–44
  range equation, 31–33
  resolution, 28–30
  target and background characteristics, 104–109

Millimeter waves, 1–21, 24, 97–98, *see also* specific components, devices, sources, and system applications
atmospheric effects on, 2, 4–5, 98–103
new technology, 5–11
rib waveguide experiments, 302–306
system applications, 11–17
Minority charge storage, in diodes, 178
Missile guidance, millimeter-wave, *see* Millimeter-wave missile guidance
Mixer, 5–6, 8–9, 34, 57, 59–61, 145, 260–262
MMIC, *see* Millimeter-wave-integrated circuit
Modal expansion coefficient, rib waveguide, 299
Model target, frequency-scaled radar measurements on, 83–84
Mode-matching technique, waveguide analysis, 218–220, 298, 301
Molecular spectroscopy of interstellar clouds, 14
Molybdenum wire, movable metal-grating coupler, 285
Monopulse antenna design, 10
Monopulse radar system, 70, 73, 84–85
MOPA radar transmitter, *see* Master oscillator, power amplifier radar transmitter
Mott diode, 9
Mottky diode, 9
Movable metal-grating coupler, 283–290, 302–304, 306–308
Movable short, in waveguide analysis, 224
Moving target indication (MTI) techniques, 7, 12, 80, 90
Missile fuzing, radar application, 12
MTI techniques, *see* Moving target indication techniques
Multipath phenomena, 42–44, 98, 123
Multistage depressed collector, traveling-wave tube, 164

N

Narrow band radar, glint noise, 131
National Radio Astronomy Observatory, Kitt Peak, Arizona, 13
National Radio Modeling Facility, U.K., 84, 86
Navigation, radiometer, 13
Near-millimeter waves, 1, 24
effect of atmosphere on, 38
imaging system, 327–352
design considerations, 333–340
image quality considerations, 344–351
sensitivity, 339–340, 342–344, 351
system limits and wavelength trade-offs, 340–344
propagation, 330–333
Nimbus 6 satellite, radiometers, 13
NMMW, *see* Near-millimeter waves
Noise
active conical scan radiometer, 118, 121
amplifier, 7, 191
diode, 189–191
FMCW sensor, 135
mixer, 5–6, 8–9, 34, 260–262
and narrow beamwidth, 98
passive conical scan radiometer, 113–114, 116–117
receiver, 12, 131, 270, 339–341
and resolution, 346
Noise power
active conical scan radiometer, 118–119, 121–122
millimeter-wave seeker, 131
total power radiometer, 114
Noise temperature, passive conical scan radiometer, 113–114
Non-separable geometrics, dielectric waveguide, 207

O

Oil smoke, effect on millimeter-wave propagation, 5
Open dielectric waveguide, 218, 233
Open-ended waveguide, 239, 258
Optical dielectric waveguide, 205–207, 233
Optical fiber, 277, 310
Optically pumped laser, 6
Optical missile seeker,98–99
Oscillator, *see also* specific oscillators
dielectric millimeter-wave integrated circuits,257–260, 269–270
Oscillator-modulator-detector module, 269–271
O-type waveguide, 277, 310–322

Oxygen, atmospheric, effect on millimeter waves, 2, 37, 99

P

Parachute-suspended munitions, guidance system, 95
Passive clutter, 111
Passive millimeter-wave seeker, 112–113, 123, 128, 132
Passive millimeter-wave sensor, 87
Passive radiometer sensor, terminal missile-seekers, 11
Periodic permanent magnet, traveling-wave tube, 160–161, 163, 167
Phase constant, dielectric waveguide, 207–277, 285, 288, 299, 302–303, 305, 308, 322
Phase-locked millimeter-wave sources, 6–7
Phase randomization, near-millimeter wave imaging system, 337
Phase shifter, 75–76, 254–257, 268
Photolithographic integrated circuit, millimeter-wave radar, 74
Pico Veleta radio astronomy facility, Spain, 13
Pierce electron gun, 161
Pierce interaction impedance, 158
Pillbox-type antenna, 70, 72
Pinhole-type probe, waveguide analysis, 226–227
PIN phase shifter, 255–257, 268
p-n junction diode, 169
Point-to-point communications, 14
Polyethylene
  insular waveguide, 203
  O-type waveguide, 310–311, 313–320, 322
  rib waveguide, 301–304, 306, 308
  slab waveguide, 277, 282–283, 285, 288, 291, 295–297, 320
Power density
  active conical scan radiometer, 119
  passive conical scan radiometer, 114
Preamplifier, pulse sensor, 136
Preselector filter, 247, 270
Prism, measurement of waveguide attenuation constant, 283
Propagation window frequencies, millimeter-wave radar, 27
Pulsed diode, 178–180, 187–188, 196
Pulsed oscillator, 62–63, 65, 187–188
Pulsed radar, range delay resolution, 29
Pulse millimeter-wave seeker, 144–145
Pulse sensor, 136–140

Q

Quasi-optical components, in millimeter-wave technology, 10, 74–75

R

Radar, 12–13, *see also* Microwave radar; Millimeter-wave radar
  all-solid-state, 7
  camouflage, 16
  frequencies, 23–24
  MTI (moving target indication), 7, 12
  range equation, 31–33
  receiver bandwidth, 29
Radar cross section, 45–49, 78, 104–107, 109, 118–120
Radial velocity of moving target, 29–30
Radiation, effect on biological materials, 17
Radiation loss, dielectric waveguide, 227, 233–236
Radio astronomy, 13–14
  precision surface antennas, 10
Radiometer, 9, 11–13, 123
  active conical span, 118–122
  effect of rain on, 101
  passive conical span, 113–117
  sensitivity, 115
  target and background characteristics, 105–106, 109–111, 116–117
Radiometric constant, 117, 119
Radiometric sky temperature, 110–112
Radiometric temperature, 105, 110
Radiometry, millimeter-wave applications, 12–13, 97–99
Radio transceiver, 14
Radome, small-diameter millimeter-wave seeker, 146
Railway communications, 14
Rain, effect of
  on millimeter waves, 2, 4, 37–42, 98–103, 125, 132

on near-millimeter wave imaging system, 330–333, 341
and radiometric sky temperature, 112
Range delay resolution, radar, 29
Range equation, 31–33
millimeter-wave missile guidance systems, 112–122
near-millimeter wave imaging system, 339
Raster scan, near-millimeter wave imaging system, 334–335, 337–338
RATSCAT radar cross-section measurement facility, Holloman AFB, 83
RCS, *see* Radar cross section
Read diode, 175–177
Receiver, 12–13, *see also* specific receivers
active conical scan radiometer, 119, 121
all-solid-state, 7, 9
FMCW sensor, 133, 135
millimeter-wave missle guidance system, 123, 131, 142, 146
near-millimeter wave imaging system, 329, 333–336, 338–340
passive conical scan radiometer, 113–114, 116–117
pulse sensor, 136
radar, 29, 31, 56–57
V-band, 270
wideband millimeter-wave, 16
Rectangular resonator, 192–193
Rectangular waveguide, 10, 201–202, 205–208, 216
leaky-wave antenna, 267
Reflected signals
clutter, 44–56
multipath effects, 42–44
Reflection coefficient, 43–44
Reflectivity, atmospheric, of millimeter waves, 36–37, 40–42
Resolution
millimeter-wave radar, 28–30
near-millimeter wave imaging system, 328–330, 340–341, 344, 349, 351
Resonant combiner, 192–193
Resonator, 192–193, 245–250, *see also* specific resonators
Rib waveguide, 276, 297–309, 320, 322
Ring resonator, 224, 245–247
Ritz-Galerkin (RG) variational technique, dielectric waveguide analysis, 239, 242
Rod antenna, 263
Rod waveguide, 201–202, 212
Rotary joint, radar, 31

S

Samarium cobalt magnet, 161
Satellite communications
atmospheric propagation effects, 2
millimeter-wave applications, 15–16
Satellite-to-satellite relay link, 15
S/C, *see* Signal-to-clutter ratio
Scale modeling radar application, 12
Scanning geodesic lens, millimeter-wave radar, 80–81
Scattering, *see also* Backscattering; Forward scattering
atmospheric, of millimeter waves, 2, 36–37, 40–42, 101
dielectric waveguide, 208
radar input signal, 104–105
Scattering center, near-millimeter wave imaging system, 330, 337, 339, 346, 350
Schottky diode, 9, 57, 59, 260
Scintillation effects, on millimeter-wave systems, 42
SD waveguide, *see* Strip dielectric waveguide
Seawater surface
clutter, 25, 53–59
reflection coefficient, 43–44
Self-oscillating mixer, 260–262
Semiactive millimeter-wave missile guidance system, 124–125
SHF band, *see* Superhigh-frequency band
Ship-to-ship communications, 14
Short-pulse, range-gated signal processing, 142–143
Side-by-side dual-frequency sensor, 103–106
Signal interception, 14, 16
Signal power
active conical scan radiometer, 119
passive conical scan radiometer, 113–114
Signal-to-clutter ratio, 78, 80
Signal-to-noise ratio
active conical scan radiometer, 118–119, 121

FMCW sensor, 135–136, 140
millimeter-wave radar,77–78, 80
millimeter-wave seeker, 132, 143
near-milloimeter wave imaging system, 328, 339, 341, 348–349
passive conical scan radiometer, 113–115
pulse sensor, 136
in radar range calculation, 32
and target recognition, 345
Silicon
antenna, 263
dielectric waveguide, 206, 271
diode, 151, 168, 173–180, 185–188
oscillator, 257
Single-diode reflection amplifier, 192
Single-drift diode, 174–175, 178, 196
Single-pulse signal-to-noise ratio, in radar range calculation, 32
Sky-warming effect, in radiometric operation, 111–112
Slab waveguide, 210–214, 219–220, 234–236, 239–240, 276–297, 319–320
Slow-wave structure, traveling-wave tubes, 153–154, 157–158
Small-diameter millimeter-wave seeker, 146–147
ıoke, effect of
ɔmparison of seeker operation in three spectral regions, 99
ı millimeter waves, 5, 37, 42, 98, 125
near-millimeter wave imaging system, 329, 331
S/N, *see* Signal-to-noise ratio
Snow, effect of
on millimeter waves, 103
on near-millimeter wave imaging system, 331
reflectivity, 52–54
Solenoid, traveling-wave tube, 160–161, 163, 167
Solid-state
millimeter-wave sources, 6–7, 168–196
portable terrestrial communications system, 14–15
radar transmitters, 58
rf amplifier, 191
STARTLE radar system, 12
Space harmonics, waves in grating structure, 248, 265
Speckle averaging, near-millimeter wave imaging system, 334, 337–339, 341, 349
Speckle effect, near-millimeter wave imaging system, 330, 337, 346–349
Specular glint, *see* Glint
Square-law detector, 114–115, 119
Stable amplifier, 191–192, 194
STARTLE radar system, 12–13, 90
Steel wire, movable metal-grating coupler, 303
Step discontinuity, dielectric millimeter-wave integrated circuit, 239
Strip dielectric (SD) waveguide, 203–204
Subharmonic mixing, 8–9
Submillimeter-wave integrated circuit, 275–276, 298, 322
Submillimeter waves, 1–2
atmospheric attenuation versus frequency, 27
dielectric waveguides, 275–326
cylindrical waveguide, 310–322
rib waveguide, 297–309
slab waveguide, 277–297
Superheterodyne radiometer, 13
Superheterodyne receiver
millimeter-wave-integrated circuit, 146
radiometer operating temperature, 114
near-millimeter wave imagining system, 329
Superhigh-frequency (SHF) band radar, 23
Superresolution, 345
Surface-wave antenna, 262–264
Surveillance and Target Acquisition Radar for Tank Location and Engagement, *see* STARTLE radar system
Surveillance radar, 35, 80–82
Switched radiometer, 115, 117

T

Tank fire-control radar system, 12–13
Tank target, 89–90, 346–348
Target-background signature data, of millimeter-wave guidance system, 102–112
Target glint, *see* Glint

Target radiometric temperature contrast, 116
Target search, detection, tracking and homing
Johnson criteria, 344–345
millimeter-wave systems, 35, 80–81, 89, 123–133, 141–143
near-millimeter wave systems, 328–330, 333–334, 341, 346, 348, 351
Target signature, near-millimeter wave imaging system, 339
Teflon, in waveguides, 203, 247
TE modes
millimeter-waves, 15, 210–212, 219, 221, 238–239
submillimeter waves, 278, 280, 282, 284, 286, 288, 290–291, 293, 299, 301–303, 311–312, 316
Terrain return fluctuation power, active conical scan radiometer, 119–122
Terrestrial communications, millimeter-wave applications, 14–15
Thermal radiation detection, 12
Thermionic devices, radar transmitters, 58
Thermography, 13, 17
Three-pole filter, 247
Time-on-target (TOT), 126–127
TM modes
millimeter waves, 210–211, 219–221, 237–239, 303–306
submillimeter waves, 278, 281–282, 284, 286, 288, 290–293, 296–297, 299, 301–303, 308–309, 320
TOT, *see* Time-on-target
Total energy seeker system, 123
Total fluctuation noise power, active conical scan radiometer, 118
Total power radiometer, 114–115
Track gate, millimeter-wave seeker, 131
Tracking radar, 35, 88–91
Track-while-scan radar techniques, 81, 85
Transit velocities, in IMPATT operation, 169
Transmitter
millimeter-wave radar, 34, 58, 61–68
near-millimeter wave imaging system, 337–338, 340
V-band, 270
Transverse broadening, submillimeter guided wave, 294–295
Transverse resonance technique, waveguide analysis, 237–238
Traveling-wave tube, 34, 151–168, 196
coupled-cavity circuits, 154–157
design, 157–160, 163–165
electron beams, 160–161
electron guns, 161–163
practical considerations and performance, 165–167
radar transmitter, 58, 65
Tumor detection, 17
Turbulence, atmospheric, effect on millimeter waves, 5, 42
Two-pole filter, 247
TWT, *see* Traveling wave tube

U

Ultrahigh-frequency (UHF) band radar, 23
Uncoupled dielectric waveguide, 215–216

V

Vacuum-tube millimeter-wave sources, 6–8
Varactor-controlled oscillator (VCO), 188–189
Varactor multiplier, 7
Variational method, waveguide analysis, 216–218
V-band
mixer, 260
receiver and transmitter, 270
VCO, *see* Varactor-controlled oscillator
Video filter, passive conical scan radiometer, 114–115
Visible imaging system, 327, 335
Voice communication system, 14–15
Voltage standing wave ratio (VSWR)
dielectric waveguide, 232
leaky-wave antenna, 259, 265, 270

W

Warfare, electronic, *see* Electronic warfare
Water vapor, effect of
on millimeter waves, 2, 37–38

on near-millimeter wave imaging system, 329–330, 332
Waveguide, 10, 15, *see also* specific waveguides
Wavelength-dependent phenomena, in missile guidance, 123
W-band balanced mixer, 260
Wideband millimeter-wave receiver, 16
Window minima, 2

Y

YIG resonator, 188

Z

Zoned dielectric lens antenna, 69–70